Advanced Energy Technologies for Green Buildings

Green buildings are designed to reduce their environmental impact over their entire life cycle. This book provides an overview of the principles of green building and energy-efficient systems including benefits of green buildings, the different types of green building rating systems, and the various energy-efficient systems that can be used in buildings. It includes case studies (India's Green Building policy) of green buildings and energy-efficient systems.

- Illustrates the green building concept and approach toward green design
- Explores the role of green buildings for society and sustainability
- Describes energy-efficient Heating, Ventilation, and Air Conditioning (HVAC) and low-power Internet of Things (IoT) devices in buildings
- Discusses a design example of green buildings and quantitative figures for measurement of green buildings
- Includes technologies, illustrations, and applications supported by examples

This book is aimed at graduate students and researchers in energy and civil engineering and architecture.

Advanced Energy Technologies for Green Buildings

Beemkumar Nagappan, Gloria Pignatta,
and A D Dhass

CRC Press
Taylor & Francis Group
Boca Raton London New York

CRC Press is an imprint of the
Taylor & Francis Group, an informa business

Designed cover image: Getty Images

First edition published 2026
by CRC Press
2385 NW Executive Center Drive, Suite 320, Boca Raton, FL 33431

and by CRC Press
4 Park Square, Milton Park, Abingdon, Oxon, OX14 4RN

CRC Press is an imprint of Taylor & Francis Group, LLC

ISBN: 9781032818320 (hbk)
ISBN: 9781041271345 (pbk)
ISBN: 9781003752837 (ebk)

DOI: 10.1201/9781003752837

Typeset in Times
by Newgen Publishing UK

Dedication

Beemkumar Nagappan

This book is dedicated to the pioneers of sustainable engineering and green building design, whose visionary work inspires a greener world. To my beloved family—my wife, daughters, and parents—whose unwavering love and support have been my foundation. To Dr. Chenraj Roychand, Chairman of JGI Group of Institutions, for his inspiring leadership and commitment to transformative education. To Dr. Jawahar Nesan, Chief mentor of JGI Group of Institutions, for his dedication to society-centric education and his profound influence on my approach to grassroots research and community empowerment. And to future innovators, may this work ignite your passion to create buildings that sustain both people and the planet.

A D Dhass

I dedicate this book to the pioneers of renewable energy and green building technologies, whose efforts have inspired sustainable solutions for a greener tomorrow. To the educators and researchers who have shaped my journey, especially the late Dr. E. Natarajan, former Dean, College of Engineering, Guindy, Anna University, Chennai, for his unwavering guidance. To my family and mentors for their constant support. To Indus University, Ahmedabad, provided this opportunity and commitment to transformative education. May this work encourage future generations to build a sustainable world through innovative energy systems.

Gloria Pignatta

This book is dedicated to the global community of scholars, practitioners, and advocates driving the transition to sustainable built environments. To the researchers and architects working to mitigate climate change and promote energy equity through innovative building solutions. To my mentors and colleagues at the University of Perugia, the Singapore-MIT Alliance for Reaseach and Technology (SMART), and the University of New South Wales (UNSW Sydney), whose collaboration and inspiration have shaped my path. And to future generations of designers and engineers, may this book fuel your creativity and commitment to crafting buildings that nurture both people and the planet.

Contents

Figures

Tables

Preface

The pressing need to mitigate environmental impact and enhance sustainability in the built environment has positioned green buildings at the forefront of architectural and engineering innovation. *Advanced Energy Technologies for Green Buildings* aims to elucidate the principles, technologies, and strategies that enable buildings to achieve superior energy efficiency, reduced environmental footprints, and enhanced occupant well-being. This book is designed to serve as a comprehensive resource for graduates, researchers, engineers, architects, and policymakers seeking to advance the design and implementation of sustainable buildings.

The book addresses the multifaceted challenges of green building design, from material selection and energy-efficient systems to the integration of renewable energy and smart technologies. It explores both established and emerging technologies, such as bioclimatic design, thermal energy storage, and Internet of Things (IoT)-enabled building management systems, which collectively contribute to reducing energy consumption and promoting sustainability. Despite the promise of these technologies, challenges such as high initial costs, scalability, and integration with existing infrastructure persist. This book not only highlights these challenges but also discusses ongoing research and innovations that pave the way for broader adoption.

Structured to provide both theoretical insights and practical applications, the book begins with an overview of green building principles, rating systems, and sustainable materials, emphasizing their economic and environmental benefits. Subsequent chapters delve into specific technologies, including insulated building envelopes, natural daylighting, renewable energy systems, and advanced energy storage solutions like pumped hydro, electrochemical, and thermal storage. The role of smart technologies, such as IoT and Building Information Modeling (BIM), in optimizing building performance is also thoroughly examined. Case studies, particularly those reflecting India's green building policies, provide real-world context and illustrate successful implementations.

A key focus of this book is the interplay between design, technology, and policy in achieving energy-efficient buildings. Topics such as efficient HVAC systems, passive and active lighting strategies, and future trends in green building innovation are discussed to equip readers with the knowledge to address current and emerging challenges. By investigating the properties, performance, and practical applications of energy-efficient systems, this book aims to inspire further research and collaboration among scholars, engineers, and industry professionals.

The journey toward sustainable buildings is both a technical and societal endeavor. We hope this book serves as a catalyst for advancing the field, encouraging readers to explore innovative solutions and contribute to a greener, more resilient future.

Acknowledgments

Beemkumar Nagappan
This book, *Advanced Energy Technologies for Green Buildings*, is the result of the collective efforts, support, and inspiration from numerous individuals and institutions. I extend my heartfelt gratitude to JAIN (Deemed-to-be University), Bengaluru, India, for providing an exceptional academic environment and resources. I am particularly grateful to Chancellor Dr. Chenraj Roychand for his visionary leadership and unwavering support for research and innovation. I also express my sincere appreciation to Dr. Jawahar Nesan, whose guidance inspired me to rethink the relationship between education, curriculum, and research with meaningful societal impact. His vision that scientific innovation should emerge from grassroots engagement and strengthen human involvement in knowledge creation has greatly influenced this work. I also acknowledge my research collaborators, co-authors, and students, whose dedication and enthusiasm have been instrumental in this endeavor.

A D Dhass
I gratefully acknowledge the support provided by Indus University, Rancharda, Ahmedabad, Gujarat, India, which has been pivotal in the completion of this book. My thanks extend to my colleagues and collaborators for their encouragement and insights. I am also appreciative of the mentors and advisors who have guided my research journey in sustainable energy and green technologies. Special thanks to my family for their unwavering support and understanding throughout this process. This book reflects the collective efforts of all who have contributed to advancing sustainable building solutions.

Gloria Pignatta
The creation of *Advanced Energy Technologies for Green Buildings* would not have been possible without the support and collaboration of many individuals and institutions. I sincerely thank the University of New South Wales (UNSW Sydney), Australia, particularly the Scientia prorogram and the School of Built Environment, for providing a dynamic and supportive research environment. My gratitude extends to my colleagues at the High Performance Architecture Research Cluster and the Material and Manufacturing Institute for their valuable insights and collaboration. I am deeply grateful to my mentors at the University of Perugia (Italy), and my collaborators across Singapore, Italy, India, Cyprus, Greece, and Australia, whose expertise has enriched this work. I also acknowledge the contributions of my students and research assistants, whose enthusiasm has been a constant source of inspiration. Finally, I extend my heartfelt thanks to my child Alexander, my family, and friends for their enduring support and encouragement. This book is a testament to our shared commitment to sustainable built environments.

Authors' Biography

Beemkumar Nagappan completed his undergraduate engineering program in the domain of Mechanical Engineering from the Coimbatore Institute of Technology (CIT), Coimbatore, India, in 2004, and then completed his post-graduation in Energy Engineering from the College of Engineering Guindy (CEG), Chennai, India. He received his PhD degree in Mechanical Engineering for innovative research in solar thermal energy storage field from Sathyabama Institute of Science and Technology, Chennai, India, in 2018. He has more than 18 years of teaching experience at various levels and he is currently working as a Professor & Deputy Dean Research in JAIN (Deemed-to-be University), Bengaluru, India. He is recipient of a research grant of INR 15.20 lakhs & faculty training grant of INR 3 Lakhs from AICTE - Delhi (GoI) and he also received a research grant of INR 40 Lakhs from VGST, Government of Karnataka, India. Since 2021, he has been recognized among Stanford University's World's Top 2% Scientists for four consecutive years. He has published more than 125 research articles in various WoS (SCI) and Scopus indexed journals with the cumulative impact factor of 450 (as per SCI, Clarivate Analytics). He received many awards for his outstanding contribution to research and academics. His current research interests include thermal energy storage with nano-encapsulated PCM, methods to reduce NO_x emission in bio-fueled CI engine, and solar-powered refrigeration systems. He is a life member in various professional bodies like the Indian Society for Technical Education (ISTE), Society of Automotive Engineers (SAE), India.

Gloria Pignatta is Scientia Senior Lecturer (Associate Professor) at the Univeristy of New South Wales (UNSW Sydney, Australia. She received a PhD degree in Energy Engineering from the University of Perugia (Italy), in 2016. In the same year, she was Postdoctoral Fellow at the Inter-University Research Center on Pollution and Environment "Mauro Felli" (CIRIAF), University of Perugia (Italy). From 2017 to 2018, she worked as Postdoctoral Associate and Building and Construction Expert at the Singapore-MIT Alliance for Research and Technology (SMART), in Singapore. Since 2018, she has been High Performance Architecture (HPA) Research Cluster expert in the School of Built Environment, Faculty of Arts, Design and Architecture at UNSW Sydney. She is also Investigator at the Material and Manufacturing Institute (MMFI) at UNSW. In 2021, she was nominated City Futures Research Centre (CFRC) Fellow.

She is an expert in building energy efficiency and sustainable solutions for the built environment. Her research touches on different fields, from engineering to architecture, material science, and urban climate, and aims to face the global problem of climate adaptation and mitigation, as well as social challenges, such as energy inequality. She has been actively engaged in research and contributed to the knowledge and literature with more than 60 research papers, 3 books, and numerous international conference papers. Her collaboration network touches different countries: Italy, Cyprus, Greece, Denmark, UK, France, Germany, Spain, Singapore, Sweden, USA, Canada, India, and Australia.

A D Dhass received his BE, ME, and PhD degrees in Electrical and Electronics Engineering, Energy Engineering, and Solar Energy in 2005, 2009, and 2018, respectively, from Anna University, Chennai, India. Since 2020, he has been working as Assistant Professor in the Indus Institute of Technology and Engineering (IITE), Indus University, Ahmedabad, Gujarat, India. His areas of research include solar, heat transfer, thermoelectric generators, Green Building and phase change materials.

1 Green Building Materials

1.1 INTRODUCTION

Environmentally friendly and resource-efficient green building includes site selection, design, construction, operation, maintenance, restoration, and deconstruction. It enhances cost-effectiveness, functionality, durability, and comfort in architectural design. Sustainable development began with the 1970s' fossil fuel energy crisis and the resulting environmental damage. The green construction movement began with energy efficiency and sustainability. Eco-friendly building benefits the environment, economy, and society. Modern sustainability programs require integrated and synergistic design of new and renovated buildings. Sustainable design aligns the building life cycle with eco-friendly activities for synergy. Green construction uses numerous methods to lessen environmental and health impacts. It promotes passive solar, active solar, photovoltaics, green roofs, rain gardens, and vegetation-based rainfall runoff reduction. Instead of concrete or asphalt, use packed gravel or permeable concrete to restore groundwater [1].

1.2 WHAT IS A GREEN BUILDING?

Green buildings use sustainable and resource-efficient methods throughout their siting, design, construction, operation, maintenance, and renovation. Green buildings use recycled materials, reduce water and energy use, and employ resource-efficient design and construction methods. They also reduce soil, air, and water pollution.

Thus, sustainable construction affects business and society. Sustainable building improves society. Sustainable building owners save money over time. They rent or sell faster, potentially increasing revenue. Green construction benefits society, the economy, and ecology. Green buildings provide environmental, social, and economic benefits; thus, construction players need incentives to adopt them. Green construction projects are encouraged through incentives. Designers, builders, consultants, and project owners must collaborate on designing green buildings.

Project owners have more power as they make important decisions. Thus, green construction incentive concerns mostly affect project owners. Motivating recipients, especially project owners, is its hallmark. To be eligible for the award,

DOI: 10.1201/9781003752837-1

green building requirements must be fulfilled. Compensation offered by the government is an example of an external incentive. Highlighting the advantages of green building can also increase interest in sustainable development among stakeholders. Despite their differences, most studies agree that both can promote green construction [2].

1.3 GREEN BUILDING CONCEPTS IN INDIA

The concept of green buildings is changing, but it reduces environmental impacts through better waste management, operation, maintenance, construction, location, and design. Green buildings are self-sufficient and energy-efficient, making them sustainable or high-performance [3]. Green buildings focus on the following five categories to reduce environmental impacts and improve tenant well-being [4]:

- Site planning for sustainability
- Water saving and efficiency
- Renewable energy, energy efficiency, and greenhouse gas reduction
- Preservation and reuse
- Better health and indoor air

India has a small green construction sector, with 5% of buildings qualified. According to the Indian Green Building Council (IGBC), India has built 7.17 billion square feet of "green" construction. Due to 6000 green projects and approximately 577,000 acres dedicated to substantial development initiatives, the government may meet the 75% green footprint goal for buildings 2 years earlier than the target.

According to Leadership in Energy and Environmental Design (LEED)–Green Building Council (GBC) India data, Maharashtra has the maximum number of green buildings, followed by Karnataka, Haryana, Tamil Nadu, and Uttar Pradesh. Higher awareness of green buildings and their long-term advantages will improve green construction.

Clay, sand, stone, and bamboo are locally produced and renewable, reducing green construction carbon emissions. Waste from development is composted or reused. Sustainable solid waste management is crucial both during and after construction. After occupation, green buildings in India segregate wet and dry trash to reduce environmental impact [5].

1.4 BARRIERS TO GREEN BUILDING ADOPTION

Davis [6] identified three main green building development barriers:

- *Builder Incentives:* Energy efficiency and workforce productivity are green building's main benefits. This benefits the owners but costs the builder. Thus, cost is the biggest barrier to green building.
- *Product Information and Sourcing:* Low-income countries' lack of understanding about green materials for high-performance building systems is the biggest hurdle to green building development. This requires developers to hire advisors.

- *Client Knowledge:* Proving the benefits of green architecture and doing longitudinal research are the best ways to overcome this obstacle [6]. The need for sustainable development (SD) in construction started the green building movement. Sustainable growth can be achieved through green building. Challenges have slowed Green Building (GB) deployment worldwide. The necessity of evaluating the challenges to GB adoption has not been emphasized.

1.5 TOP FIVE BARRIERS

Green building adoption remains constrained by five structural barriers that limit scalability, investment, and market maturity:

1. Insufficient information, education, research, knowledge, awareness, and expertise
2. Expense
3. Absence of incentives and support
4. A deficiency in interest and demand
5. Absence of green building codes and regulations [7].

1.5.1 Insufficient Information, Education and Research, Knowledge, Awareness, and Expertise

The goals of "green" building are to improve tenant quality of life and lessen the built environment's social and environmental impacts. The data underpinning green building design is vast. Green buildings solve several environmental problems. Many modern frameworks organize and quantify green building. This tool organizes green building themes and classes.

The evidence for green building design is overwhelming. Green buildings solve several environmental challenges. Many frameworks organize and quantify green building. This tool organizes green building themes and classes. Table 1.1 presents prior Green Building Literacy (GBL) frameworks, green building grading systems, green school awards, and focus group comments.

1.5.2 Expense

While green buildings have many advantages, their high cost limits their widespread implementation. Economic constraints and social expectations, like cost and value, limit technical innovation and design development in the complex and slow-adapting building industry. Changing building design processes to meet environmental issues requires challenging many design criteria and a new perception of building construction, use, and disposal. Innovative designs may become commonplace. During transitional periods, specification, interpretation, and implementation are ambiguous, causing conflict and resistance. Individuals can accelerate change despite organizational reluctance.

Buildings must all satisfy the economic feasibility test and provide justification for variations in cost and value. Construction costs for green buildings are usually higher. Green buildings may cost extra. Their experience shows 1% with eco-friendly

TABLE 1.1
Green Building Knowledge Categories [8]

Category	Sample Topics (Factual and Conceptual Knowledge)	Sample Competencies (Procedural Knowledge)
Sites are sustainable	Management of stormwater, impacts of building on light pollution, and the urban heat island phenomenon.	Evaluate the water quality at the building site. Implement water-conserving landscape elements, such as rain barrels.
Shape of the building	Dimensions and density of the structure (e.g., occupants per square foot/meter); placement of doors and windows; alignment of the building on the site concerning solar and wind exposure	Assess and contrast solar gain through windows orientated towards different cardinal directions.
Energy and atmosphere	Reduction of energy consumption, enhancement of energy efficiency, and the carbon footprint generated by construction and operational activities	Compute kilowatt-hours and identify energy-saving alternatives.
Water	Conserving water in the home with fixtures like low-flow toilets and sinks; addressing water challenges in the outdoors with measures like rainwater collection and drought-tolerant plants.	Landscape features that reduce water usage should be used. Select plumbing fixtures that reduce water usage.
Materials and waste management	Building material considerations include recycled content, recyclable, fast regenerated, sustainable wood products, locally produced, repurposed, and material use reduction.	Build furniture with recycled materials. Consider the environmental, social, and economic impacts of building materials including stone, wood, and carpet.
Indoor environmental quality	Maintaining interior humidity and temperature; connecting residents to nature with vistas, lighting, and biophilic design.	Assess radon and mold levels. Decorate to improve air quality. Consider daylight's psychological impacts.
Social justice	Affordableness; accessibility; local community impacts	Encourage equitable green building solutions. Utilize volunteer expertise in neglected regions
Beauty and inspiration	According to the International Living Future Institute, designs that "elevate our spirits and inspire us to be better than we currently are" are innovative.	Make a building site history mural. Comparing old and new construction methods
Economics	Cost-cutting features; long-term budgeting; performance-versus-cost trade-offs	Determine energy savings and green building material costs.

TABLE 1.1 (Continued)
Green Building Knowledge Categories [8

Category	Sample Topics (Factual and Conceptual Knowledge)	Sample Competencies (Procedural Knowledge)
Life cycle assessment	environmental impacts of construction materials from manufacture to disposal; Furniture and building materials' inherent energy	Adhere to "cradle to cradle" or "cradle to grave" construction products.
Operations and metrics	BIM, performance monitoring, green cleaning, and occupant training.	Evaluate cleaning products' environmental impact on air and water quality. Trail HVAC system performance.
Local and healthy food	Promote healthy food production and consumption infrastructure (e.g., greenhouses, edible landscapes).	Plan and build a greenhouse and a wheelchair-accessible vegetable garden.
Policy	Green building and politics, including local/regional/ national laws, ordinances, guidelines, and strategic plans.	Review local residential building codes for eco-home design alternatives and limitations.

designs. Due to this misconception, developers, clients, and designers overlook higher environmental performance goals [9].

1.5.3 Absence of Incentives and Support

Green building technology (GBT) adoption differs from other industrial non-green innovations. GBTs are eco-friendly technologies; their adoption and spread should address technological, economic, environmental, social, and political factors. Politics and the environment spread eco-innovation. Developers, designers, general contractors, subcontractors, suppliers, governments, and tenants or end-users utilize GBTs in construction projects. Building projects use GBT with numerous adopters. According to the market, GBTs create green buildings, which are end-user commercial products. Developers might decide not to construct GBTs if there is little social demand for them [10].

Eco-friendly buildings use sustainable design. This is called "green design" or "environmentally sustainable design". Global sustainability standards require local governments to identify buildings as green or non-green. Green and traditional construction approaches vary greatly. Sustainable property investments appeal to investors for several reasons. Tenant wellness and product/service sustainability credentials are excluded [11].

1.5.4 A Deficiency in Interest and Demand

Short-term solutions and non-green material, technological, and management strategies are used in most building projects. Construction management-at-risk, design–bid–build, and design–build delivery methods offer strengths and cons for conventional projects. Design–build is best for green building because team members interact. As the architect and contractor work together, the contractor can advise on green building design. Design changes and reworks delay and cost more without contractor experience. Green building project's essential personnel should be included from the beginning. These experts know about green certification and green building. Despite the risks involved, many developers opt for conventional methods when creating green. Misinformation about the advantages of green buildings, such as their effects on productivity, health, and interior environmental quality, has resulted from a lack of study [12].

1.5.5 Absence of Green Building Codes and Regulations

Urban green building legislation, codes, regulations, and awareness are lacking, making green construction methods difficult to implement. Industry stakeholders lack legislation, standards, and understanding of sustainable construction, complicating the situation. Data pertaining to sustainable and energy-efficient construction is sometimes restricted, imprecise, expensive, and challenging to verify.

High perceived cost hinders green construction. Insufficient technical expertise and practical knowledge hinder decision-making in areas such as energy efficiency, sustainable construction, and energy-saving devices and appliances. Green building data is scarce at the municipal level, making tracking, implementing, and scaling sustainable urban construction difficult. Cities cannot create effective evidence-based policies or track net-zero targets without centralised information on precertified and certified structures, built-up area typologies, pilot programs, and exemplary practices.

Effective decision-making requires end-user information when we buy or invest in green versus non-green businesses. Green buildings, their benefits, how daily behaviors affect energy and water use, and the relationship between energy, water, resource usage, and the environment are unfamiliar to end users, making it difficult for the end-users to make educated decisions [13].

1.6 GREEN BUILDINGS' CHALLENGES

The Green Building Challenge Framework and Tool increased knowledge transfer during development and testing. Each team found Environment Protection Agency (EPA)-accelerated learning helpful. Each national team applied this differently. Local information distributed by the team raised industry awareness. However, future architecture changes may optimize process value [14].

Success factors in the Project Delivery Attributes (PDAs), Critical Success Factors (CSFs) are situations, circumstances, or events that affect project outcomes without prescribing success. Success criteria are used to evaluate project performance. The association between "success factors" and "success criteria" includes causal (e.g., the owner's commitment to sustainability) and consequent (e.g., sustainability performance) aspects, leading to project success (Figure 1.1).

The framework presented in Figure 1.2 emphasizes multiple paradigms in GB development research. This framework shows the paradigm's uniqueness and

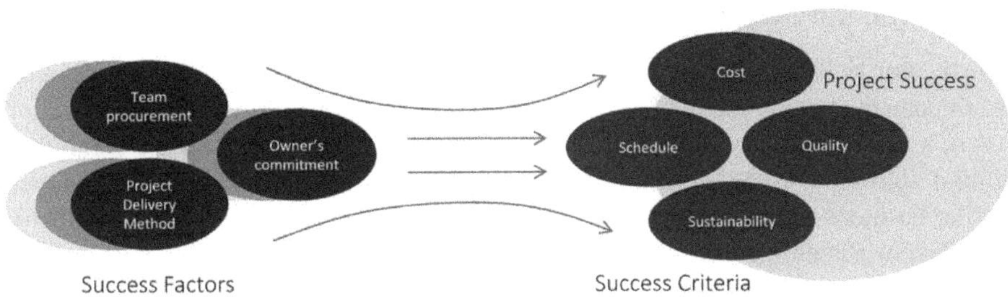

FIGURE 1.1 Project success correlates with success elements and criteria [15].

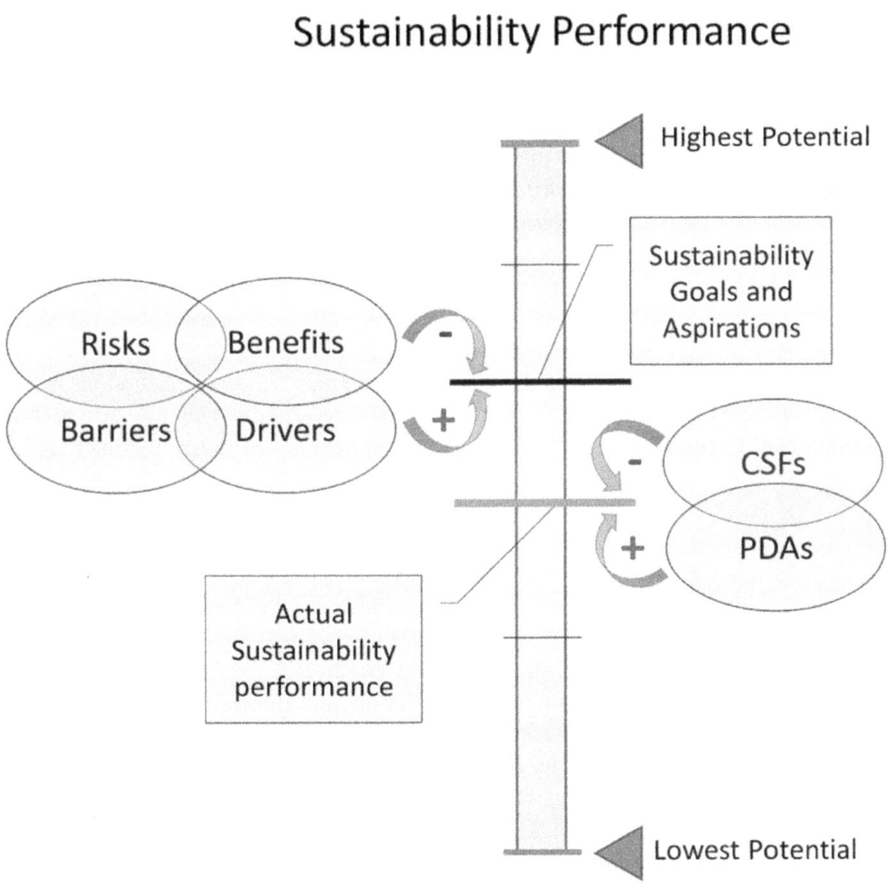

FIGURE 1.2 Research concepts for green building development [15].

importance. GB is assessed on a sustainability continuum. Sustainability potential is lowest at the bottom and highest at the top. Both project goals and actual project sustainability are measured by two horizontal bars on the scale. The sustainability scale's measuring bars are positioned this way because projects often underperform their planned sustainability performance.

1.7 GREEN BUILDING RATINGS

Global green building rating systems assess construction sustainability. Classifications and standards are updated to follow sustainable construction progress. Sustainable construction methods are promoted by "going green" and "environmental sustainability". Official statistics show that buildings use a lot of energy. Building professionals may be uninterested in sustainability. Due to rising energy costs and environmental concerns, sustainable construction characteristics are desired. To reduce pollution and conserve resources, governments created green building rating systems.

Energy-efficient buildings with grading systems improve living conditions and property value. Globally, there are about 600 green rating systems. This was the first tool to assign goal-level grades to building performance for a variety of variables. Energy expenses drop by 6–30% for Building Research Establishment Environmental Assessment Method (BREEAM)-certified buildings and 18–39% for LEED-certified ones. LEED certifications ignore sustainability's entirety.

Therefore, all major green building rating systems update their criteria. Mid-2016 saw LEED v4 rating tool updates after its 2013 release. The main BREEAM UK New Construction change occurred in mid-2014. Green Star Australia and New Zealand released their latest versions in mid-2015 and mid-2016. International sustainable construction standards exist. After two decades of research on green building rating certificates, no systematic examination of each rating system's criteria and upgraded methods has been done. Despite trend and credit analysis of individual rating instruments, a complete comparison has yet to be done.

Green Star NZ is a new method that has increased green building registrations in New Zealand, while BREEAM, LEED, and Comprehensive Assessment System for Built Environment Efficiency (CASBEE) were chosen for their international relevance [16].

1.7.1 BREEAM

BREEAM began in 1990 in Britain. BREEAM International for New Construction 2016 is the latest international project certification. Success percentages total points: Pass: 30%, Good: 45%, Very Good: 55%, Excellent: 70%, Outstanding: 85%. New building project areas include Health and Wellbeing, Energy, Transport, Water, Materials, Waste, Land Use and Ecology, Pollution, and Innovation [17].

The "Environmental Assessment Method" was created to evaluate and certify constructions. Its main goal is to inform the public and real estate market about a building's environmental status. The Building Research Establishment (BRE) and politicians have promoted it as a design tool for 10 years [18].

There are 76 criteria in BREEAM, and each category has its own set of requirements that can be disregarded and vary depending on the type of building and certification procedure. Certain categories have minimum performance standards that must be met regardless of Code level. The necessary criterion is not considered during calculation [19].

1.7.2 LEED

The US Green Building Council (USGBC) introduced LEED in 2000. While there are many different systems for rating "green buildings," this is the most popular. Since its founding, LEED buildings have been associated with energy efficiency. Until recently, energy usage information was scarce to support this claim. The USGBC asked the New Buildings Institute (NBI) to study LEED-certified commercial building energy use in 2006. The March 2008 final analysis found that "average LEED energy use is 25–30% superior to the national average." The NBI is also concerned about LEED energy performance heterogeneity. The NBI report has been hotly criticized since its release.

According to critics, the National Bureau of Investigation's voluntary building owner data gathering may have caused biased sampling. The 2003 Commercial Buildings Energy Consumption Survey (CBECS)' mean energy use intensity (EUI) for all US commercial buildings was compared to the median EUI of LEED-certified buildings, which they say affected statistics. NBI is criticized for focusing primarily on "medium energy" LEED buildings and comparing their median EUI to the mean of all CBECS [20].

Green buildings have been popular in recent years as a solution to environmental challenges in architecture and construction. Resource efficiency and ecological design are used to create and manage a sustainably built environment. LEED is the most popular green building grading system worldwide. The six LEED categories have corresponding rating credits. A project can receive LEED Certified, Silver, Gold, or Platinum certification based on credits earned. Historical LEED projects show how sustainable development technology has changed credit allocation patterns. Decision-makers can use LEED project performance, particularly credit frequency indicators, to gain historical data on awarded credits and insights into credit implementation for future projects with similar goals.

Progressive Building Information Modeling (BIM) and sustainable construction companies have found synergies between these two market developments. Green BIM refers to using BIM tools to achieve sustainability and building performance goals, combining two independent movements into a new practice. BIM's information richness and parametric modeling provide immediate insights into design decisions' effects on building performance. BIM demands an integrated design process (IDP) that incorporates a diversified and dedicated project team to manage high-performance project complexity and sophistication through collaborative and interoperable information sharing. However, little research has been done to examine the business process of BIM deployment or how BIM execution planning in green construction projects may affect project outcomes.

Even though BIM software may not fulfill industry requirements, some significant organisations have implemented green BIM using a comparable set of standard BIM tools, highlighting the epistemic gap in current knowledge. In green construction projects, BIM success is not solely determined by technology. Construction projects may have an impact on the goal of "sustainable building," which they contend is "not impeded by a deficiency of technologies and assessment methods, but rather confronted with organizational and procedural challenges associated with the implementation of new methods" [21].

1.7.3 CASBEE

CASBEE was founded in 2001 by academics, businesses, and local governments in Japan. Certified constructions are scarce because of their Japanese focus. The rating

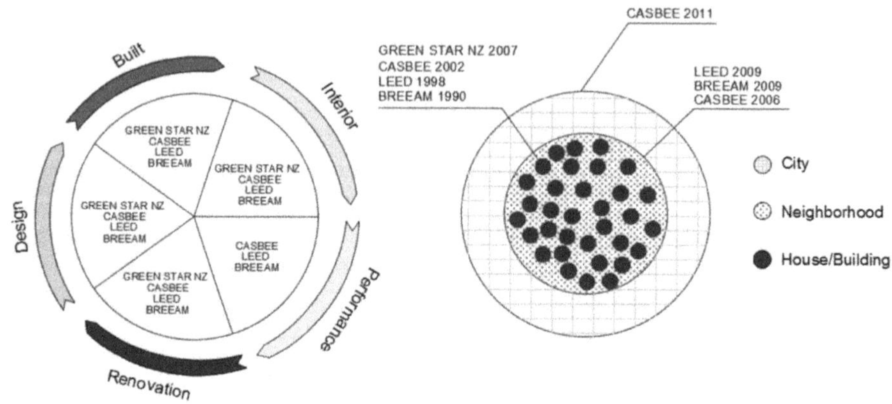

FIGURE 1.3 Overview of green rating tools [22].

with the most experience released a global pilot version in 2015. CASBEE evaluates buildings from design to maintenance using its Buildings, Commercial Interiors, and Temporary Construction guidelines. The CASBEE for Urban Development and Cities manuals evaluate buildings.

1.7.4 Green Star NZ

In 2007, the New Zealand Green Building Council (NZGBC) introduced the Green Star NZ grading system, inspired by the Australian Green Star. The latest grading system is Green Star NZ, the only institution without an operational building evaluation handbook. There are not many certified constructions because the product has only been available for 10 years. However, it has done well, getting 125 certifications. The overview of green rating tools details is presented in Figure 1.3.

1.8 COMPARISON OF GREEN BUILDING RATINGS

Green building rating agencies operate globally and regionally. This organization certifies green buildings through voluntary rating systems. Green building rating techniques vary for new, core and shell, existing, and major changes. Green building rating systems can certify infrastructure, community, and public transportation projects. LEED and BREEAM are available worldwide. Multiple functionalities are built into green building grading systems. It may be used to organize building design, construction, operation, and maintenance in addition to certification. Quality green building rating systems are recognized. These methods help project stakeholders understand sustainable development goals. Plan-based criteria are used to priorities projects during pre- and post-construction. Comparative assessments begin with a credit framework analysis of green building grading systems. Table 1.2 describes scheme credit structures in a conventional way. Because plans must fit local climate, culture, construction traditions, and geography, they are unlikely to be uniform. Each scheme has a distinctive structure for subjective and statistical sustainability target assessment.

TABLE 1.2
Structure of Green Building Rating Schemes [23]

Layers →	Layer 01	Layer 02	Layer 03	Layer 04	Layer 05
Layer naming →	Category Group	Category	Indicator		
Purpose →	Comprehensive domains for delineating sustainable development	A subset of the Category Group	Variables for quantitative and qualitative evaluation		
Scheme ↓					
1. Green Globes	—	Areas	Subareas	—	
2. LEED USA	—	Quality Sections	Indicator	—	
3. BREEAM	—	Environ. Sections	Indicator	—	
4. DGNB System	Topic	Criteria Group	Criteria	—	
5. GRIHA	—	Category	Indicator	—	
6. IGBC System	—	Category	Indicator	—	
7. BEAM Plus	—	Category	Credit requirement	—	
8. Green Mark	Sections	Criteria	Indicator	—	
9. CASBEE	Theme	Category	Indicator 1	Indicator 1.1	Indicator 1.1.1
10. Green Star	—	Category	Sub-category	Credit Criteria	

Comparative research has been used by many researchers worldwide. Comparison levels include general, category group, category, indicator, key category, and triple-bottom-line. A brief description of each comparison type follows:

Level 1: A comprehensive comparison encompasses the background, history, aims, certification levels, and scope of coverage.

Level 2: Category group comparison: "Category group" encompasses sustainable development's environment, economy, and society.

Level 3: Category comparison: Sustainability encompasses site, energy, water, environmental quality, and human health and well-being.

Level 4: Indicator comparison: Green building rating methods conclude with metrics that quantify and evaluate built environment sustainability.

Level 5: Comparison of key categories: Differences in green construction plan finance structures make direct comparison impossible. Green building ratings recognize energy, human considerations, and management. Each primary category's measurements are then determined. Aggregating all indicator scores in a key category determines their importance. All schemes use this method, and key category findings are compared.

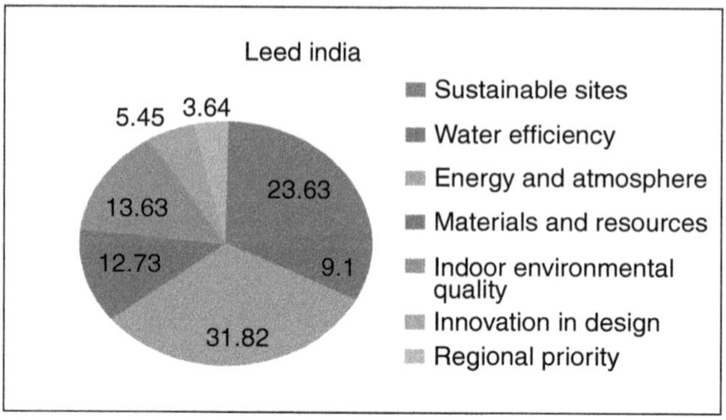

- Credit points under different categories

1. 100 possible points under the five core categories – SS, WE, EA, MR and IEQ
2. 6 possible points under 'Innovation in design'
3. 4 possible points under 'Regional priority'
4. Total possible points achieved are 110.

Credit point for different levels of certification

Certified	40–49 points
Silver	50–59 points
Gold	60–79 points
Platinum	80 points and above

FIGURE 1.4 LEED India evaluation criteria [24].

Level 6: Comparison of the triple bottom line: The overall score is divided between ecology, economics, and society across schemes. The purpose of the credit determines how each indicator's score is distributed across the sustainability pillars. Upon evaluating the indicators' intents, it is clear that they often served various goals with different relevance. Three objectives are ranked using the Analytical Hierarchy Process (AHP). The AHP pairwise comparison method is used [23]. Figure 1.4 shows LEED India evaluation criteria. LEED for India's commercial and institutional buildings promotes high-performance, healthful, durable, cost-effective, and sustainable buildings.

Green Rating for Integrated Habitat Assessment (GRIHA) improves building design to reduce energy consumption and increase energy efficiency while maintaining comfort. Figure 1.5 lists GRIHA evaluation criteria. Building performance is considered throughout its life cycle.

1.9 SUSTAINABLE MATERIALS

Construction uses mostly wood, concrete, and steel. Their performance must be assessed during project building. In all construction projects, these materials are used extensively. All buildings need these materials. Informing designers during the selection process is the main goal of evaluating these materials. They can find alternatives or minimize use wherever possible [25].

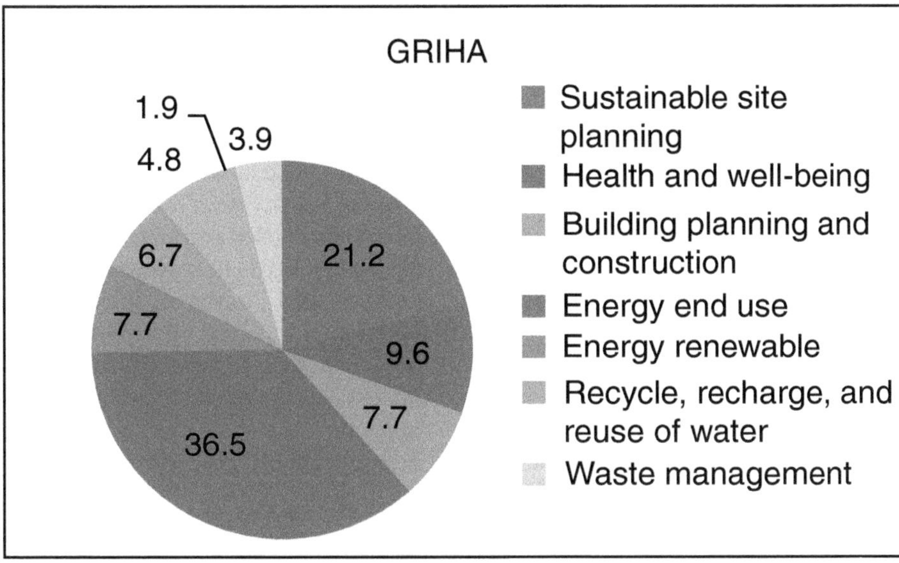

- Credit points under different categories

5. 100 possible points under the eight categories
6. 4 possible points under 'Innovation in design'
7. Total possible points achieved are 104.

- Points achieved GRIHA rating

One star	50–60
Two stars	61–70
Three stars	71–80
Four stars	81–90
Five stars	91 points and above

FIGURE 1.5 GRIHA evaluation criteria [24].

1.10 GREEN BUILDING MATERIALS

Green building materials reduce environmental impact and help builders achieve sustainability. Green building materials are classified into four categories [26]:

1. Materials chosen for their vast coverage of big surfaces such as walls, floors, and ceilings.
2. Material composition representation.
3. Expectation of high ozone reactivity and product yields, and
4. Recent small-scale chamber studies using comparable components.

Renewable, natural, or abundantly available natural materials requiring minimal processing—such as fast-growing renewable resources, and sustainably sourced materials approved by an independent organization like the Forest Stewardship Council (FSC)—include bamboo, cork, and FSC-certified wood.

Local products reduce transportation and greenhouse gas emissions. Rammed earth and crushed earth bricks use earth. Bamboo is a great building material, but it deteriorates with time. A common example is bamboo, which comes from Asia. Enviro board, papercrete, and wood–plastic composite are recycled construction materials. Despite not being recycled, metals, wood, plastic, and glass can be recycled or reused.

Salvage, deconstructed, remanufactured, or refurbished—retrieving goods before disposal reduces the need to make furniture and fittings, including cupboards, doors, windows, and flooring. Durable materials need less replacement. Stone, copper roofing, hardwood floors, and fine furniture and cabinets are durable, although other materials consider plastic "green" due to its longevity.

Efficient producers use less energy, emit less greenhouse gas, and produce less trash than typical manufacturers [27].

1.11 CEMENT-BASED ELECTROMAGNETIC FUNCTIONAL MATERIALS

As people priorities health and environmental sustainability, eco-friendly construction materials that absorb and filter electromagnetic waves will become essential. Due to its availability, environmental flexibility, and ease of preparation, cement is a versatile construction material.

Urbanization and population expansion drive cement construction demand. Sustainable architecture requires cement-based construction materials that absorb and shelter electromagnetic waves.

Conventional cement is made by pulverizing limestone, sandstone, silicate minerals, iron powder, and certain slags in a specific ratio. Pure cement absorbs and shields electromagnetic radiation poorly. Additives improve cement absorption and protection. Figure 1.6 shows bulk ferrite, nano-Fe_3O_4 magnetic fluid, and flake carbon as green cement-based electromagnetic functional materials.

Renewable energy and resources should be used where possible. The effects are assessed throughout product life. Sustainable building materials must have few environmental and health risks. The following criteria must also be met:

- minimal embodied energy
- energy efficiency
- waste generation reduction or elimination
- low toxicity
- water conservation
- cost-effectiveness

Sustainable building materials provide distinct advantages to building owners, including [29]:

- diminished maintenance and replacement expenses
- energy efficiency

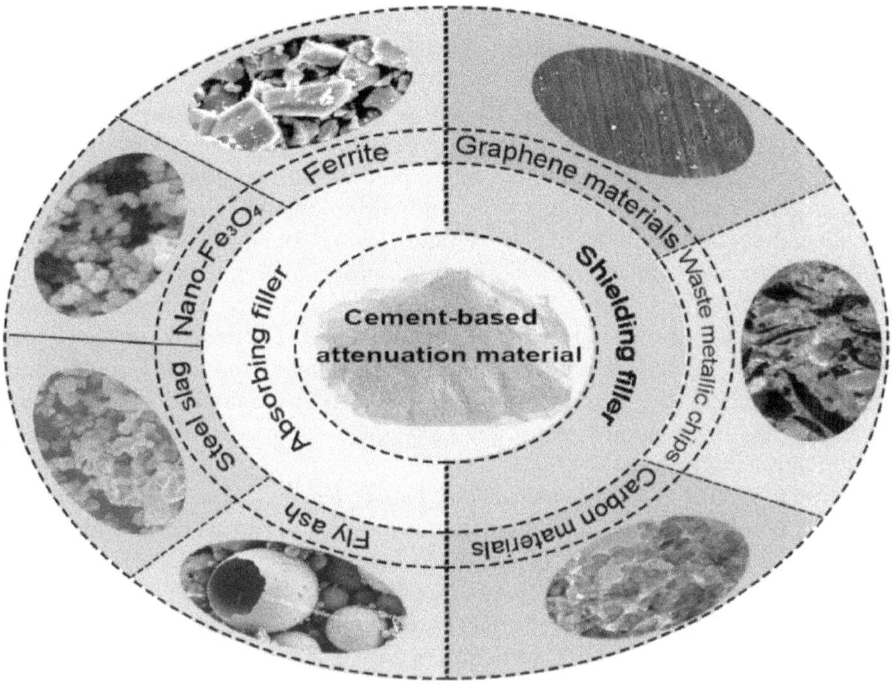

FIGURE 1.6 Typical cement-based electromagnetic functional materials' classification and microstructure [28].

- enhanced occupant health and productivity
- reduced expenditures related to alterations in space layouts
- enhanced design flexibility

Lack of processing limits the strength and durability of natural materials. Carbon emissions result from energy use in natural material processing and transportation. The development of energy-efficient building materials and goods will be necessary to reduce carbon emissions. Here are summaries of building materials and methods [30]:

1. Composite cements
2. Stabilized mud blocks for construction
3. Compacted fly ash blocks
4. Compressed earth walls
5. Green roofing systems.

1.11.1 Composite Cements

Use non-hazardous waste to partially replace clinker creates sustainable cements. Recycled rubbish can become cement components and sustainable binders with environmental resilience. The production of cement is energy-intensive and environmentally harmful. Clinker, the main component of most cement products, requires a high-temperature process (approximately 1500°C), a lot of natural solid fuels, and a lot of CO_2 emissions.

Based on comprehensive research, the goal is to reduce clinker content and develop viable substitutes for sustainable cements. There is growing interest in waste-derived cement ingredients, often called additional cementing materials. Conventional cements are subject to limitations and chemical, physical, and mechanical requirements specified by international standards. Thus, novel sustainable cements, which replace clinker with supplementary cementitious materials (SCMs) in amounts of 20–35%, must be classified by their mechanical strength class, setting time, soundness, and presence of chemical substances that may damage the hydration process. The testing of novel binders is too frequently limited to mechanical properties, which leads to inadequate characterization. Nevertheless, this does not ensure that new sustainable cements would be used safely in construction [31].

Extra-cementitious concrete provides better mechanical and durability than control concrete. In many cases, carbonated concrete performs poorly. Figure 1.3 shows the carbonation depth of Ordinary Portland Cement (OPC) and mixed concrete under natural and accelerated carbonation settings. Figure 1.7 shows that replacing OPC with different pozzolanic components makes concrete more carbonation-prone, increasing carbonation depth. Under regulated settings of 65% relative humidity, 20°C, and 5% CO_2, replacing 70% OPC with Fly Ash (FA) increases carbonation depth by 74%. Another study found that replacing OPC with 70% FA led to an increase in carbonation depth by 4.8 mm at 6.5% CO_2, 65% relative humidity, and 30°C. Decalcification reduces C–S–H gel and alkalinity, thus increasing carbonation depth. Silicon fume (SF) replaced 10% OPC to reduce carbonation depth by 18%. The decrease is due to SF concrete changing all large pores into smaller ones. A denser concrete is generated, increasing carbonation resistance [32–41].

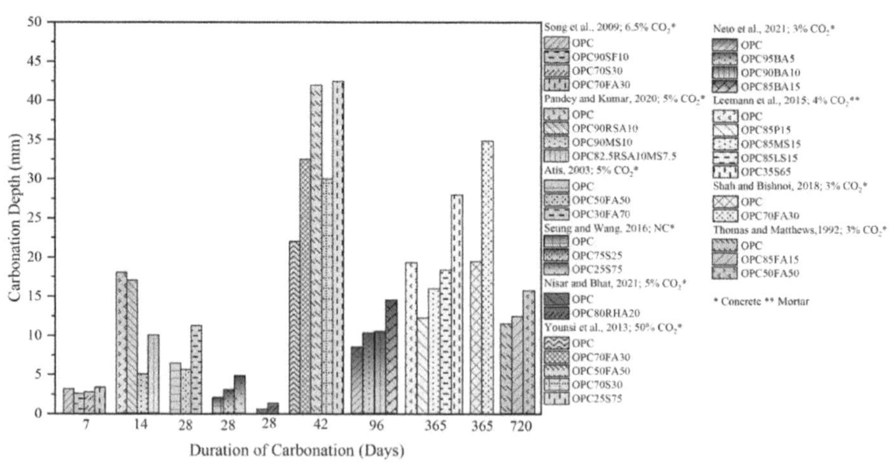

FIGURE 1.7 Carbonation depth of OPC and blended concrete [42].

1.11.2 Stabilized Mud Blocks for Construction

Build load-bearing masonry with these blocks. Cement–soil mortar is used in stabilized mud blocks (SMB) masonry. Manual block-making machines made 305,143,100-mm blocks. The block density was regulated by mixing soil, sand, and cement in a mechanical mixer and pouring a precise weight into the machine mold. These blocks cured for 28 days under damp burlap. For block making, red loamy soil, sand, and Portland cement were used. The soil had liquid limit (LL) of 45% and a plasticity index (PI) of 24. Studies show that cement-stabilized soil mixture blocks work best with sandy soils. Due to the weight ratio of 1:2 for red loamy soil and sand, a sandy soil mixture with 9% clay, 17.7% silt, and 73.3% sand was generated [43].

A manual or mechanized press can compact the soil-stabilizer mixture. The press must be strong enough to compact the block. Using a manual press reduces electricity and diesel needs. Fly ash, quarry dust from crushers, granite fines, stone dust from stone cutting and polishing firms, and other mining or industrial wastes can be used with natural soil [44].

1.11.3 Compacted Fly Ash Blocks

Stabilizers have been studied extensively to improve mud block compressive strength. Cement stabilizes, increasing compressive strength. Use mud blocks to reduce or eliminate cement. Lime stabilization has not led to strength gains above 5%. Stabilization requires more cement. Many studies have used FA and Ground Granulated Blast Furnace Slag (GGBS) as partial mud substitutes. In typical building construction, IS 1725–1982 standard specifies soil-based blocks.

Fly ash decreases pore interconnectivity by creating effective cementitious materials with lime, which affects water absorption rates in lime-stabilized mud blocks more than GGBS. Stepwise fly ash addition increases compressive strength while retaining 10% GGBS. Stepwise addition of GGBS while retaining fly ash at 10% increases compressive strength. The pozzolanic activity of GGBS-lime may not have affected water absorption, but it may have affected strength growth after 28 days.

On days 7, 14, and 28, wet and dry strength tests were performed. 7-, 14-, and 28-day wet compressive strength blocks were submerged in potable water for 48 hours. Compression strength testing followed. Results for the control mix are shown in Figure 1.8 [45].

1.11.4 Compressed Earth Walls

Stabilizing earth with cement and ramming it at appropriate moisture content creates cement-stabilized rammed earth (CSRE), a strong, durable building material with minimal embodied energy. CSRE has the highest dry density at an ideal moisture level, increasing compressive strength.

The ratios of gravel, sand, silt, and clay affect the strength of laterite soil-based CSRE. To determine their suitability for CSRE construction, sandy, clayey, and gravelly laterite soils were tested. Laboratory sieve analysis and an on-site jar test estimated primary element distribution for soil selection [1]. Three types of

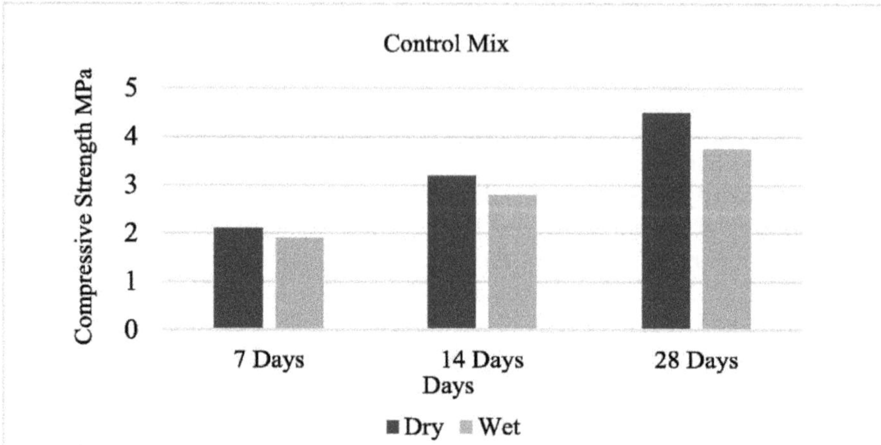

FIGURE 1.8 Compression strength test result of control mix [45].

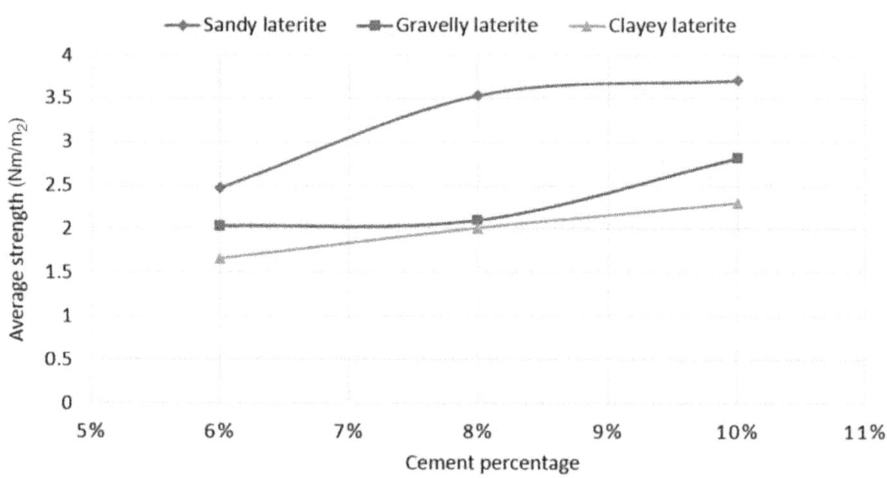

FIGURE 1.9 Variation of strength of CSRE with soil type [46].

lateritic soil were used to make wall panels and tested for compressive strength 28 days later. Each soil type's mean compressive strength was calculated from duplicate wall panels exposed to failure loading. Figure 1.9 shows CSRE wall panel strength with different laterite soils. Gravelly and clayey laterite had 60% gravel and 40% fines, respectively. To maximize CSRE strength, soil particles (clay and silt) should be 25–35%.

Rammed-earth buildings have many structural and functional advantages. It is simple and inexpensive to build. Rammed-earth buildings are suitable for cold and warm climates since they require less maintenance. However, water saturation reduces strength, wind and rain erosion, and dimensional stability is poor.

Dimensional stability measures wind and water-induced linear dimensional changes. Stabilization—physical (aggregate grading), mechanical (compaction), or chemical (cement)—can reduce the above drawbacks [47].

1.11.5 Green Roofing Systems

Green roofs improve building energy efficiency and the environment. Green roofs are a major solution to building and metropolitan environmental challenges. Green roofs have several benefits beyond a pleasant atmosphere. This improves stormwater management and reduces air pollution and noise. By photosynthesizing carbon dioxide into oxygen, green roofs increase urban plant and animal biodiversity and reduce carbon emissions. Green roofs improve heat transfer through roofing systems, enhancing energy efficiency. Green roofs reduce summer temperatures, effectively chilling fluid before it returns to the chiller, improving HVAC system performance. The reduced temperature improves nearby photovoltaic cells. Green roofs reduce heat stress, making roofing membranes more durable.

Green roofs reduce urban heat islands at the municipal level. The soil depth of extensive green roofs is less than 10–15 cm, whereas intensive green roofs are deeper than 15–20 cm. Extensive green roofs can be retrofitted without reinforcement because they carry no extra weight. Figure 1.10 compares green and conventional roofs.

The heating and cooling demand (Q_h and Q_c, respectively).

The temperature of the exterior surface of the roof slab (T_r).

The indoor air temperature (T_i).

T_g and T_f are the soil and foliage temperatures (kelvin), respectively,

The heat flux inside of the building (q_{ri}'').

Green roof materials must be carefully selected to be environmentally friendly. Plant diversity and coverage were important; therefore, substrate materials and vegetation must be chosen. Layer composition and quality affect green roof performance: the appropriate combination can improve winter efficiency, while the right plant can promote cooling. Their choice may affect stormwater discharge quality. On vegetation, evapotranspiration is essential. Evapotranspiration turns surface heat flux into latent heat emissions for green roofs. The green roof's thermal performance improved with irrigation, promising to increase building energy efficiency and be cheaper than air conditioners. Additional green roof layers add thermal inertia and insulation, reducing summer heat transfer inside [49].

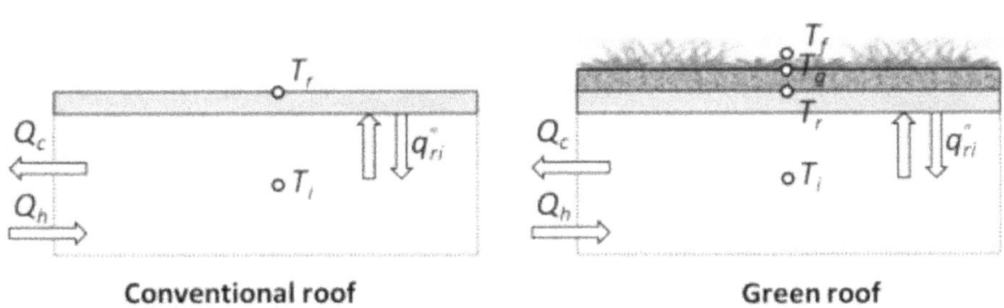

FIGURE 1.10 The comparison of conventional and green roofs [48].

1.12 ECONOMIC BENEFITS

Green buildings are becoming crucial to a sustainable society. Promoting green building is essential to sustainable development, which is a global priority due to rising environmental concerns. Technology improves material well-being and cultural wealth, while pollution, global warming, and the depletion of natural resources continue. Resource conservation and environmental protection are essential for sustainable human progress and quality of life.

Energy, water, land, material, and environmental quality are the main environmental benefits of green building. Energy-efficient green construction reduces coal use and acid rain emissions. In accordance with local contextual conditions, green buildings maximize natural light and ventilation, exploit solar, geothermal, and other renewable energy sources, and reduce reliance on artificial lighting and air conditioning. Green building exteriors use thermal insulation and improved sealing for multi-story windows, roofs, and walls to reduce energy usage. Sustainable buildings may reduce energy use by 70–75%, while ideal designs can cut it by 80%. Green buildings conserve more water and reduce sewage discharge than typical construction by using water-saving appliances, water system upgrades, rainwater harvesting, and purification systems to recycle water features.

Based on the "Unified Standard for Green Building," green buildings must promote intensive land use, optimize land usage, and use industrial byproducts like fly ash and slag to conserve land and materials. Reduce fresh adobe consumption, utilize green building materials wisely, and protect land resources. Environmental quality green buildings enhance indoor air quality, temperature, lighting, and acoustics, enhancing occupant comfort, productivity, and reducing sickness [50].

Economic gains are only directly related to school buildings' energy and water use. Numerous measurement and assessment tools require technological criteria to reduce energy and water usage and reduce environmental impacts. Cutting down on operations and maintenance costs, which are much higher than building costs, may be a good reason to build in a sustainable way. A large initial investment in educational buildings may be motivated by low operating costs [51].

REFERENCES

1. Yousif, Y., Misnan, M. S., & Ismail, M. Z. (2023, December). The influence of labelled green building materials on the performance of green construction projects. In *IOP Conference Series: Earth and Environmental Science* (Vol. 1274, No. 1, p. 012028). IOP Publishing.
2. Olubunmi, O. A., Xia, P. B., & Skitmore, M. (2016). Green building incentives: A review. *Renewable and Sustainable Energy Reviews*, *59*, 1611–1621.
3. Pramanik, P. K. D., Mukherjee, B., Pal, S., Pal, T., & Singh, S. P. (2021). Green smart building: Requisites, architecture, challenges, and use cases. In *Research anthology on environmental and societal well-being considerations in buildings and architecture* (pp. 25–72). IGI Global.
4. Samari, M., Godrati, N., Esmaeilifar, R., Olfat, P., & Shafiei, M. W. M. (2013). The investigation of the barriers in developing green building in Malaysia. *Modern Applied Science*, *7*(2), 1.

5. Alpha Sand. www.alphasand.in/blog/green-building-concept-in-india
6. Davis. A. (2001). Barriers to building green. Retrieved from www.greenbiz.com/blog/2001/11/01/barriers-building-green
7. Darko, A., & Chan, A. P. (2017). Review of barriers to green building adoption. *Sustainable Development*, *25*(3), 167–179.
8. Cole, L. B. (2019). Green building literacy: A framework for advancing green building education. *International Journal of STEM Education*, *6*, 1–13.
9. Cole, R. J. (2000). Cost and value in building green. *Building Research & Information*, *28*(5–6), 304–309.
10. Leontev, M. (2021). Analysis of obstacles to green building projects: the experience of Russia and Europe. In *E3S Web of Conferences* (Vol. 258, p. 09069). EDP Sciences.
11. Leskinen, N., Vimpari, J., & Junnila, S. (2020). A review of the impact of green building certification on the cash flows and values of commercial properties. *Sustainability*, *12*(7), 2729.
12. Hwang, B. G., & Tan, J. S. (2012). Green building project management: obstacles and solutions for sustainable development. *Sustainable Development*, *20*(5), 335–349.
13. Mohanty, B. (2012). *Low Carbon Green Growth Roadmap for Asia and the Pacific*. United Nation. Retrieved From https://repository.unescap.org/bitstream/handle/20.500.12870/4013/ESCAP-2012-PB-Buildings-policyrecommendations.pdf?sequence=1& isAllowed=y
14. Mackley, C. J., & Milonas, S. (2001). Knowledge transfer and green building challenge. *Building Research & Information*, *29*(5), 346–354.
15. Ahmad, T., Aibinu, A. A., & Stephan, A. (2019). Managing green building development–A review of current state of research and future directions. *Building and Environment*, *155*, 83–104.
16. Doan, D. T., Ghaffarianhoseini, A., Naismith, N., Zhang, T., Ghaffarianhoseini, A., & Tookey, J. (2017). A critical comparison of green building rating systems. *Building and Environment*, *123*, 243–260.
17. Awadh, O. (2017). Sustainability and green building rating systems: LEED, BREEAM, GSAS and Estidama critical analysis. *Journal of Building Engineering*, *11*, 25–29.
18. Schweber, L., & Haroglu, H. (2014). Comparing the fit between BREEAM assessment and design processes. *Building Research & Information*, *42*(3), 300–317.
19. Rezaallah, A., Bolognesi, C., & Khoraskani, R. A. (2012, May). LEED and BREEAM; Comparison between policies, assessment criteria and calculation methods. In Rogerio Amoeda (Ed.), *Proceedings of the 1st International Conference on Building Sustainability Assessment (BSA 2012), Porto, Portugal* (pp. 23–25).
20. Scofield, J. H. (2009). Do LEED-certified buildings save energy? Not really… . *Energy and Buildings*, *41*(12), 1386–1390.
21. Wu, W., & Issa, R. R. (2015). BIM execution planning in green building projects: LEED as a use case. *Journal of Management in Engineering*, *31*(1), A4014007.
22. Doan, D. T., Ghaffarianhoseini, A., Naismith, N., Zhang, T., Ghaffarianhoseini, A., & Tookey, J. (2017). A critical comparison of green building rating systems. *Building and Environment*, *123*, 243–260.
23. Varma, C. S., & Palaniappan, S. (2019). Comparision of green building rating schemes used in North America, Europe and Asia. *Habitat International*, *89*, 101989.
24. Arya, A., & Sharma, R. L. (2022). Strategies for green building rating in India: A comparison of LEED and GRIHA criteria. *Materials Today: Proceedings*, *57*, 2311–2316.
25. Sahlol, D. G., Elbeltagi, E., Elzoughiby, M., & Abd Elrahman, M. (2021). Sustainable building materials assessment and selection using system dynamics. *Journal of Building Engineering*, *35*, 101978.

26. Gall, E., Darling, E., Siegel, J. A., Morrison, G. C., & Corsi, R. L. (2013). Evaluation of three common green building materials for ozone removal, and primary and secondary emissions of aldehydes. *Atmospheric environment*, *77*, 910–918.
27. Green Building Canada. https://greenbuildingcanada.ca/green-building-guide/green-building-materials/
28. Cao, M. Q., Liu, T. T., Zhu, Y. H., Shu, J. C., & Cao, M. S. (2022). Developing electromagnetic functional materials for green building. *Journal of Building Engineering*, *45*, 103496.
29. Hussain, A., & Kamal, M. A. (2015). Energy efficient sustainable building materials: An overview. *Key Engineering Materials*, *650*, 38–50.
30. Venkatarama Reddy, B. V. (2009). Sustainable materials for low carbon buildings. *International Journal of Low-Carbon Technologies*, *4*(3), 175–181.
31. Bignozzi, M. C. (2011). Sustainable cements for green buildings construction. *Procedia Engineering*, *21*, 915–921.
32. Song, H., Niu, D., & Li, C. (2009). Carbonation test of concrete containing mineral admixtures, Kuei Suan Jen Hsueh Pao. *Journal Chinese Ceramics Society, 37,* 2066–2070.
33. Pandey, A., & Kumar, B. (2020). Investigation on the effects of acidic environment and accelerated carbonation on concrete admixed with rice straw ash and microsilica, *Journal of Building Engineering*, *29*, 101125. https://doi.org/ 10.1016/j.jobe.2019.101125.
34. Atis, C. D. (2003). Accelerated carbonation and testing of concrete made with fly ash. *Construction Building Materials*, *17*, 147–152, https://doi.org/10.1016/S0950-0618(02)00116-2.
35. Han-Seung, L., & Wang, X. Y. (2016). Evaluation of compressive strength development and carbonation depth of high volume slag-blended concrete. *Construction Building Materials*, *124*, 45–54, https://doi.org/10.1016/j.conbuildmat.2016.07.070.
36. Nisar, N., & Bhat, J. A. (2021). Effect of rice husk ash on the carbonation depth of concrete under different curing ages and humidity levels. *Journal of Materials Civil Engineering*, *33*, 04020427. https://doi.org/10.1061/(asce)mt.1943-5533.0003520.
37. Younsi, A., Turcry, P., Aït-Mokhtar, A., & Staquet, S. (2013). Accelerated carbonation of concrete with high content of mineral additions: effect of interactions between hydration and drying. *Cement Concrete Research*, *43,* 25–33, https://doi.org/10.1016/j.cemconres.2012.10.008.
38. da J., Neto, S.,. de França, M. J. S., de Amorim Júnior, N. S., & Ribeiro, D. V. (2021). Effects of adding sugarcane bagasse ash on the properties and durability of concrete. *Constructions Building Materials*, *266.* https://doi.org/10.1016/j.conbuildmat.2020.120959.
39. Leemann, A., Nygaard, P., Kaufmann, J., & Loser, R. (2015). Relation between carbonation resistance, mix design and exposure of mortar and concrete. *Cement Concrete Composition*, *62*, 33–43, https://doi.org/10.1016/j.cemconcomp.2015.04.020.
40. Shah, V., & Bishnoi, S. (2018). Carbonation resistance of cements containing supplementary cementitious materials and its relation to various parameters of concrete. *Construction Building Materials*, *178*, 219–232, https://doi.org/10.1016/j.conbuildmat.2018.05.162.
41. Thomas, M. D. A., & Matthews, J. D. (1992). Carbonation of fly ash concrete. *Magnite Concrete Research*, *44*, 217–228. https://doi.org/10.1680/macr.1992.44.160.217.
42. Charitha, V., Athira, G., Bahurudeen, A., & Shekhar, S. (2022). Carbonation of alkali activated binders and comparison with the performance of ordinary Portland cement and blended cement binders. *Journal of Building Engineering*, *53*, 104513.

43. Venkatarama Reddy, B. V., & Gupta, A. (2006). Strength and elastic properties of stabilized mud block masonry using cement-soil mortars. *Journal of Materials in Civil Engineering*, *18*(3), 472–476.
44. Kamal, M. A. (2022). Potential of stabilized mud blocks as a sustainable material for building construction. *Journal of Construction and Built Environment*, *2*(1), 1–15.
45. Soundarya, N. (2021). Effect of fly ash and GGBS on lime stabilized mud block. *Materials Today: Proceedings*, *47*, 4636–4640.
46. Kariyawasam, K. K. G. K. D., & Jayasinghe, C. (2016). Cement stabilized rammed earth as a sustainable construction material. *Construction and Building Materials*, *105*, 519–527.
47. Jayasinghe, C., Fonseka, W. M. C. D. J., & Abeygunawardhene, Y. M. (2016). Load bearing properties of composite masonry constructed with recycled building demolition waste and cement stabilized rammed earth. *Construction and Building Materials*, *102*, 471–477.
48. Jaffal, I., Ouldboukhitine, S. E., & Belarbi, R. (2012). A comprehensive study of the impact of green roofs on building energy performance. *Renewable Energy*, *43*, 157–164.
49. Bevilacqua, P. (2021). The effectiveness of green roofs in reducing building energy consumptions across different climates. A summary of literature results. *Renewable and Sustainable Energy Reviews*, *151*, 111523.
50. Ding, W. X., Zhao, Y. H., & Ji, X. (2013). Analysis of the environmental benefits of green building. *Applied Mechanics and Materials*, *368*, 1135–1138.
51. Meron, N., & Meir, I. A. (2017). Building green schools in Israel. Costs, economic benefits and teacher satisfaction. *Energy and Buildings*, *154*, 12–18.

2 Bioclimatic Design

2.1 STRATEGIES OF DESIGN

Bioclimatic design leverages local climate conditions, such as sunlight, wind patterns, vegetation, and natural terrain, to optimize building heating, cooling, and lighting, reducing energy consumption and enhancing occupant comfort. Effective bioclimatic strategies significantly lower energy use and carbon emissions, making them vital components of sustainable architecture practice. Energy savings and carbon footprint reduction depend on building design and climate. Better building envelopes, energy-efficient appliances, bioclimatic techniques, and efficient materials for building design strategies have been important for net-zero energy in the previous two decades. Local renewable resources should be used in sustainable buildings to reduce utility-scale energy demand.

Bioclimatic design gives building occupants visual and thermal comfort from the climate. Thus, bioclimatic design improves indoor air quality (IAQ), thermal comfort, energy efficiency, and carbon footprint through solar light and heat, building orientation and compactness, envelope materials, ventilation systems, water and plants, and renewable energies. Climate factors affect bioclimatic design methodologies; therefore, buildings must be climate-adaptive to increase energy efficiency without compromising current living standards [1].

Building sustainability has improved with bioclimatic design. Due to energy accessibility, their economic benefits, which are the main reasons building owners utilize such approaches, are questionable.

Key bioclimatic design elements [2]:

1. Building envelope thermal insulation and airtightness
2. Solar winter heating. This is feasible with passive solar heating.
3. Natural ventilation (NV) and passive cooling reduce building heat.
4. Shading and plants protecting the building from summer warmth.
5. Year-round solar day lighting.
6. Ensuring inhabitants' indoor and outdoor comfort.

 DOI: 10.1201/9781003752837-2

2.1.1 Proper Thermal Insulation of the External Envelope of the Building and Air-Tight Structures

Monitoring was limited to the as-built solution. Energy Plus dynamic software simulated thermal behavior and calibrated a virtual model using measured values. The generated model approximates observed data effectively as interior surface temperatures deviated by 1.3% and outdoor surface temperatures by 2.7%.

The measured and simulated ground-floor south wall surface temperatures are shown in Figure 2.1.

The two building levels' external, interior, and air temperatures during the summer testing campaign are shown in Figure 2.2. On the ground and first floors, there are solid and semisolid brick walls.

Minimum external surface temperatures occur at 5:00 a.m. for both walls, with the first-floor lightweight wall having a lower value of about 2°C. As sun radiation increases, both walls' external surface temperatures rise in the following hours. They have identical plaster finishes. Peak temperatures occur at 11:00 a.m., with the low inertia wall having values around 2°C higher than the other wall. The difference is due to different outgoing heat flux values for the two walls.

As a result of the entrance of outside air, opening the windows equalizes the temperature of the interior air on two levels: 25°C in the morning and 31°C at the peak hours. Values change when July 30th is a day with closed windows. Under the open layout, the temperature at night reaches 29°C on the top floor and 28°C on the ground

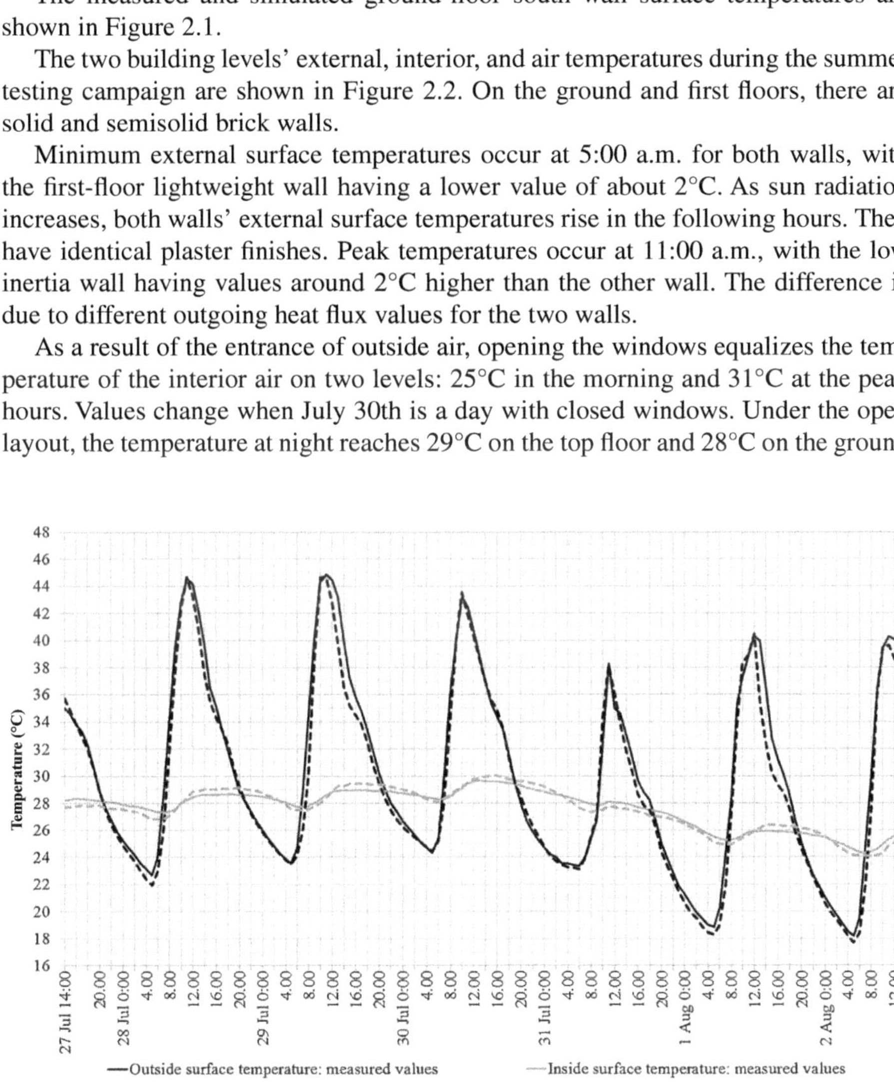

FIGURE 2.1 Comparison between measured and simulated temperatures demonstrating the accuracy of the calibrated energy simulation model [3].

floor. Daytime lows of 29°C are sustained, a 2°C decrease when contrasted with a vented ground floor. On the other hand, when compared to the open layout, the upper level gets around 1°C too hot.

It is understood that the disparity occurs because the irradiative contribution of extra constructive elements reduces when windows are open. The ground floor gets too cold when the windows are closed, while the first floor gets too hot. A two-floor dataset was utilized in a dynamic simulation together with a calibrated model. For determining if the observed discrepancies are caused by boundary conditions, the two walls were positioned at the same building level (ground floor).

An adiabatic layer removed ground-floor heat flow in a later variant. This change made the results independent of the story and highlighted the envelope mass contribution. Standard recommendations were followed for internal load programs [3].

2.1.2 The Use of Solar Energy for Heating in Winter Time

This can be achieved by a proper passive solar heating system. Passive solar heating and NV use similar methods. The buoyancy effect, caused by air temperature differential and density difference at the input and outflow, is the main factor affecting airflow rate. Facades are usually built to trap or store heat or allow airflow to cool the building.

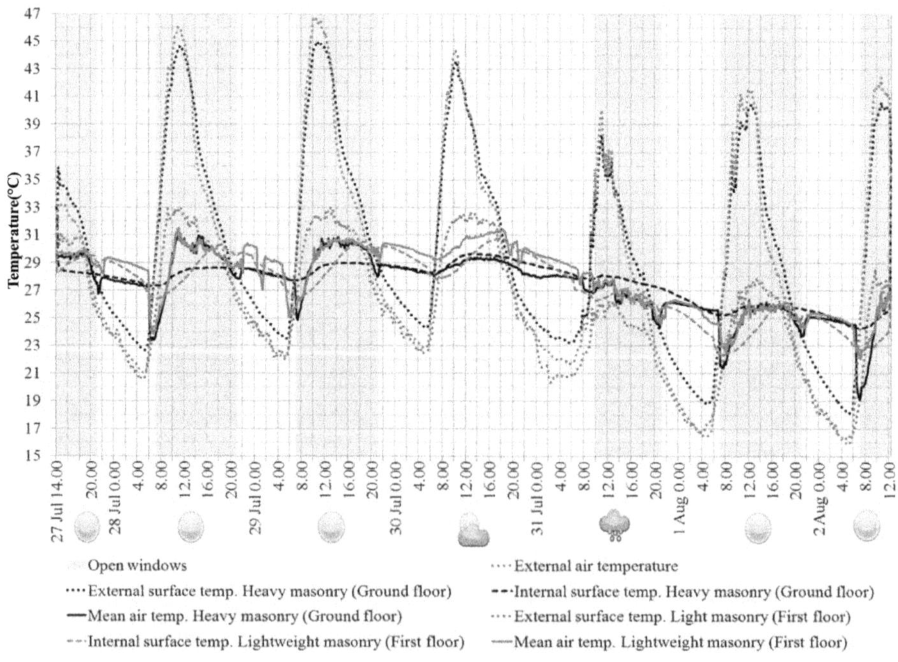

FIGURE 2.2 Internal and external wall surface temperature profiles for different floor configurations and varying ventilation conditions [3].

2.1.2.1 Trombe Wall

The typical Trombe wall has a large wall covered in exterior glazing and an air channel in between, excluding dampers A and B (Figure 2.3). Glazing on the huge wall preserves solar energy. Conduction transfers energy into the room through the wall. The lower wall vent brings cooler room air into the channel. The buoyancy effect raises this air after the wall heats it. Through the higher wall vent, hot air returns to the room.

2.1.2.2 Solar Chimney

By converting heat energy into airflow kinetic energy, the solar chimney helps buildings circulate air. The air density differential between the chimney entrance and outflow controls airflow through the solar chimney. Fan-directed hot air into the building allows cooling and heating.

A solar chimney attached to a wall works like a Trombe wall. The technology passively heats the space by distributing solar collector-heated air. Solar chimneys use NV to cool indoors in cold or mild temperatures.

2.1.2.3 Unglazed Transpired Solar Façade

A façade design made of metal sheets with perforations that collect heat, heats the air without a heat storage facility. This solar collector schematic is shown in Figure 2.4.

2.1.2.4 Solar Roof

Water firms, roof ponds, roof gardens, and thermal insulation are passive roof cooling solutions. In climates with high solar altitude, solar roof ventilation may outperform Trombe walls. Solar energy collection increases with roof collectors' bigger surface area, raising the air exit temperature. With a roof solar collector system alone,

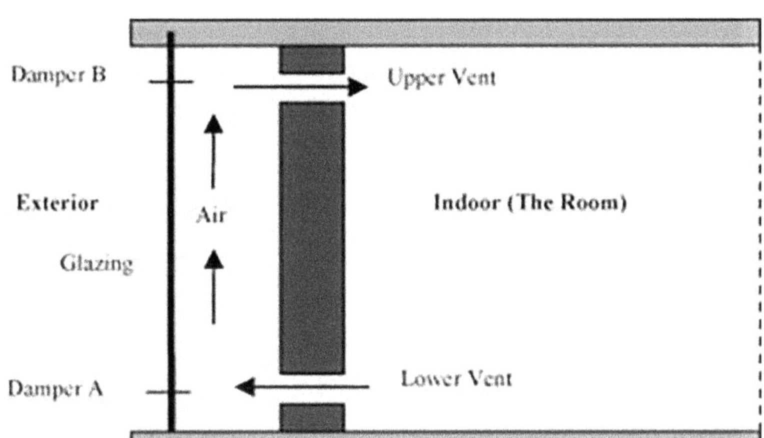

FIGURE 2.3 Schematic representation of a classical Trombe wall configuration (excluding dampers) [4].

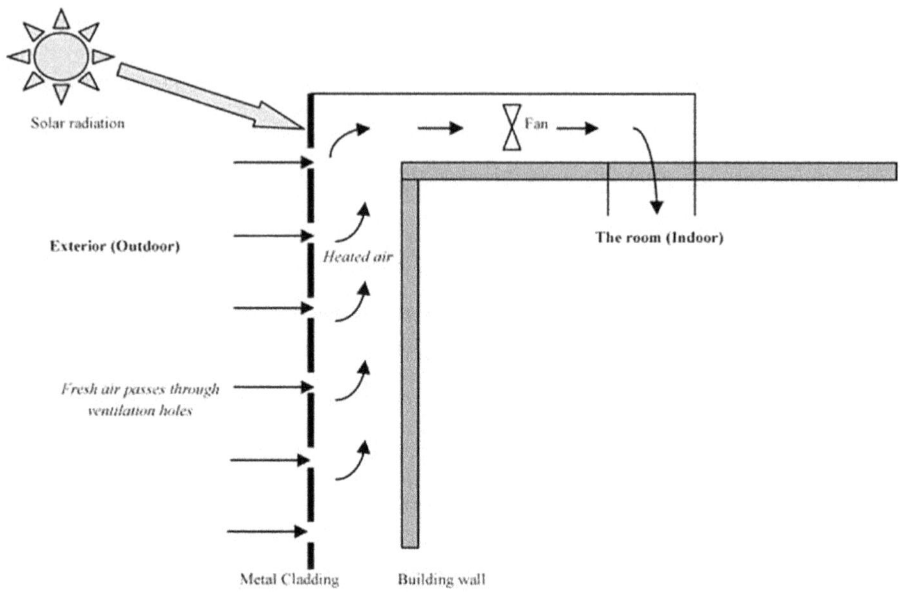

FIGURE 2.4 Schematic diagram of an unglazed transpired solar façade [4].

a room's thermal comfort is limited. Adding a Trombe wall to a roof solar collector may increase cooling, especially in hot areas. In the ventilated roof component, a reinforced concrete slab and insulation layer were conventional, but the insulation and higher prefabricated slab had an air gap.

In summer, the building's insulation reduces solar gains, which is the main benefit.

2.1.3 Removal of the Heat from the Building by Passive Cooling Systems and Natural Ventilation

Passive cooling uses energy-efficient strategies to lower a building's temperature without using energy. Phase change material (PCM), a latent energy storage technology used in building envelopes, is a promising passive cooling option. It acts as a thermal regulator in new construction and thermal energy retrofitting of inefficient and uncomfortable structures. High energy storage density allows PCM to store energy across a small temperature variation. PCMs can be included into building materials by direct immersion, encapsulation, or shape stabilization.

Buildings with PCMs had lower peak indoor temperatures and less cooling and heating electricity. PCM must be fully charged and discharged each cycle to maximize its potential. The PCM must solidify overnight to absorb heat the next day. PCMs are usually placed on the building envelope's interior to regulate inside temperatures. As it reduces heat loss to the outside, insulation may prevent PCM from solidifying overnight. The heat trapped in the PCM is released indoors, resulting in little energy savings and maybe increased space cooling energy use.

Night ventilation, or free cooling, can help PCM-enhanced buildings charge during the night. Night ventilation cools buildings by dissipating heat from PCM with colder external air. Night ventilation performance depends on ventilation control systems, the building's location, kind, and shape, and climate characteristics, including external temperature, wind speed, and wind direction. The use of shape-stabilized phase change materials (SSPCM) plates during night ventilation improved indoor thermal comfort and reduced daytime electricity consumption by 76%.

NV enhanced PCM's potency. Indoor low night temperatures solidified PCM in insulation. To reduce peak cooling loads the next day, the PCM was charged. Figure 2.5 shows the office building's inside and outdoor temperatures during a Madrid weather simulation for one day. The night and early morning outdoor temperature was lower than the inside temperature. At below 25°C, the PCM's surface temperature matches its melting temperature in Madrid. It appears that external cool air charged the PCM overnight. Combining the two methods with smart control increased energy savings.

2.1.4 Protection of the Building from Overheating by Proper Shading Solutions and Vegetation in Summer Time

Escalating temperature concerns, climate change awareness, and the urgent need to cut construction carbon emissions are escalating. Climate change dangers include heat stress, harsh weather, bad air, and water- and vector-borne diseases [6].

Shade devices are mostly used to block direct sunlight and solar radiation for cooling and allow solar gains for heating. Heating and cooling require uniform interior

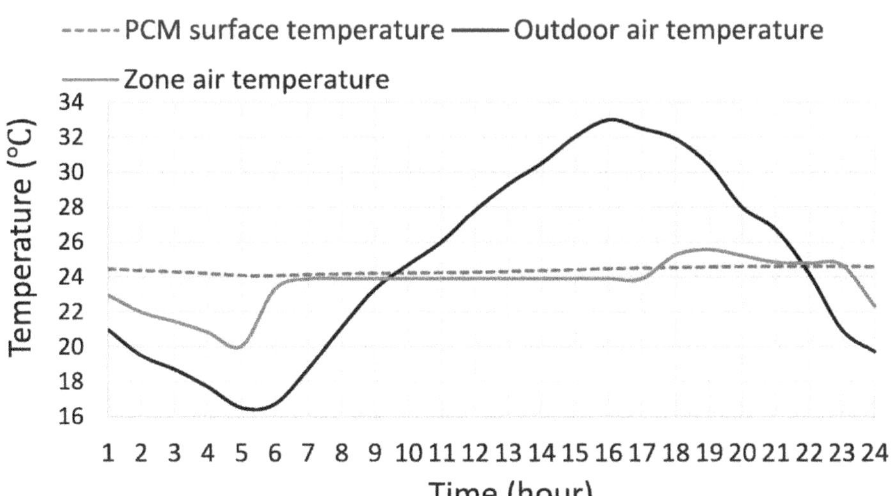

FIGURE 2.5 Daily air temperature profiles comparing simulated indoor and outdoor environments, showing the PCM surface temperature profile [5].

lighting dispersion. Designing glass facades with shading devices in any building requires considering building type, natural light, and latitude. Building design and orientation affect shading devices. Shade devices alter daylight, thermal, and visual comfort.

Design issues should be cost-effective. Similar studies found that solar shading and nighttime ventilation can sustain comfort without cooling. Protections are cheaper than cooling equipment installation.

Exterior shading systems effectively filter solar radiation before window glass. The glass blocks far-infrared radiation from interior surfaces when fitted indoors. It increases cooling loads during overheating.

Green façade blocks solar radiation, improving indoor thermal comfort. It lets enough sunlight into buildings. The leaves absorb most solar radiation for photosynthesis and evapotranspiration. Reduces building overheating. More leaves on the façade improve thermal performance and energy efficiency in warmer regions.

Lowering global warming is a priority worldwide. EU plans aim to cut greenhouse gas emissions 20% below 1990 levels by 2050. The building sector is vast, thus current and new structures must be energy efficient. Shades are a precaution. Shade makes cooling the structure hardest [7].

Framework (Figure 2.6) uses measured hourly indoor temperatures over a limited period to assess overheating in an existing building during the entire summer, extreme future years, and various mitigation strategies for existing buildings.

Building sector adaptation is urgently needed to handle future growth and climatic sensitivity as climate change and economic growth will raise worldwide cooling energy consumption in the world's warmest regions by 2100 [9].

2.1.5 The Use of Solar Energy for Daylighting All Year Round

Sunlight entering interior rooms affects their quality and function. Architectural daylighting decreases energy costs and creates renewable energy. Artificial light, electricity, and heating, ventilation, and air conditioning (HVAC) costs decrease with daylighting. Designers chose daylighting for energy efficiency, sustainability, and low upkeep. Buildings save 15–40% energy with daylighting. Daylighting improves student, employee, and retail customer well-being, productivity, and satisfaction because people need it. User psychology, eyesight, and eye health gain from daylighting. Daylighting energy generation, transit, and distribution are pollution-free. Glare control is crucial. Direct sunlight in office buildings glares on work surfaces, making computer screen viewing difficult. Window and skylight orientation maximizes direct and diffused daylight into the building, boosting light levels and minimizing glare. Light needs, climate, and architectural design should determine window and skylight placement.

Daylighting requires building heat regulation. Natural lighting can cause significant heat gains, especially in warmer climates, if not planned properly. Adding light without heat might be difficult. National energy savings can surpass 40%

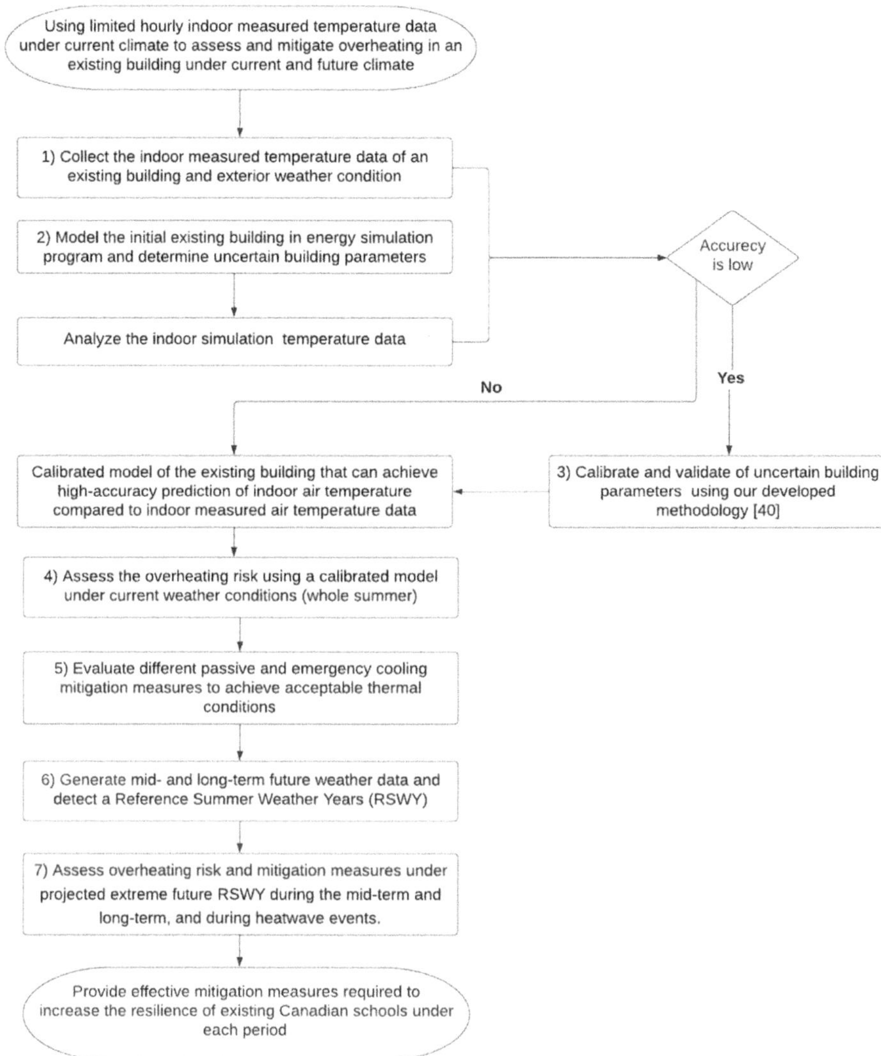

FIGURE 2.6 Framework outlining the assessment of overheating risk in existing buildings, recommending mitigation solutions based on limited indoor hourly data [8].

with appropriate daylighting. National energy conservation reduces pollutants and improves the environment [10].

The side window is the easiest sidelighting method. The window opening area determines daylight intake. The best way to increase daylight exposure is to add windows, especially side windows and clerestory. Figure 2.7 shows how a vertical clerestory and side window affect daylight penetration. Many restrictions limit their use.

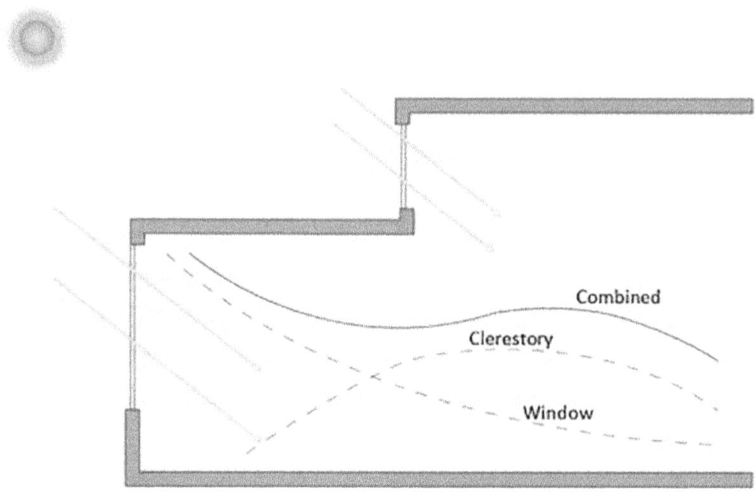

FIGURE 2.7 Daylight penetration achieved through the integration of a vertical clerestory and a lateral window [11].

1. Daylight depends on window or clerestory size. Some structures have insufficient window apertures.
2. Daylight is concentrated around windows and decreases.
3. Windows cannot automatically control daylight, making occupants uncomfortable under bright sunlight.

A light shelf uses ceiling reflections to redirect sunlight to the back of a room. The light shelf controls sunlight intensity dependent on solar elevation angles. The light shelf can tilt to reflect sunshine onto the ceiling in low-light rooms. This ceiling adds diffuse light for uniform illumination and deeper sunlight penetration. The light ledge blocks direct sunlight and decreases glare. Figure 2.8 shows daylight penetration in an oblique light shelf room.

2.1.6 Providing a Pleasant and Comfortable Indoor and Outdoor Environment for the Inhabitants

Due to global warming, several countries have reduced greenhouse gas emissions, improved energy efficiency, and promoted renewable energy. Outdoor thermal comfort affects people's activity in streets, retail centers, playgrounds, and urban parks. The amount and intensity of these activities depend on how uncomfortable residents feel in outdoor climates. Outdoor wind velocity exceeds inside air speeds. Summer wind can be enjoyable up to a certain velocity, but winter wind is typically annoying.

Relative humidity greatly affects outdoor temperature. Evaluation of subjective outdoor environment responses must include these factors. Outdoor activities

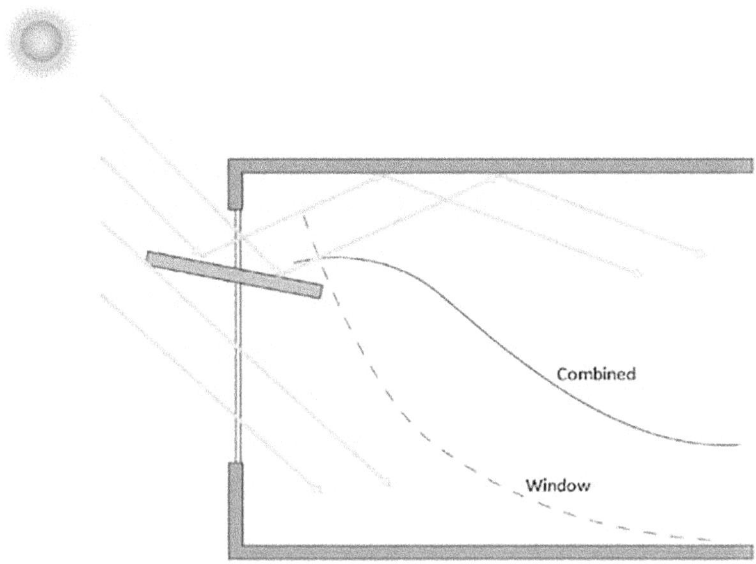

FIGURE 2.8 Daylight infiltration in a space featuring an angled light shelf [11].

often include dressing for the season and the atmosphere. Despite subject perception and alteration of biological processes, metabolic rate is similar. Few studies have addressed outdoor thermal comfort and its factors, unlike interior thermal comfort.

Urban paving also lowers air temperature. Walkways can be treated with appropriate paving materials and arena cooling technologies to lower surface temperature. Paving materials quickly absorb heat, causing glare and thermal radiation in metropolitan environments. Pavement absorbs diffuse sky and wall rays in the morning. The ground distributes heat to the cooler air layer as the sun rises. The ascending warm air is replaced by cooler air until the courtyard temperature matches the outside. Paving is exposed to powerful radiation longer than vertical walls, emphasizing the significance of surface treatment.

Physiological and psychological considerations, human behavior, outdoor element design, and outdoor environment variables affect outdoor comfort. Understanding how these components interact is crucial to comfort. Climate chamber experiments and outdoor research are used to determine indoor thermal comfort. Climate chamber experiments are done in a lab, but field research can be done anywhere with the right equipment. Comfort temperatures in real-world investigations vary and can be difficult to match with those from climatic chamber tests [12].

Modern housing has a lack of hierarchical open spaces and movement patterns, enclosure and size, non-suitability of open spaces, informal character, indifference to climatic conditions, visual relief, and spatial character. Many housing projects employ neighborhood and outdoor spaces as leftover negative spaces rather than activity or interaction zones. Luxury homes and low-income dwellings are less habitable as they cost more. Thus, most Indian contemporary housing projects lack healthy neighborhoods.

Mental health is also prioritized in residential settings. Neighborhood connections require privacy and physical separation. Roderick Lawrence defined privacy as "controlling access to spaces," meaning the resident is alone, and "controlling access to information," implying oversight.

Even though neighbors and social connections are important for survival, privacy is also important for mental and emotional health. Interference from neighbors, residents, or passersby can destroy a good environment. Progressive public–private evolution is promoted. Psychologically, home neighborhood safety is essential.

Child and senior protection from automobiles and other risks, while safe and cheerful neighborhood pedestrian movement, are safety concerns. Thus, interpersonal safety, especially outside, is psychologically necessary. "Designing for surveillance and privacy must be balanced delicately" [13].

2.1.7 Understanding Bioclimatic Architecture

Bioclimatic architecture produces climate-appropriate structures. Studying natural variables optimizes comfort, energy efficiency, and sustainability.

Bioclimatic architecture was applied in Greek and Roman buildings. Many civilizations have modified this building strategy to their climates.

2.1.8 Bioclimatic Architecture Strategies

2.1.8.1 Solar Gain Optimization

A building is heated by window sunlight. Window direction affects solar gain. Solar gain occurs in south-facing windows due to the low winter sun. North windows are inefficient and get little winter daylight. Summer afternoon solar gains warm south and west-facing windows.

Window shade affects solar gain. The roof overhangs and well-placed deciduous trees in front of south and west-facing windows minimize summer overheating but allow winter solar gain.

Curtains and drapes protect from sunlight and heat. In winter, open curtains optimize sunlight gain, while closing them at night or on rainy days reduces heat loss [14].

2.1.8.2 Bioclimatic Zoning

Building energy limits require climate zoning. Building energy efficiency is recommended for India's five climate zones. Indian climate diversity influences these strategies' efficacy. Build-centered bioclimatic zoning incorporates climate-responsive solutions. Climate, building envelope, and people are thermally related. Building envelopes govern human thermal contact with meso- or microclimates. Agricultural, ecological, and architectural climate zones exist. Climate classifications use independent and dependent environmental factors [15] such as air temperature, relative humidity, rainfall, soil type, etc.

A building should respond consistently across climates within the same category. Because climate indicator sensitivity varies per system, a single climate classification may not work for all needs. Solar radiation is the fundamental indicator of photovoltaic (PV) energy generation in PV zoning. Climate zoning building thermal performance depends more on outdoor temperatures. A climate zone may be useful for some uses but not for others [16].

2.1.8.3 Smart Shading

Cold areas need significant energy for building heating. In cold climates, blinds and window shades for heating season energy savings are as crucial as summer use. Thermally resistant blinds and window shades, and heating season-specific management systems lower building heating load. Research on solar shading system automation has focused on building cooling load and artificial illumination reduction. Their use during the cooling season saves energy in many areas. Motorized blinds and window shades save cooling energy in the study's ASHRAE climatic zone. Automatic shading systems saved cooling energy in all temperature zones. Studies on non-insulative solar shading systems showed that heating season cooling load reduction management did not apply. These shading techniques can cost the building energy during the heating season. Summer and winter energy savings conflict with manual and automatic shading systems.

Non-insulative shading techniques save heating season energy in different climes. Medium and cool temperatures result in negative energy savings or penalties from the technology. Non-insulative solar shade saves energy in cooling-dominant climates with warmer winters. Non-insulative shade installation may raise the heating season energy penalty in particularly cold and subarctic areas due to decreased cooling energy demand. Thus, cold-climate buildings cannot employ cooling season energy savings to offset heating season penalties. Seasonally improper automation control increases heating season penalties induced by non-insulative window shades reducing glazing unit thermal resistance. Customized heating season control helps save energy costs in cold climates. For heat transfer prevention, a heating season control system should retract window shades during peak solar irradiance and deploy them during low solar intensity [17].

2.1.8.4 Thermal Mass

A building's thermal mass impacts internal temperatures, electricity use, and occupant comfort by retaining sensible or latent heat. The phrase thermal capacitance is common because big heat capacity components may delay temperature swings, like electrical circuits. The bulk of buildings worldwide employ transitory heating and cooling, with few attaining steady-state temperatures. Figure 2.9 demonstrates that the dynamic evaluation of the two obstacles produces very divergent results. From transitory impacts, the dynamic model demonstrates that the high-mass blockwork wall moderates temperatures. Solar flux, air heat loss, and wall heat flow comprise the outside energy balance. The blockwork wall's high heat capacity minimizes

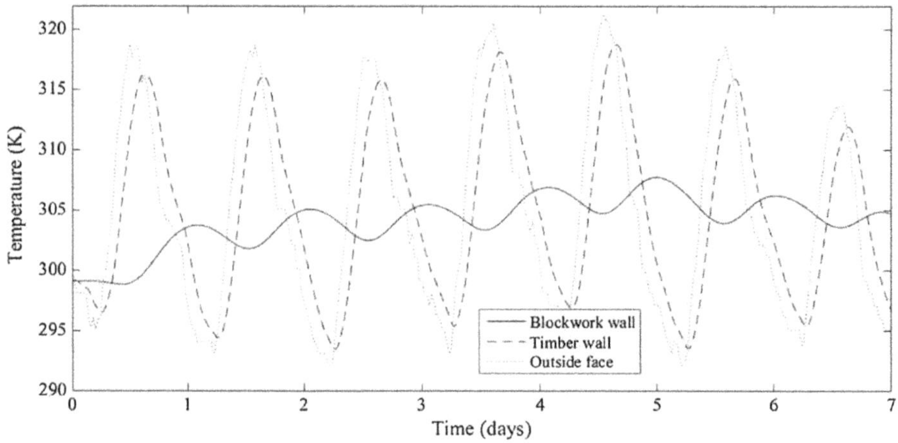

FIGURE 2.9 Indoor wall temperatures in lightweight timber versus high-thermal mass solid-wall (blockwork) in buildings [18].

interior surface temperature rise compared to static analysis (Figures 2.9 and 2.10). Different from the blockwork wall, the lightweight hardwood wall's dynamic analysis moderates temperature.

2.1.8.5 Natural Ventilation

Free space cooling/heating in buildings currently uses NV. NV is energy-efficient compared to air-conditioning. Depending on the weather, NV can provide thermal conditioning and IAQ without mechanical devices. These systems are being advocated

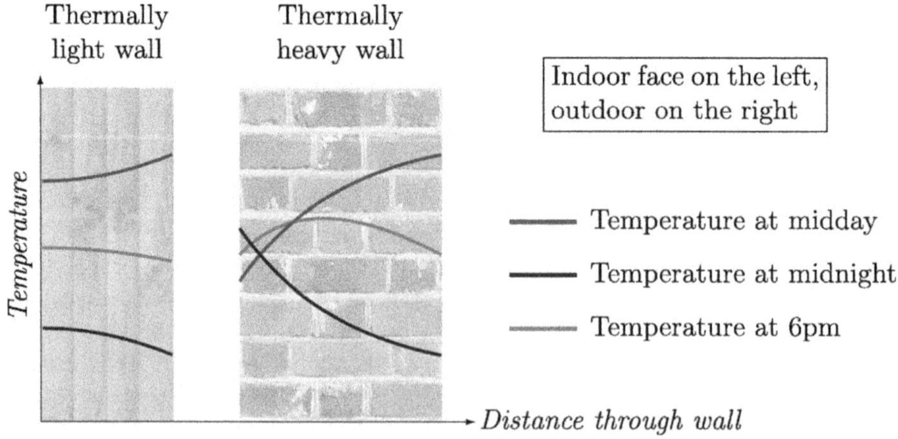

FIGURE 2.10 Representation of temperature fluctuations across lightweight and heavyweight walls [18].

as an alternative to mechanical ventilation systems due to their energy savings, durability, low noise, and carbon dioxide reduction. NV systems conserve energy and improve health despite their flaws: they require natural processes and are affected by temperature and wind. NV systems can provide respiratory health benefits and a comfortable interior with correct design and maintenance. Along with sound IAQ, residents may demand outdoor access [19].

Night ventilation in naturally ventilated office buildings may depend on a shared security strategy. Building design and insurance terms might affect security policies. Ground-floor or other easily accessible rooms may need better burglary prevention than higher-floor ones. The size and placement of façade apertures may also deter burglaries. Weather protection during resident absence may require midnight window closure. Because of the ventilation policy and façade security, occupants may have less effect on night ventilation than during the day [20].

2.1.8.6 Use of Local Materials

Many developing nations' housing crises are deteriorating despite new public and private sector policies, programs, and solutions. Most homeless people in Africa, Asia, and South America are low-income and need housing programs. Numerous nations have provided public housing for years, as market solutions and financial resources may be inadequate for this demographic, and housing is crucial for socioeconomic and political advancement. Industrial technology, such as concrete blocks and slabs have supplanted traditional components and processes, prompting wattle-and-daub homeowners to seek replacement with the new material, despite concrete blocks offering worse thermal comfort. The indigenous replaced their comfortable, low-cost, and ecological homes with contemporary fashion, which represented wealth and status [21].

Nanoscale material attributes may affect concrete's strength, ductility, early age rheology, creep and shrinkage, fracture behavior, durability, etc. The most important hydration product in cement paste, calcium–silicate–hydrate (C–S–H), has an unknown nano- and micro-scale structure. One must understand nanoscale structure and how it influences local mechanical properties to improve cement and concrete. Fast imaging and Young's modulus determination of a nanoscale feature or region are needed. Many cementitious materials have been imaged using electron beams; however, mechanical characteristics are not available. Atomic force microscopy (AFM) has been used to study cement clinker surface changes in saturated calcium hydroxide solution, followed by water and sucrose solution; humidity-induced changes in hydrated cement paste microstructure; calcium hydroxide carbonation; and cement paste microstructure densification with silica fume or fly ash. The diamond indenter probe of AFM can extract nanoscale local mechanical properties and give high-resolution pictures. This method works for soft materials but only gives qualitative elastic modulus values and proportions. Quantifying mechanical properties locally with nanoindentation is reliable [22].

2.1.8.7 Courtyards

The architects built a comfortable location for inhabitants to spend summer months having breakfast, lunch and supper by the pool and garden by putting a pool in the middle and growing trees (typically fruit trees) in the yard. Courtyard-built quarters were protected from hot, dusty air by tall walls. The courtyards' trees, flowers, bushes, and pool created a microenvironment that lowered air temperature and increased humidity. Tall courtyard trees shaded the walls and ground from summer's direct sunlight. Patio reduced building cooling energy slightly. It increased building heating energy needs [23].

Early courtyard design requires alignment and positioning. North, South, East, West, and central courtyards are shown in Figure 2.11. Mapping or emphasizing on previous characteristics in courtyard centers studied how position and orientation affect microclimate parameters such ambient temperature, relative humidity, wind speed, mean radiant temperature, and temperature change over time.

From vernacular architecture until today, Mediterranean buildings have courtyards. Only in recent decades has its tempering potential been assessed. Courtyards soften severe microclimates and encourage outdoor use in hot and cold countries. Research shows that courtyards can increase thermal comfort by over 17°C on hot days in warm cities, especially when external temperatures are greater. Such morphology reduces built-up area in warm climate city cores by enhancing urban fabric porosity. Courtyards reduce urban heat island (UHI) by creating a suitable environment. Courtyards enhance surrounding rooms, improving daily living [25].

2.1.8.8 Breathable Façade

Indoor humidity management benefits building and occupant health. Traditional desiccant dehumidifiers with vapor compression devices offer cool, dry air and dilute indoor moisture to decouple latent and perceptual load. Such methods are energy-intensive in hot, humid areas with significant latent load spaces. Directly eliminating internal moisture through naturally ventilated façades saved energy. Water vapor diffusion and direct water removal were recently achieved with a "breathable" building envelope. This material naturally transfers moisture from indoors to outdoors, lowering indoor RH peaks. Allow NV of external air into the envelop to promote water vapor permeability and transport moisture from the interior.

Arid regions favor ventilated building facades. Indoor humidity decreases are restricted by outdoor humidity. RH may rise in the cooled environment, reducing thermal comfort. Higher outside humidity may reverse moisture flow to the interior, impacting façade operation.

External humidity limits passive ventilated façades in humid climates. By pumping indoor moisture outdoors, passive ventilated façade performance and application in humid places can be improved.

A renewable energy-regenerating desiccant dehumidifier brings dry air to the façade. High water vapor capture, fast sorption kinetics, and low regeneration temperatures make MOF desiccants popular [26].

2.1.8.9 Operable Windows

Window operation lets people control local settings and access outside spaces. Operable window operation affects energy use and internal climate. Operable

FIGURE 2.11 Square courtyard models orientated differently [24].

windows use NV to adjust the local environment, improve thermal comfort, and reduce cooling demands when correctly installed and operated. Operable windows offer thermal comfort, lower operation costs, and better IAQ, making them a popular alternative to air-conditioned buildings [27].

In simulation-based investigations, non-stochastic representations of occupant presence and control actions are considered "dated," meaning they are no longer relevant.

Prior research in this topic has highlighted the lack of inter-comparison, ambiguity surrounding the validity range of established models, and the lack of trustworthy algorithms for using these models in building performance modelling. Building performance modelling using occupant behavior models and their prediction capacities across contexts creates major uncertainty that could be harmful.

Previous research on window operation models has ignored their feedback mechanisms. This condition reduces the complexity of verifying behavioral models by ignoring the effect of outputs (like window states) on inputs (like interior temperature). It is irrational to expect empirical evidence to match every possible behavioral model action sequence. Thus, virtual building performance models must imitate the building's responsiveness to behavioral impulses. Numerous attempts have addressed over-parameterized building simulation model calibration difficulties in recent years. The calibration approach uses heuristics to define influential parameters and schedules. This involves a coarse grid to identify a few potential parameters with decreased variability bounds, followed by a guided grid search to refine these parameter vector solutions. The technique focusses using this set of solutions rather than a calibrated one.

Building performance models are calibrated using an evidence-based process that links parameter values to the source of information. The availability of monitored building data has improved calibration.

The test bed from previous research provides environmental and occupancy data for window operation model execution and evaluation. Its biggest drawback is that it ignores model feedback. A calibrated office simulation model and all observed data are used to evaluate window operating model predicted performance. The calibrated building performance model allows us to compare window operation models' effects on simulation-based building heating energy consumption and occupant thermal comfort.

Ecological satisfaction affects productivity and health. There are two main thermal comfort models: static Predicted Mean Vote and Predicted Percentage of Dissatisfied (PMV/PPD) and adaptive. PMV/PPD was invented by developed by P.O. Fanger. PMV is a renowned thermal comfort model. A regulated climate chamber under steady state circumstances provided experimental data for the heat balance-based development. The adaptive model emphasizes dynamic occupant-environment interactions based on substantial field study data. Occupants regulate their temperature with clothes, movable windows, fans, heaters, and sunshades. Passive solar heating solutions include direct daylighting through tall south-facing windows (Figure 2.12a), Trombe walls (Figure 2.12b), glass chambers used to create a greenhouse effect (Figure 2.12c), atriums (Figure 2.12d), and double-skin facades (DSF) (Figure 2.12e) [28].

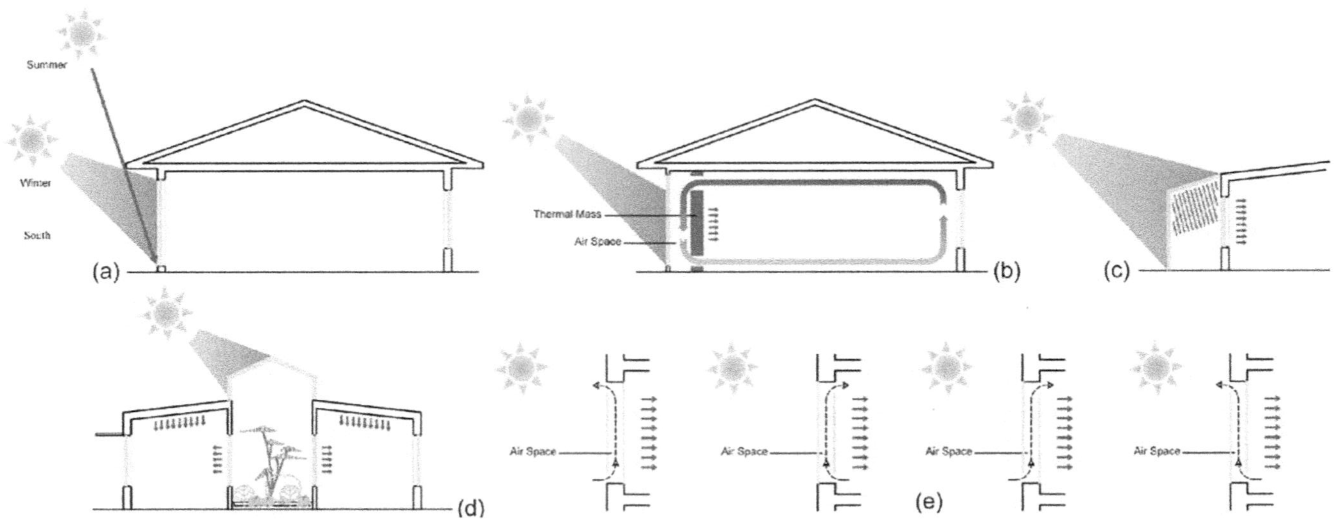

FIGURE 2.12 Examples of passive solar heating strategies: (a) tall south-side windows for direct daylighting; (b) Trombe wall system; (c) greenhouse effect space; (d) atrium design; and (e) double-skin façade (DSF) [28].

2.1.8.10 Rooftop Solar Array

Many structures in the country are low, giving them more horizontal space. Thus, rooftop solar PV electricity generation in India appears promising. The National Solar Mission of India (JNNSM) prioritizes Mega Watt (MW)-scale grid-connected solar PV facilities in India. Rooftop solar PV potential must be estimated and quantified to support its large-scale adoption [29].

Figure 2.13 shows a hybrid solar PV design. To determine its north-south orientation and major physical parameters, the PV system rooftop was surveyed. The data was used to construct a two-dimensional (2D) rooftop model. They also validated that the specified area could create enough power. Revising the basic model with AutoCAD software produced the final 2D model. A Solid Works or SketchUp 3D model shows the solar PV system's post-installation appearance.

2.1.8.11 Rainwater Harvesting

Rainwater harvesting (RWH) technologies are becoming essential to sustainable stormwater management. Increased attention to these systems' environmental and life cycle consequences requires more accurate system designs. In addition to lowering financial costs, the goal is also to lower the costs and resources of making the system, installing it, and keeping it running [31].

Rainfall is the main variable for a Rainwater Harvesting System (RWHS) system, and its temporal fluctuation determines its performance. A rainwater collecting system's design emphasizes on tank size to ensure enough water for usage. Large tanks waste energy, time, and money. Conversely, a small tank cannot supply enough water. Thus, when designing an RWHS, household needs and local characteristics must be considered [32].

Since ancient times, humans have harvested rainwater. The need to improve water sources for humans in arid locations has refined it. Many early rainwater gathering studies focused on size and economics because rainwater is scarce and requires enormous, expensive water tanks. Many water managers regard RWH as a way to collect drinking water from rooftops or store irrigation water in rural tanks, despite its environmental importance. Changes in the paradigm have expanded rainwater-gathering perspectives. Sustainable economic growth and rainfed agriculture, ecosystem balance, urban flood control, surface urban runoff quality management during rainfall events, watershed management, climate change impact mitigation, and green infrastructure development are among its environmental contexts. This caused a global environmental crisis. Drinking water consumption in many places worldwide is expected to rise due to economic expansion and population growth.

Some cities raised awareness of water's importance and encouraged its wise use to minimize societal turmoil. Harvested rainwater can be used for crop irrigation and drinking, providing environmental benefits. As cities grow quickly around the world, food needs must be met by sustainable intensification of farming systems. Managing water well can increase crop yields while lowering damage to the environment. Water collection has reduced hazards, increased yields, and benefited ecosystems.

Drinking and agricultural irrigation water in a heavily populated metropolitan region must be reliable. Alternatives include building enormous dams to collect

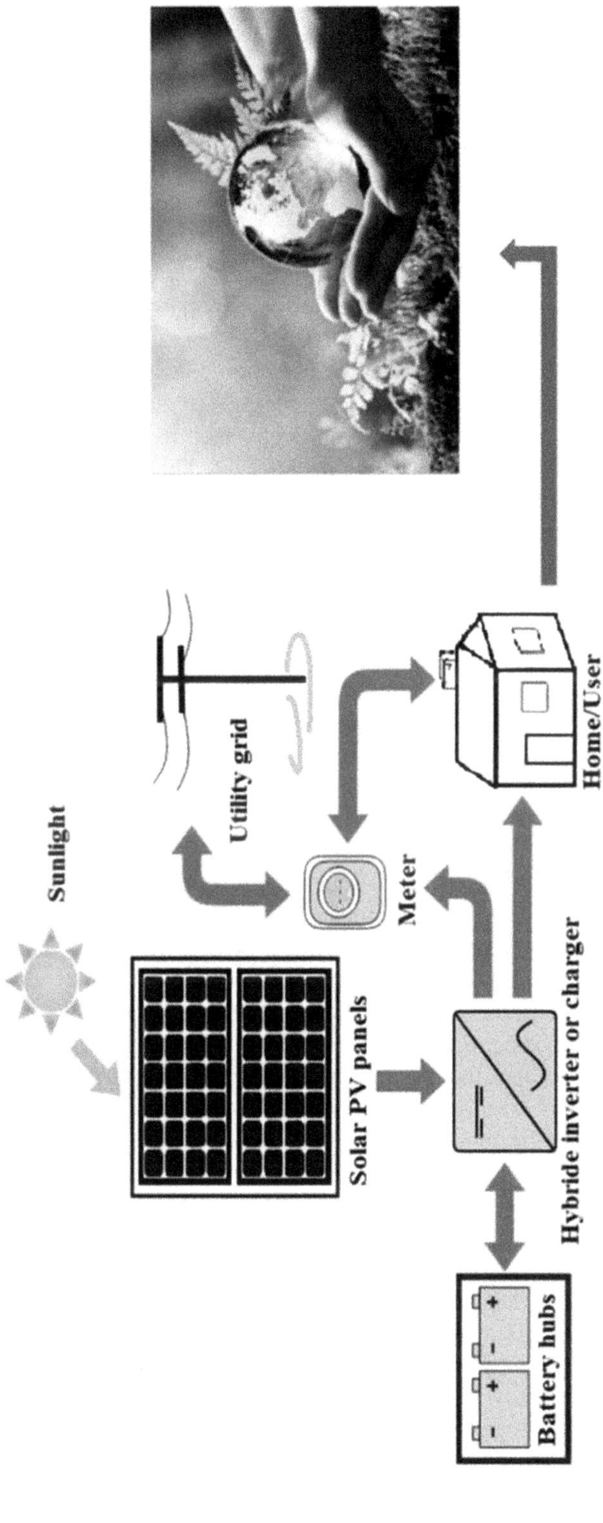

FIGURE 2.13 Detailed schematic of a standard solar PV system [30].

water from manmade lakes or overusing groundwater. These extractions could harm adjacent landscape habitats and their functions, reducing the ecosystem's ability to provide downstream water, biodiversity habitats, and livelihoods. Urban areas' less infiltration alters downstream flow, which may increase flooding. Urban RWH reduces these effects. Rainwater collecting tanks significantly reduce downstream flooding by shifting flows over time. Collecting rainwater reduces the need for other water sources [33].

2.1.8.12 Innovative Waste Management

The issue of waste production continues to be a persistent challenge, prompting numerous countries worldwide to prioritize its management. Transforming waste materials into valuable products enhances sustainable development and fosters efficient waste management. In recent years, extensive research has focused on the conversion of waste materials into valuable products, such as struvite synthesized from urine and biodiesel produced from waste cooking oil. The development of strategies for producing sustainable materials and manufacturing processes across various applications is essential for advancing sustainable development [34].

For a long time, municipal managers have recognized and adopted Municipal Solid Waste Management (MSWM). However, towns in different countries have adopted different solutions due to different social and economic settings, consumption habits, and technological advancements. Using landfills and encouraging recycling and recovery are the main goals. Due to environmental concerns and limited landfill space, municipal solid waste avoidance and reuse, recycling, and recovery are becoming more important, especially in highly populated countries [35].

Digital technologies and equipment applications can improve solid waste management throughout the Industrial Revolution 4.0 (IR 4.0), eliminating, recovering, and reusing trash. The introduction of IR 4.0 technology could improve solid waste management and efficiency. Machine learning, artificial intelligence, and image recognition can automate waste segregation, reducing worker exposure to dangerous items. Radio frequency identification (RFID) and wireless communications enable material traceability, helping to understand circular economy (CE) potential.

Computers led to substantial technological advancement during the 3rd Industrial Revolution, which enabled the production of automated systems, technologies, and electronics. These powerful systems needed to communicate through data transfer and reception to become more dynamic and advanced. IR 4.0 promotes computer device interconnection to enable intelligent automation using AI and ML. This time is marked by autonomous systems, especially cyber–physical systems that rely heavily on the Internet of Things. The promise of IR 4.0 is shown by systems that create and send information wirelessly.

This breakthrough technology is used in automated farming and global communications. Process automation, improvement, and optimization are IR 4.0's benefits. Real-time data improves management and decision-making. These benefits can improve SWM efficacy when used to various applications. The Internet of Things is a new framework with networked gadgets. These gadgets use electronic sensors and

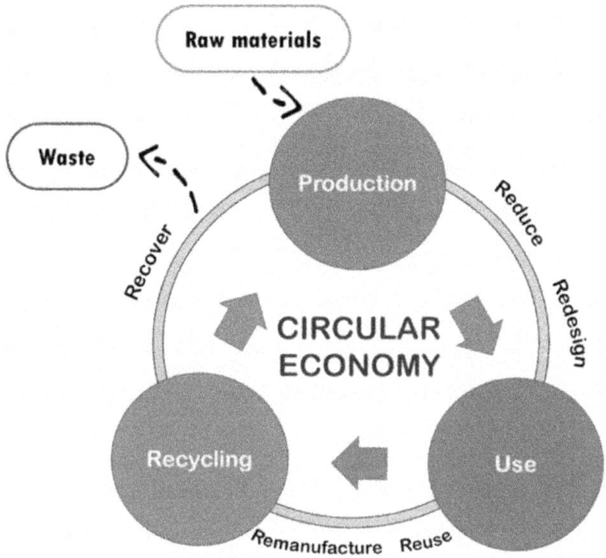

FIGURE 2.14 CE concept diagram [36, 37].

components to process data and communicate via the internet, efficiently conveying information. Communications technology and electronic engineering have helped build IoT infrastructure.

Smart transport uses mobile phones as Internet of Things (IoT) devices to predict traffic patterns using real-time geographical data. This allows optimal route calculation, saving time, energy, and carbon emissions. Recently, additive manufacturing has gained popularity in Industry 4.0. The linear economy paradigm predated recycling and reusing resources. This paradigm extracts natural resources, produces items, consumes them, and discards them. Rapid growth in emerging nations and recognition of unsustainable behaviors make the CE important.

Unlike the linear economy, the CE emphasizes the 3Rs and 6Rs. This method optimizes systems by reducing natural resource waste from product production. Reusing, repurposing, recycling, remanufacturing, redesigning, recovering used products, and minimizing waste while adding value and prolonging product lives are stressed. Figure 2.14 explains CE thoroughly [36, 37].

2.2 BIOCLIMATIC ANALYSIS

Bioclimatic Analysis Tool is a spreadsheet that calculates, visualizes, and evaluates parameters and criteria. It uses formulae and graphics from Auliciems, Szokolay, Givoni, ASHRAE, Olgyay, and Mahoney, among others. Based on spreadsheet climatic data, this technique is automated. Bioclimatic Architecture uses environmental resources to create worthy, pleasant, and healthy spaces for users' body, mind, and spirit [38].

Energy conservation and natural integration are stressed in bioclimatic design. It uses passive ways to improve a building's energy efficiency and indoor comfort without using more energy. For passive energy reduction and minimal active system use, this method assesses form layout, orientation, internal spatial arrangement, façade design, vegetation usage, NV, and building mass utilization [39].

Architects need bioclimatic research and tools to visualize and understand local thermal comfort. Bioclimatic tools mostly use psychrometric charts to assess a location's climate's thermal comfort potential. The National Building Code of India classifies India into bioclimatic zones (Figure 2.15).

The chart's comfort zone was expanded to include passive design zones based on passive heating and cooling calculations. The instrument proves NV, direct evaporative cooling, and passive solar heating work. Depending on the region, natural comfort potential ranges from 25% to 46% of the annual duration [40].

Bioclimatic architecture reduces building carbon emissions and non-renewable energy use. This strategy may reduce building heating and cooling costs by using natural resources. By creating comfortable and healthful indoor settings, bioclimatic design improves building occupants' quality of life [41].

2.2.1 Key Principles of Bioclimatic Architecture

2.2.1.1 Site Analysis

A site study checks to see how well the suggested RFID system works with the current system on the site. The process evaluates electrical disturbances, noise, and physical barriers. Finding interrogation zones that work with the current infrastructure is the goal. Installing is possible after design selection and location analysis. The system is installed using design and site analysis data [42].

Designers study topography, utilities, hydrology, and other aspects to understand the site's physiographic setting. Additional evaluations include programming, site selection, land use, infrastructure, regulatory, biological, cultural, and historical settings.

Evidence indicates that designers seldom possess complete knowledge at the commencement of a project. A designer may lack the funds to acquire all the necessary information. Owing to their complexity, procedural frameworks can present challenges for designers. An architect may investigate several approaches to address difficulties and enhance requirements and information. Iterating through solutions, analysis, and solutions may be necessary. Architectural professionals write about how a site is examined during early building design [43].

2.2.2 Orientation

Building orientation is vital for managing solar radiation and wind movement. Skylight design, along with building orientation, maximizes solar energy use. Skylights are essential to atriums' structural envelopes because they connect indoor and outdoor spaces. Large skylights let sun radiation into cool places, reducing daytime heat demand. However, poor thermal performance may cause significant nighttime heat loss.

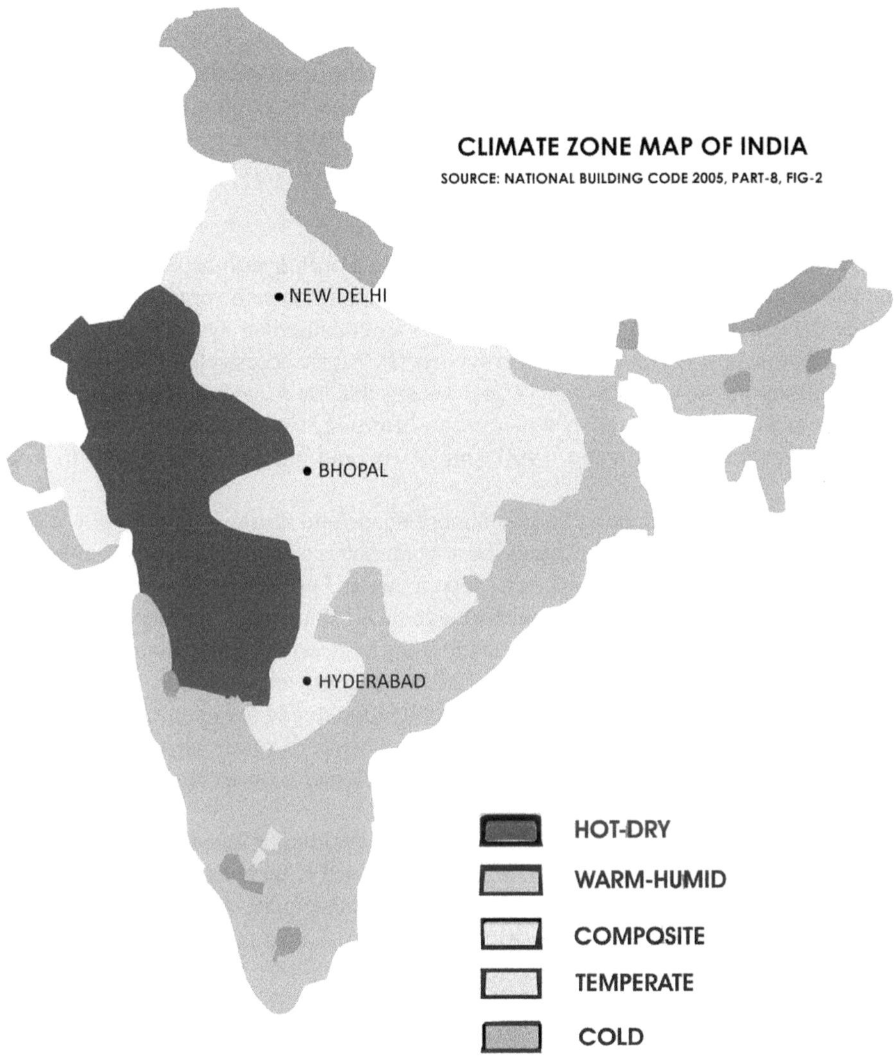

FIGURE 2.15 Bioclimatic zone map of India.

Source: National Building Code of India-2005 [40].

Building thermal performance is affected by the area ratio (AR), which is the ratio of skylight to roof area. Maintaining atrium thermal comfort requires the right skylight area. Simulation studies on energy usage imply that reducing skylight area improves thermal comfort and daylighting at a 1/4 aspect ratio. Further research suggests that increasing the AR may improve daylighting. The ratio of skylight to floor area also affects building energy usage. Skylight-to-floor ratios between 5.5% and 6% reduce energy demand by 19% [44].

An optimal orientation maximizes a building's energy performance over its life cycle by improving heating, lighting, and cooling efficiency. The ideal building orientation depends on the demand for natural light, the benefits of solar radiation for heating, and the need to prevent overheating. This determination also considers winter wind protection and summer humidity benefits [45].

2.2.3 Passive Solar Design

After considering the project location, building orientation is determined. The local climate and site conditions must be assessed to enhance natural light and ventilation and reduce energy demand. Although active (mechanical) air-conditioning systems are becoming more common, post-occupancy comfort surveys (PCSs) are accessible worldwide. Active air-conditioning is newer than architectural designs that lower indoor temperatures.

As heat events increase, so will air conditioning use. Global air-conditioning system sales tripled between 1990 and 2016, and the trend is likely to continue [46].

Energy savings can depend on greenhouse shape and design. According to a five-shape comparison, northern China's best greenhouse shape is a single-gable roof with an opaque wall. Large changes in design factors between single- and multi-span passive greenhouses affect energy performance. Due to its larger surface area per unit volume, single-span greenhouses exchange more heat, which can cause energy losses in colder areas or gains in warmer ones. Multi-span greenhouses may have stronger shadow effects and reduced solar radiation collection due to lower surface-to-volume ratios. Multi-span greenhouses have a larger footprint, therefore air flow, solar radiation dispersion, and cold patches from structural elements obstructing direct sunlight must be considered.

Thermal storage media like water, soil, rock, brick, and PCM ensure production in cold climates in many passive agricultural greenhouses. Inclement or overcast weather may render passive agricultural applications inadequate for short-term storage and capable of managing only a restricted heat load. A heat pump is required for the seasonal storage of summer heat for winter use [47].

2.2.4 Natural Ventilation

Due to climate change, extreme events, including heat waves, overheating, and UHIs, affect residential and tertiary structures with high occupancy, such as hospitals, schools, commercial centers, and workplaces. The bulk of low-carbon plans ignore ventilation as a mitigating method. Indeed, NV and mixed-mode methods are understudied, especially in warm or hot-and-humid conditions. Ventilation can help with both. To improve IAQ, ventilation is often used. The role of ventilation in reducing energy-cooling demand has only been highlighted recently. External pressure gradients and the stack effect define NV. Because of the variety of cases, NV) can be categorized as unilateral, cross, or stack ventilation. These NV categories' key principles are shown in Figure 2.16.

Passive building cooling technologies reduce electric air conditioner loads and improve energy efficiency. The building's façade apertures and accessories allow natural ventilation and appropriate airflow within. The balcony, a semi-open transitional

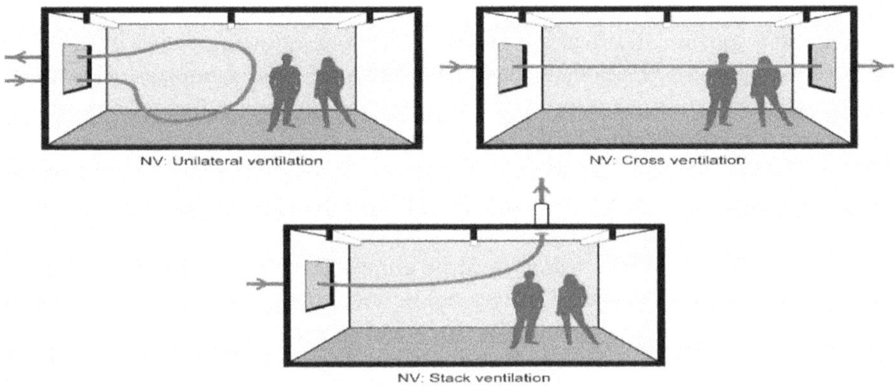

FIGURE 2.16 Strategies for natural ventilation [48].

space between the building's interior and exterior, is significant. Over 2019–2023, only 7% of research on single-sided versus cross-ventilations was published, according to Scopus and Web of Science. Individual-sided natural ventilation is suggested as an alternative to wind-driven urban building natural ventilation.

Semi-open places like balconies have been disregarded in national construction. Limited research in this area has led professionals to overlook this crucial part of architectural design [49].

2.2.5 Materials

Modern architecture uses modern materials and technologies to produce environmentally friendly constructions with 21st-century esthetics and utility. Architects' material choices have long-term effects. The choices can greatly impact the project's esthetics, style, environmental impact, durability, care, expense, and practicality. Construction materials are as follows:

1. Steel
2. Glass
3. Plastic
4. Stone
5. Concrete

2.3 SOLAR ASPECTS AND DIDACTIC-RELATED TOOLS

Creating a design that fits a location requires understanding the sun's course and seasonal changes. Developing a Permaculture design requires understanding solar factors and microclimates. When mapping your design site, the sun is important due to your proximity to the equator. In temperate zones, especially those distant from the equator, mapping the sun sector and designing for solar exposure is essential. The video below explains solar sector mapping. In a 2D screen like this, understanding the sectors can be difficult! Watching the sun move across the sky across time and seasons reveals its patterns and habits. The goal is to learn the sector compass so you can arrive at a spot and discover the cardinal directions, allowing you to instinctively comprehend the sun's positions throughout the solstices and equinoxes. This allows speedy site evaluation, measuring conditions, and understanding crucial design aspects in a brief observation [50].

As our lives become more digital, Science Education must prepare future instructors and develop techniques for integrating digital media into schools. Media competences include using digital tools and media for collaboration, learning, work, and problem-solving, which must be emphasized on. "DigitalPakt Schule," a German government initiative, offers public incentives to improve schools with information and communication technology and integrate specialized knowledge. Federal and state governments are allocating at least 5.55 billion euros for this endeavor. These actions are essential to digitalizing education and should be thoroughly investigated. Electronic books and animations can help integrate cutting-edge subjects into chemistry curriculum when used properly. Along with established methods, these multimedia forms offer a modern approach to autonomous learning.

An important study and technology inquiry is finding a sustainable global energy supply. PVs will impact this sector, requiring decades of interdisciplinary research into solar cells. Thus, PVs must be substantially integrated into school curricula, both in physics and chemistry, to support Education for Sustainable Development [51].

In education, museums, zoos, and botanical gardens are gaining importance. In contrast to schools and universities, people can learn from their surroundings and artefacts, which may not be possible in a more structured context. Digital media allow information to be conveyed visually, aurally, or in virtual and actual worlds. Adaptive and interactive functionalities can also use the learner's prior knowledge to encourage active learning. Digital media combines leisure with study with an entertaining, interactive, and educational experience. Thus, museums increasingly curate material for learners using digital media, allowing visitors to interact with exhibitions more independently than passively [52].

REFERENCES

1. Elaouzy, Y., & El Fadar, A. (2022). Impact of key bioclimatic design strategies on buildings' performance in dominant climates worldwide. *Energy for Sustainable Development, 68*, 532–549.
2. Smart Cites Marletplace. Guide for Bioclamitic Design. https://smart-cities-marketplace.ec.europa.eu/sites/default/files/pimes_guide_for_bioclimatic_design.pdf (Accessed on 05th September, 2024)
3. Stazi, F., Bonfigli, C., Tomassoni, E., Di Perna, C., & Munafò, P. (2015). The effect of high thermal insulation on high thermal mass: Is the dynamic behaviour of traditional envelopes in Mediterranean climates still possible?. *Energy and Buildings, 88*, 367–383.
4. Chan, H. Y., Riffat, S. B., & Zhu, J. (2010). Review of passive solar heating and cooling technologies. *Renewable and Sustainable Energy Reviews, 14*(2), 781–789.
5. Prabhakar, M., Saffari, M., de Gracia, A., & Cabeza, L. F. (2020). Improving the energy efficiency of passive PCM system using controlled natural ventilation. *Energy and Buildings, 228*, 110483.

6. Heracleous, C., & Michael, A. (2018). Assessment of overheating risk and the impact of natural ventilation in educational buildings of Southern Europe under current and future climatic conditions. *Energy*, *165*, 1228–1239.
7. Kirimtat, A., Koyunbaba, B. K., Chatzikonstantinou, I., & Sariyildiz, S. (2016). Review of simulation modeling for shading devices in buildings. *Renewable and Sustainable Energy Reviews*, *53*, 23–49.
8. Baba, F. M., Ge, H., Wang, L. L., & Zmeureanu, R. (2023). Assessing and mitigating overheating risk in existing Canadian school buildings under extreme current and future climates. *Energy and Buildings*, *279*, 112710.
9. Gamero-Salinas, J., Monge-Barrio, A., Kishnani, N., López-Fidalgo, J., & Sánchez-Ostiz, A. (2021). Passive cooling design strategies as adaptation measures for lowering the indoor overheating risk in tropical climates. *Energy and Buildings*, *252*, 111417.
10. Sharaf, F. M. (2014). Daylighting: An alternative approach to lighting buildings. *Journal of American Science*, *10*(4), 1–5.
11. Shi, L., & Chew, M. Y. L. (2012). A review on sustainable design of renewable energy systems. *Renewable and Sustainable Energy Reviews*, *16*(1), 192–207.
12. Heidari, S., & Azizi, M. (2020). Evaluation of thermal comfort in urban areas. *International Journal of Urban Management and Energy Sustainability*, *1*(2), 24–33.
13. Gulati, R. (2020). Neighborhood spaces in residential environments: Lessons for contemporary Indian context. *Frontiers of Architectural Research*, *9*(1), 20–33.
14. Sill to Sash. Solar Gain. www.silltosash.com/windows/window-performance/solar-gain/ (Accessed on 05 th September, 2024)
15. Upadhyay, K., Sundaram, M. K., Parvanendu, G., & Elangovan, R. (2023, September). A new bioclimatic zoning for India. In *Building Simulation 2023* (Vol. 18, pp. 1718–1723). IBPSA.
16. Bre, F., Mazzaferro, L., Melo, A. P., & Lamberts, R. (2024). Bioclimatic zoning for building performance using tailored clustering method and high-resolution climate data. *Energy and Buildings*, *311*, 114157.
17. Shum, C., & Zhong, L. (2023). A review of smart solar shading systems and their applications: Opportunities in cold climate zones. *Journal of Building Engineering*, *64*, 105583.
18. Reilly, A., & Kinnane, O. (2017). The impact of thermal mass on building energy consumption. *Applied Energy*, *198*, 108–121.
19. Zhang, H., Yang, D., Tam, V. W., Tao, Y., Zhang, G., Setunge, S., & Shi, L. (2021). A critical review of combined natural ventilation techniques in sustainable buildings. *Renewable and Sustainable Energy Reviews*, *141*, 110795.
20. Roetzel, A., Tsangrassoulis, A., Dietrich, U., & Busching, S. (2010). A review of occupant control on natural ventilation. *Renewable and Sustainable Energy Reviews*, *14*(3), 1001–1013.
21. Ugochukwu, I. B., & Chioma, M. I. B. (2015). Local building materials: affordable strategy for housing the urban poor in Nigeria. *Procedia Engineering*, *118*, 42–49.
22. Mondal, P., Shah, S. P., & Marks, L. (2007). A reliable technique to determine the local mechanical properties at the nanoscale for cementitious materials. *Cement and Concrete Research*, *37*(10), 1440–1444.
23. Safarzadeh, H., & Bahadori, M. N. (2005). Passive cooling effects of courtyards. *Building and Environment*, *40*(1), 89–104.
24. Ghaffarianhoseini, A., Berardi, U., & Ghaffarianhoseini, A. (2015). Thermal performance characteristics of unshaded courtyards in hot and humid climates. *Building and Environment*, *87*, 154–168.

25. Diz-Mellado, E., López-Cabeza, V. P., Rivera-Gómez, C., Galán-Marín, C., Rojas-Fernández, J., & Nikolopoulou, M. (2021). Extending the adaptive thermal comfort models for courtyards. *Building and Environment*, *203*, 108094.
26. Harrouz, J. P., Ghali, K., Hmadeh, M., Slim, R., Katramiz, E., & Ghaddar, N. (2022). Design and control of MOFs-based indoor humidity pump integrated into the building's ventilated façade in hot and humid climates. *Energy Conversion and Management*, *268*, 115983.
27. Wang, L., & Greenberg, S. (2015). Window operation and impacts on building energy consumption. *Energy and Buildings*, *92*, 313–321.
28. Daemei, A. B., Eghbali, S. R., & Khotbehsara, E. M. (2019). Bioclimatic design strategies: A guideline to enhance human thermal comfort in Cfa climate zones. *Journal of Building Engineering*, *25*, 100758.
29. Singh, R., & Banerjee, R. (2015). Estimation of rooftop solar photovoltaic potential of a city. *Solar Energy*, *115*, 589–602.
30. Behura, A. K., Kumar, A., Rajak, D. K., Pruncu, C. I., & Lamberti, L. (2021). Towards better performances for a novel rooftop solar PV system. *Solar Energy*, *216*, 518–529.
31. Ward, S., Memon, F. A., & Butler, D. (2012). Performance of a large building rainwater harvesting system. *Water Research*, *46*(16), 5127–5134.
32. Haque, M. M., Rahman, A., & Samali, B. (2016). Evaluation of climate change impacts on rainwater harvesting. *Journal of Cleaner Production*, *137*, 60–69.
33. Fernandes, L. F. S., Terêncio, D. P., & Pacheco, F. A. (2015). Rainwater harvesting systems for low demanding applications. *Science of the Total Environment*, *529*, 91–100.
34. Kannan, M. B., & Ronan, K. (2017). Conversion of biowastes to biomaterial: An innovative waste management approach. *Waste Management*, *67*, 67–72.
35. Geng, Y., Tsuyoshi, F., & Chen, X. (2010). Evaluation of innovative municipal solid waste management through urban symbiosis: A case study of Kawasaki. *Journal of Cleaner Production*, *18*(10–11), 993–1000.
36. Cheah, C. G., Chia, W. Y., Lai, S. F., Chew, K. W., Chia, S. R., & Show, P. L. (2022). Innovation designs of industry 4.0 based solid waste management: Machinery and digital circular economy. *Environmental Research*, *213*, 113619.
37. Where Is the North. www.whereisthenorth.com/article/12-bioclimatic-architecture-strategies-with-examples (Accessed on 10th September, 2024)
38. Martínez, J. C. R., & Freixanet, V. A. F. (2014). Bioclimatic analysis tool: An alternative to facilitate and streamline preliminary studies. *Energy Procedia*, *57*, 1374–1382.
39. El-Sayed, R. M. (2023). Review and analysis of bioclimatic design approach. *Engineering Research Journal (Shoubra)*, *52*(4), 111–121.
40. Khambadkone, N. K., & Jain, R. (2017). A bioclimatic analysis tool for investigation of the potential of passive cooling and heating strategies in a composite Indian climate. *Building and Environment*, *123*, 469–493.
41. Ugreen. The Benefits of Integrating Bioclimatic Architecture into Your Construction Projects https://ugreen.io/the-benefits-of-integrating-bioclimatic-architecture-into-your-construction-projects/ (Accessed on 15th September, 2024).
42. Behera, S. K., & Mishra, D. P. (2024). *Chipless RFID Printing Technologies*. Artech House.
43. Osmólska, D., & Lewis, A. (2023). Architects' use of intuition in site analysis: Information gathering in solution development. *Design Studies*, *87*, 101189.
44. Wang, Y., Qin, H., Wang, Y., Chen, J., Hou, X., Rui, P., ... & Song, H. (2024). Effect of orientation and Skylight area ratio on building energy efficiency in the Qinghai–Tibet Plateau. *Buildings*, *14*(3), 755.

45. Haseeb, Q. S., Yunus, S. M., Shoshan, A. A. A., & Aziz, A. I. (2023). A study of the optimal form and orientation for more energy efficiency to mass model multi-storey buildings of Kirkuk city, Iraq. *Alexandria Engineering Journal*, *71*, 731–741.
46. Hu, M., Zhang, K., Nguyen, Q., & Tasdizen, T. (2023). The effects of passive design on indoor thermal comfort and energy savings for residential buildings in hot climates: A systematic review. *Urban Climate*, *49*, 101466.
47. Ding, D. (2024). Design strategies of passive solar greenhouses: A bibliometric and systematic review. *Ain Shams Engineering Journal*, *15*(5), 102680.
48. Bienvenido-Huertas, D., de la Hoz-Torres, M. L., Aguilar, A. J., Tejedor, B., & Sánchez-García, D. (2023). Holistic overview of natural ventilation and mixed mode in built environment of warm climate zones and hot seasons. *Building and Environment*, *245*(1), 110942.
49. Karimimoshaver, M., Sadathosseini, M., Aram, F., Ahmadi, J., & Mosavi, A. (2023). The effect of geometry and location of balconies on single-sided natural ventilation in high-rise buildings. *Energy Reports*, *10*, 2174–2193.
50. https://open.oregonstate.education/permaculture/chapter/solar-aspect/
51. Zeller, D., Bohrmann-Linde, C., & Kläger, S. (2020). Digital learning tools for teaching–alternative solar cells with titanium dioxide (ALSO-TiO2)–A contribution to sustainable development education. *World Journal of Chemical Education*, *8*(1), 29–39.
52. Degner, M., Moser, S., & Lewalter, D. (2022). Digital media in institutional informal learning places: A systematic literature review. *Computers and Education Open*, *3*, 100068.

3 Insulated and Airtight Building Envelope

In developed nations, modern buildings require a lot of energy. Efficiency can reduce energy use in many of these constructions. Because of rising energy prices and qualms about the environment, energy efficiency has become more fashionable. Inside and outside areas are defined by the building's envelope. Regardless of the weather, the quality and indoor circumstances are contingent upon it. Numerous international studies have examined how alterations to building exteriors impact energy use. Environmental factors including temperature, humidity, air movement, rain, snow, solar radiation, and other factors, affect building envelopes. Maintenance of the building envelope prevents structure failure and maintains optimal living/working/industrial environments. Routine maintenance encompasses building envelope component inspections, repairs, and replacements, whereas response maintenance covers emergencies.

Protect building envelopes from windborne debris and pressure fluctuations. An analysis found that hurricane winds mostly affect high-rise structure fenestration. Building regulations for windborne debris and changing stresses on the building envelope are also examined. Researching, categorizing, and documenting distress symptoms rather than the fault is typical. As symptoms, water leaks, efflorescence, spalled brick, etc., are sometimes mistaken for problems. This technology helps investigators uncover distress-causing faults, not suffering itself. The fault tree contains two main branches:

1. left for building construction issues,
2. right for operating issues.

Several faults may cause building envelope deterioration in a fault tree. After identifying issues, repair or replacement may commence depending on operational, financial, and technological limits to reduce suffering [1].

The building envelope is shaped by size, material, culture, and appearance. It shields the building from cold, heat, rain, wind, and solar radiation while offering beauty and representation. Indoor environmental quality and energy efficiency criteria have evolved simultaneously, making the building envelope a more complex climate regulator. As the contact between the ambient and internal climate, the building

DOI: 10.1201/9781003752837-3

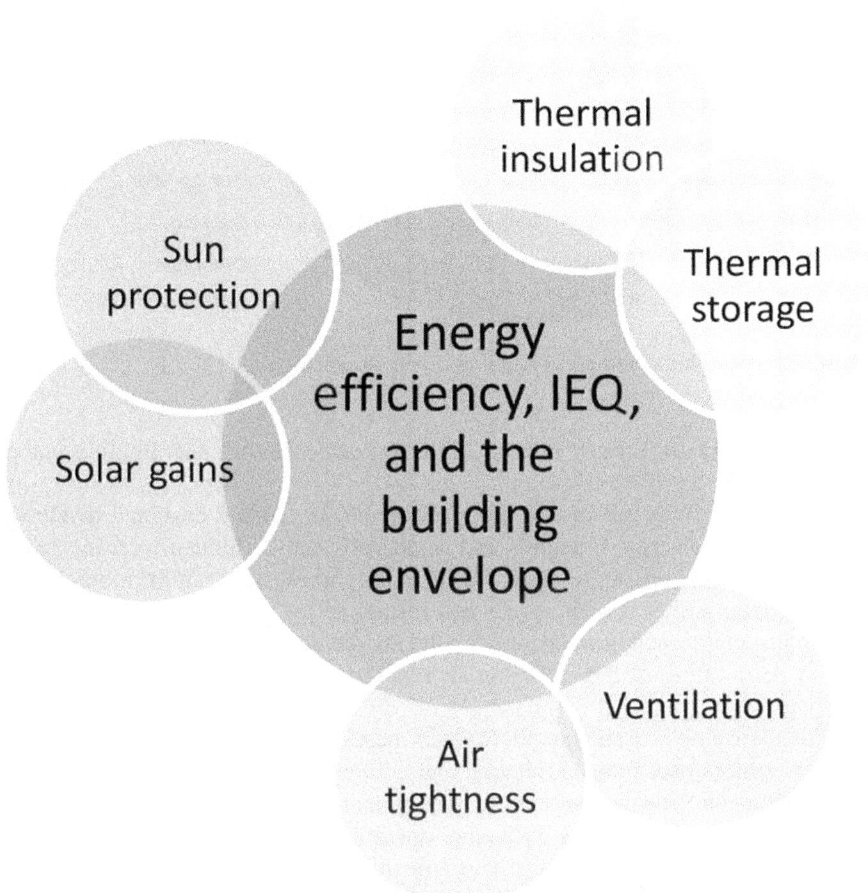

FIGURE 3.1 Envelope factors affect energy efficiency and indoor air quality [2].

Note: IEQ = Indoor Environmental Quality.

envelope is under pressure. Building envelope function and optimization rules should reflect this idea. Figure 3.1 illustrates arranging influential parameters. Building design and construction follow two rules:

1. Energy-specific regulations.
2. Indirectly, through construction laws that address structural, fire, electrical, acoustic, and comfort factors.

High energy variability in buildings. Building energy use depends on orientation, location, construction materials, classifications, interior equipment, and energy systems. Commercial buildings use primary energy (Figure 3.2). Heating, ventilation, and air conditioning HVAC consumes most energy in a building. The United

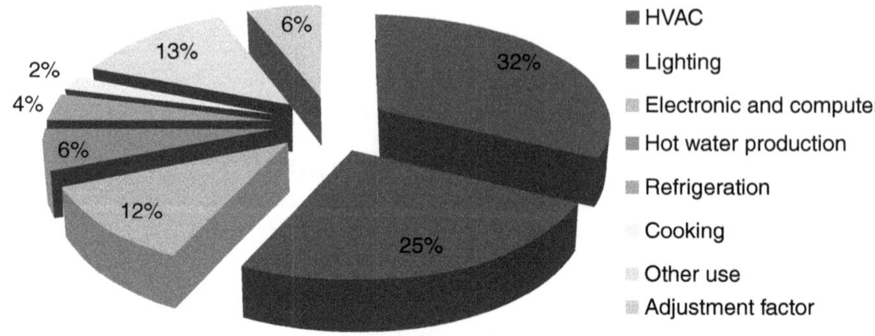

FIGURE 3.2 Standard commercial building energy use distribution [3].

States Department of Energy (US DOE) encouraged HVAC and lighting energy conservation.

Unairtight buildings are insufficiently insulated. In loosely designed dwellings, infiltration wastes energy. Fractures and small gaps can be hidden with insulation. Height, shielding, temperature, wind speed, and tightness affect infiltration. A well-insulated, energy-efficient home needs less insulation for thermal comfort. Avoiding heated or air-conditioned internal air from fleeing the building shell reduces air penetration. A perm rating of 5.0 or higher air retarder should stop air and let moisture escape [4].

Building envelopes transmit 50–60% of heat, affecting thermal performance. Insulation reduces heat transfer, heating and cooling demand, and thermal comfort. Noise and fire concerns can be reduced with proper insulation. Insulation materials' embodied energy matters. By reducing operating energy more than insulating material energy, insulated buildings have a positive net energy balance and decrease greenhouse gas (GHG) emissions. Building insulation materials might be historic, modern, or sustainable depending on origin, chemical composition, and availability (Figure 3.3).

However, multi-objective optimization requires all Figure 3.4. Insulation materials can measure thermal conductivity but not heat capacity, water vapor resistance, embodied energy, or carbon. Elevated energy varies from cradle to grave and gate. Local energy sources (renewables emit zero) embody carbon. Local energy sources and boundary conditions should determine insulating materials' embodied energy and carbon. Utilize the variables from Figure 3.4 to characterize insulating materials and create an optimization database.

3.1 THERMAL ENVELOPE DEFECTS

Construction and architecture prioritize thermal performance. The thermal performance of buildings can be assessed in several ways. Construction utilizes the most energy worldwide. Buildings consume 40% of energy and carbon. Policymakers and scientists recognize that only considerable building energy efficiency improvements can reduce carbon emissions. Construction has improved energy efficiency through

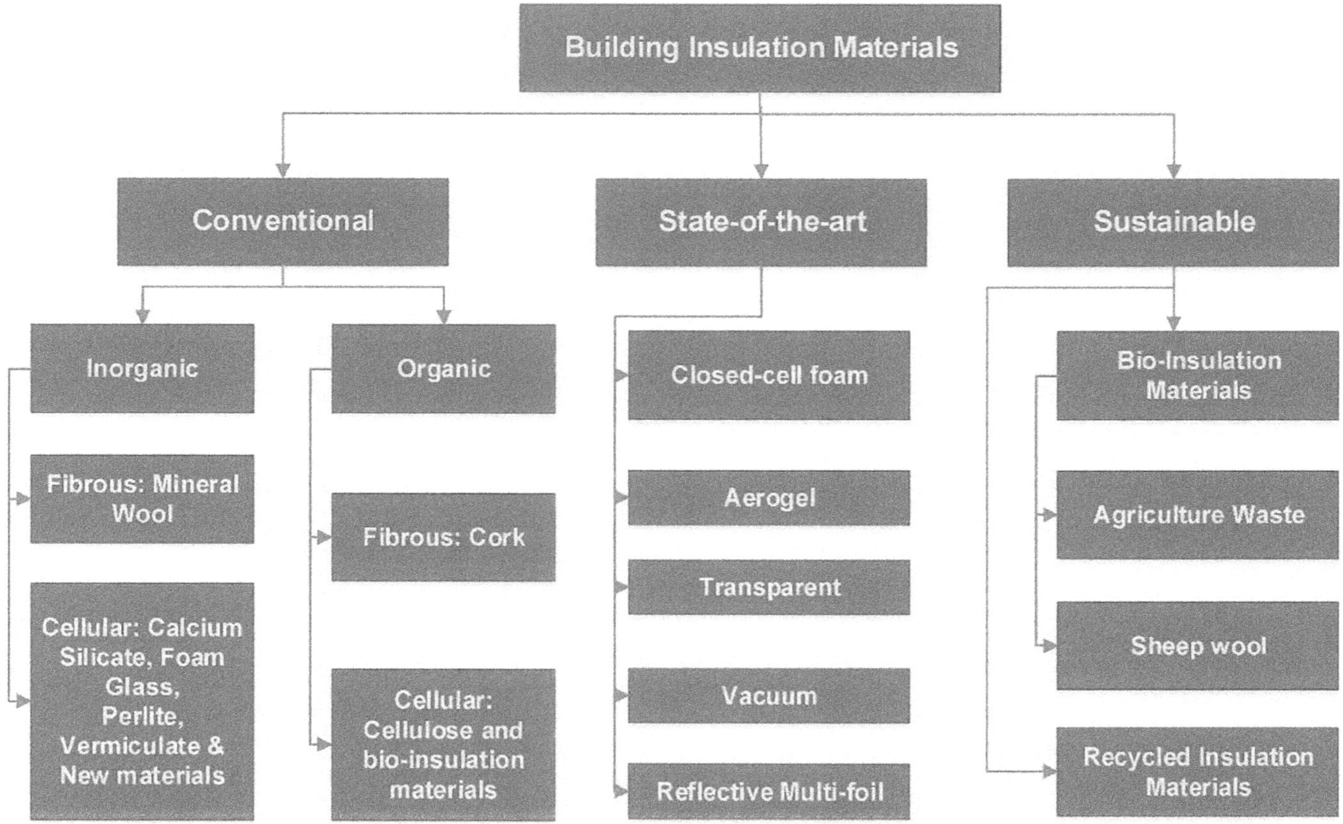

FIGURE 3.3 Classifying building insulations [5].

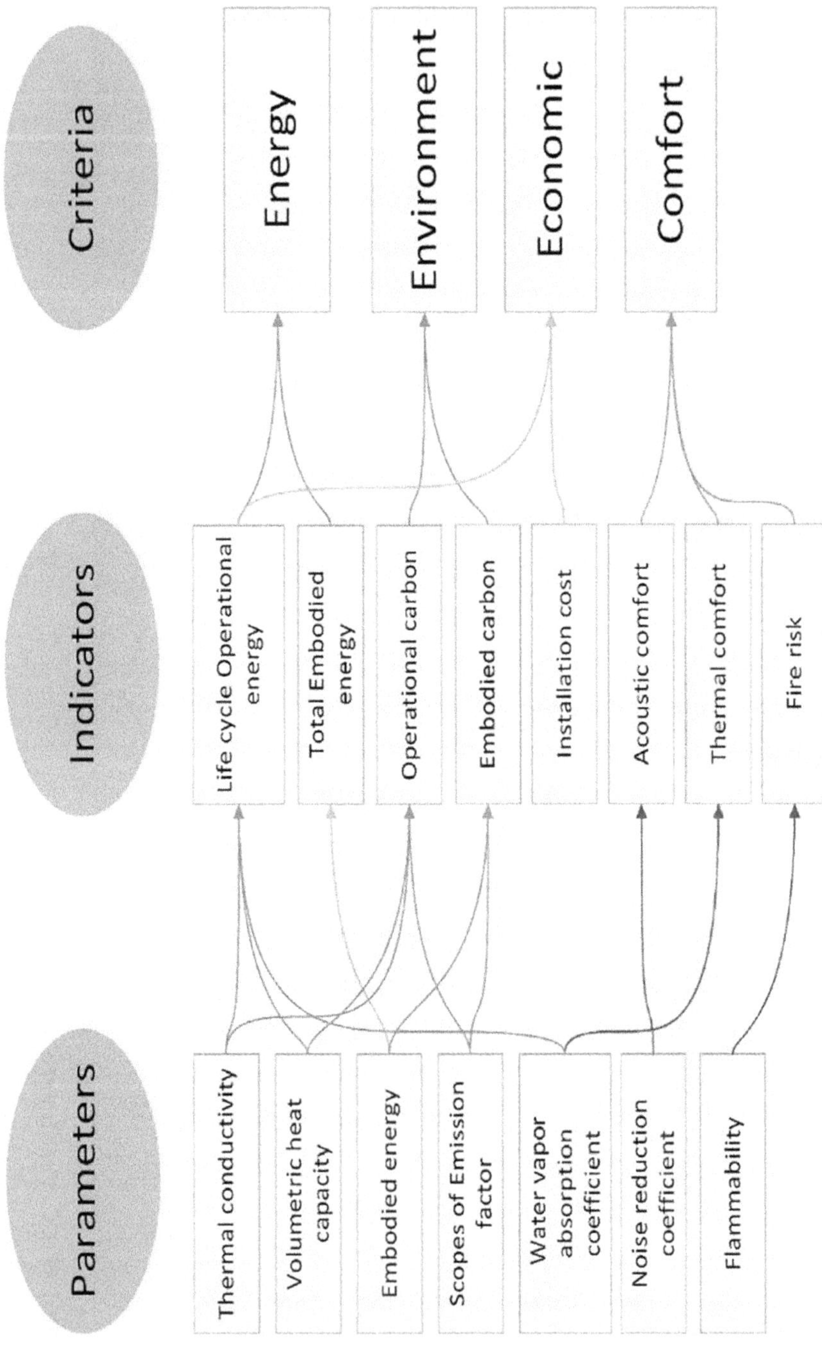

FIGURE 3.4 An approach to choosing building insulation [5].

thermal renovations and low-energy building construction. Studies suggest energy reductions are failing despite efforts [6].

For envelope retrofitting, a Nearly Zero Energy Building (nZEB) requires a thorough building performance investigation to calculate heat gain and loss. Infrared thermography (IRT) reduces the risk of failure in potential energy savings and speeds building inspection before performance-based optimization processes to achieve optimal design options. IRT can only find defects in a certain depth of target solid structures as a boundary technique. Thermographic accuracy and calibration depend on the target structure's emissivity and reflective temperature. The accuracy of standard cameras in measuring temperature is ±2%, whereas advanced models have ±1% accuracy. Emissivity, air particles, ambient temperature, wind speed, and target distance are further IRT constraints that affect picture data interpretation and measurement. Multiple measurement methods, analysis systems, and analysis types improve IRT building envelope performance evaluation. Figure 3.5 summarizes the criteria, and heat loss, air leakage, and moisture commonly cause envelope problems. These issues are detected using quantitative and qualitative IRT for defect root causes. Problems are identified through active or passive study using existing conditions or more heat.

To accurately assess building envelope performance, IRT uses many measuring methods, analysis systems, and analysis kinds. Most envelope issues are heat loss, air leakage, and dampness. Fault root causes are found using quantitative and qualitative IRT. The problem is found through active or passive study of current conditions or surplus heat [7].

Health, agriculture, and building science are diagnosed with thermal imaging. Infrared imaging diagnoses materials utilizing thermodynamics by detecting blackbody radiation. Thermal bridge (TB) detection, moisture characterization, and envelope problem diagnosis are all IR imaging uses in constructed environments. According to thermodynamics, thermal processes are dynamic; watching behavior over time helps improve the building exterior thermal performance rating. Single-shot IR readings are limited by the handled camera vibrations. Envelope performance

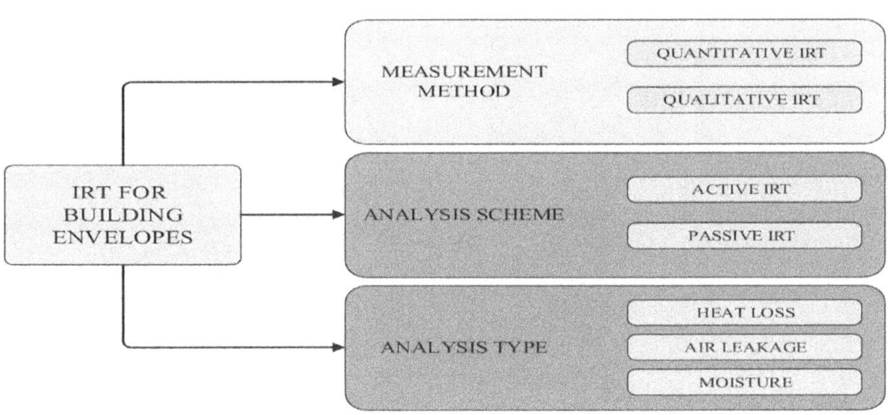

FIGURE 3.5 Classification of building envelope IRT detection criteria [7].

conclusions demand more than almost steady-state data. Thermal imaging can identify architectural issues producing energy efficiency gaps on the building envelope [8].

To achieve interior comfort with less complexity and minimize energy consumption, especially in low-income housing, residential buildings must have excellent thermal performance. Figure 3.6 shows over-wound braided (OWB) thermography from the wall sample's outside. When using three 1 cm^2 orifices with a total area of 3 cm^2, the frame-junction section saw a significant temperature increase (~4.0°C) compared to the rainscreen membrane surface temperature (Figure 3.6b). Lower fluctuation (~1.5°C) in the orifice-insulated section. Construction tape for rain-screen membrane attachment was cooler. As it is above the extruded polystyrene (XPS) rigid insulation panels' linear junction, the frame-junction orifice raises the surface temperature. OWB flaws above insulation panel joints can increase wall heat losses, lowering envelope hygrothermal performance in cold winters. Figure 3.6c's 9 cm^2 orifices (three 3 cm^2) and 12 Figure 3.6d's 18 cm^2 (three 6 cm^2) were tested. As orifice surface area increased, surface temperature rise's influence increased.

In cold winters, OWB faults above insulation panel joints increase wall heat losses and impair envelope hygrothermal performance. This was done for Figure 3.6c's 9 cm^2 (three 3 cm^2) and Figure 3.6d's 18 cm^2 (three 6 cm^2). More orifice surface area increased the surface temperature rise's effect. Temperature imaging is the finest nondestructive method for qualitative building evaluation in operation. The analysis sometimes includes numerical heat transfer models. Temperaturograms calculate the

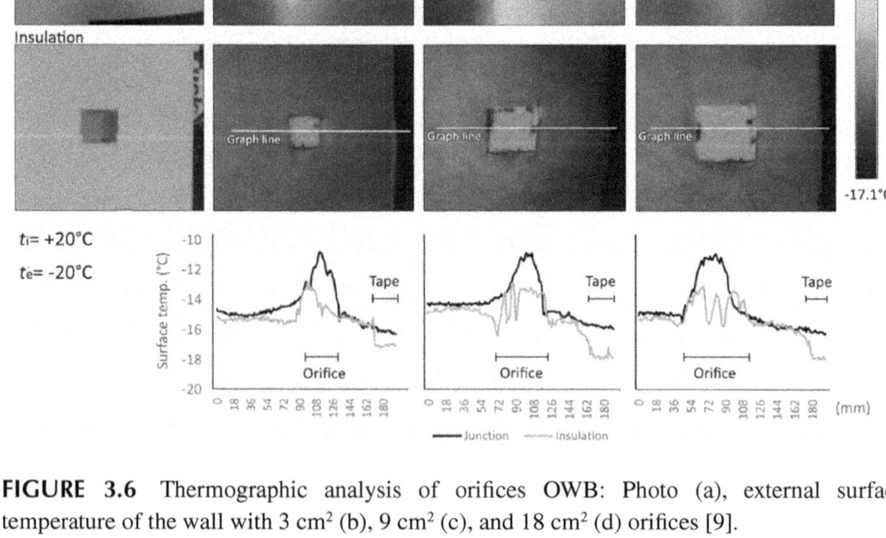

FIGURE 3.6 Thermographic analysis of orifices OWB: Photo (a), external surface temperature of the wall with 3 cm^2 (b), 9 cm^2 (c), and 18 cm^2 (d) orifices [9].

wall's heat transfer coefficient using the building's energy audit and internal and outdoor air temperatures. Thermal imaging determined TB incidence.

Common thermal imaging challenges include collision-free access from each side of inhabited structures for perpendicular measuring. Dynamic infrared analysis is used in thermal imaging literature to detect hidden details or flaws in a surface's thermal image under controlled heating. The method was used to map plaster delamination or moisture infiltration or discover plaster-covered wall paintings in heritage buildings without damaging the fabrics [10].

3.2 THERMAL BRIDGES

Envelope thermal bridging saves building energy. Energy performance is affected by 70% of heat loss from the building envelope (walls, floors, roofs, and windows). Building heat loss is mostly caused by door and window openings, although 19.87% of fabric building heat loss comes from external walls. Airtightness and insulation are increased to reduce building energy demand. Improving building envelope insulation without addressing TBs may not reduce heat losses. As opaque wall surfaces make up 20–50% of building envelopes, TBs may be inevitable [11].

Winter heat losses and summer heat gains from TBs affect building energy consumption. They also increase the interior mold risk from winter condensation. The applicable technical standards give thorough and simplified procedures to compute thermal losses under steady-state circumstances for TB analysis.

Linear thermal transmittance is important for analyzing heat exchange via building envelope TBs because most are linear.

An infrared camera's thermal sensor calculates each pixel's temperature depending on the subject's radiative radiation. This defines the entire thermal field of the camera lenses' region. Buildings with TBs, such as joints between structural elements or composition changes, experience two effects: altered heat flow and varying internal and external surface temperatures [12].

Figure 3.7 shows the thermal insulation plaster (TIP) benefits. TIP mitigates thermal bridging in structures across various climates. Cold-climate countries have a greater temperature difference between the interior and exterior, resulting in more thermal waste. TIP can save energy, money, and greenhouse gas emissions by reducing thermal bridging. TIP from the hawk board can be troweled to the ceiling to ensure proper application at the wall-ceiling joints, where thermal bridging is greatest. The plaster is then tightly applied with broad, sweeping strokes, the trowel slightly slanted away from the ceiling.

This begins at the wall's edge and moves to the ceiling. Plaster is applied in 15-mm layers to avoid trowel marks and flatten the surface. To fill gaps between the wall and ceiling, the process is repeated after the initial layer of plaster. For exact corners, use a corner trowel. TIP is eco-friendly as it may use trash and leftovers. Businesses and customers value energy-efficient TIP buildings. TIP works for energy-efficient house renovations. TIP enhances thermal comfort, energy efficiency, and sustainability.

TBs lose 30% heat, and inertia and mass distribution differ between wood frame and masonry walls. Thus, it differs from a homogeneous wall with continuous layers.

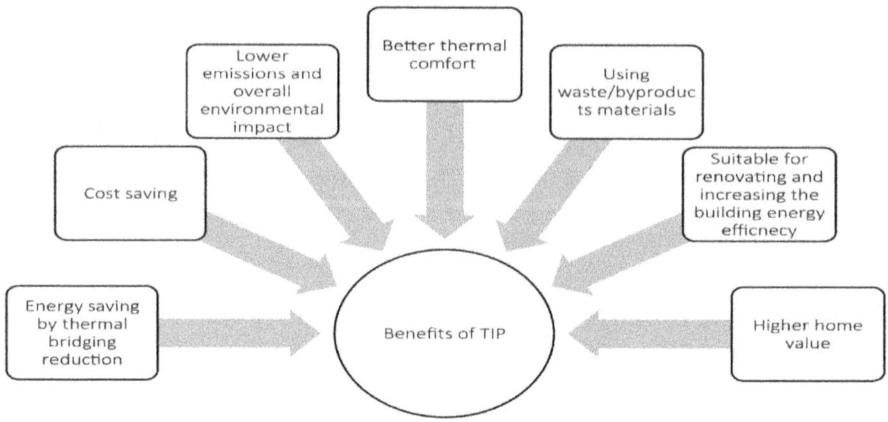

FIGURE 3.7 TIP benefits [13].

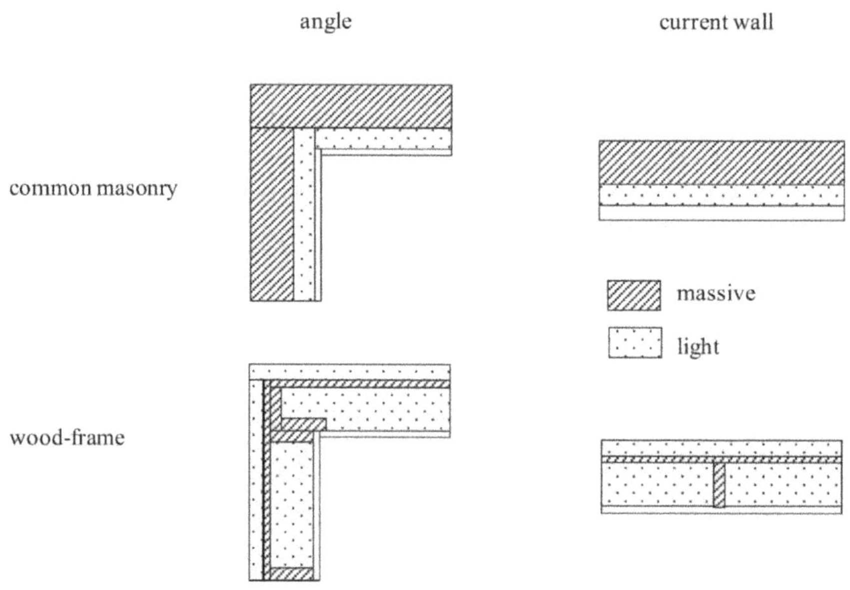

FIGURE 3.8 Wood-frame structures [14].

It generates many TBs, yet makes up most of the wall inertia. Traditional masonry walls spread mass (Figure 3.8).

Wooden frame construction impacts thermal performance; thus, it must be examined. Due to the surge in wooden frame construction in France, TBs must be examined. Presupposed transmission coefficients do not match reality. Designers must precisely predict losses and inertia, which affect comfort. Twenty years have been spent studying dynamic TB issues. The model uses the two-dimensional (2D) conduction equation. TBs were easy to add to steady-state computational tools. Numerous numerical methods exist to characterize 2D or 3D parts' thermal behavior. Figure 3.9 shows model comparisons by amplitude and latency discrepancy with the detailed scenario.

Thermal analysis shows that the thermal break alters bracket heat flow, which is strongly connected to thermal conductivity and thickness. Mechanical strength restricts thick, low-thermal-conductivity materials. High strength and good thermal insulation are typically incompatible. Typically, a thermal break with 1 W/(m·K) thermal conductivity and a maximum thickness of 6mm is the best choice. Finally, the thermal break barrier only inhibits heat transfer through bracket-wall contact, not the anchor that enters the wall [15].

Steel structure in panel cavity insulation has no effect on thermal performance. Neglecting heat conduction through metal profiles can overestimate thermal resistance by 50%. The EnergyPlus application can model thermal bridging effects using enclosure element material composition from its database (Composite Wall Constructions). This database records fabricated insulating materials with reduced widths and enhanced thermal conductivity values to accommodate thermal bridging effects when wall panel data is submitted.

Load-bearing and temporary bridges. Figure 3.10 juxtaposes the monthly thermal load peaks for CTP and WS (frameless). Methods vary from 7.13% to 22.94%. Heat loads were mitigated through sun orientation, reduced opaque panels, and west-facing translucent panels. The absence of residual loads resulted in constrained thermal load peaks due to the continuous operation of the air conditioning system throughout the day.

Modeling thermal load sizes with or without thermal bridging differs 10%. The primary explanation is the temperature difference between indoor (T = 24°C) and outdoor (T_{max} = 29.7°C) conditions. Adding 25 mm expanded polystyrene (EPS) to external panels almost eliminates simulation discrepancy.

3.3 INSULATION SYSTEM DEFECTS

Increasing envelope thermal insulation by insulating vertical walls is standard. Manufacturing defects may reduce insulation and damaged materials, faulty installation, or aged or weathered components might interfere with vertical or sloped envelope heat transfer [17].

The thin cell walls of closed-cell foam insulation decrease heat transfer. Insulation explains why water enhances thermal conductivity and how it varies by type. The cross-sectional heat transfer routes of fibrous (e.g., mineral- and glass-wool) and foam-based (EPS or polyisocyanurate (PIR) insulation materials are shown in Figure 3.11.

Wall protection is essential, and efficient thermal insulation reduces wall moisture and maintains thermal comfort, preventing wall damage. Construction materials influence thermal insulation, which regulates heat flow between structures. Due to temperature, building moisture affects the relative humidity and the indoor atmosphere. Air temperature change is 80% due to surface temperature. Visual evaluation of building exteriors with thermal images identifies energy audit issues. Pressure tests and thermal imaging detect leaks, but they are expensive and need expertise.

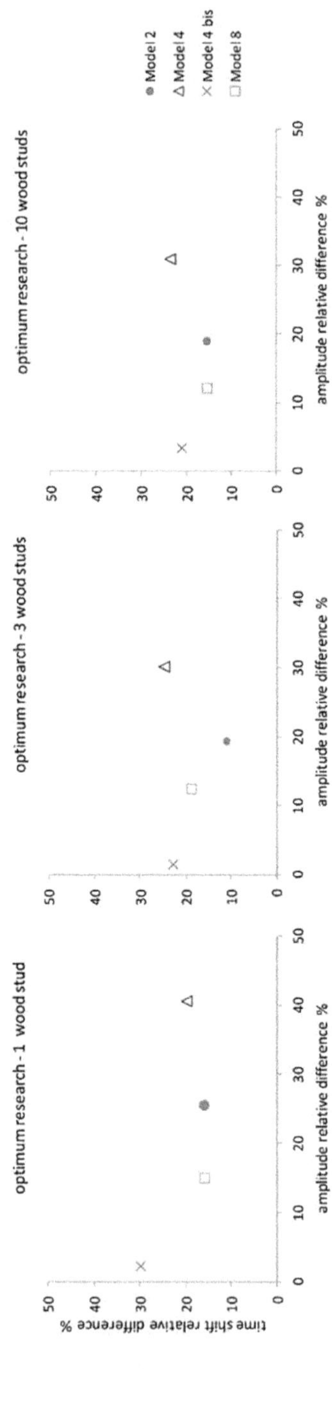

FIGURE 3.9 Comparison of models for integrated TBs [14].

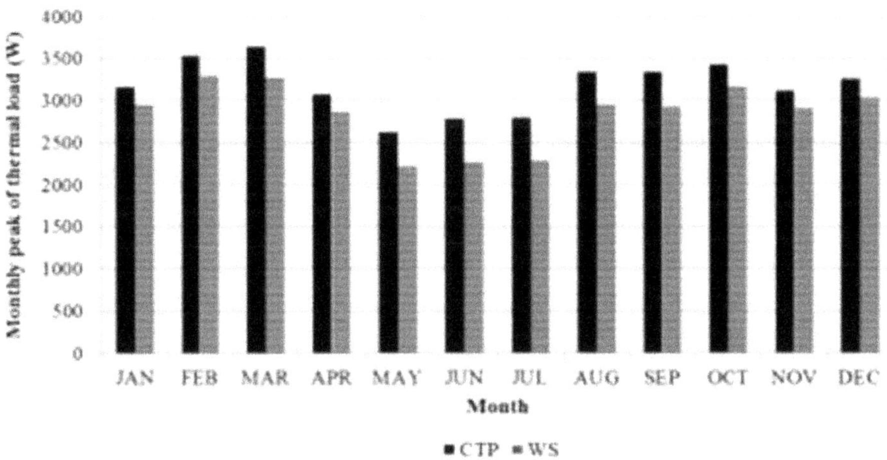

FIGURE 3.10 Monthly peak of thermal load [16].

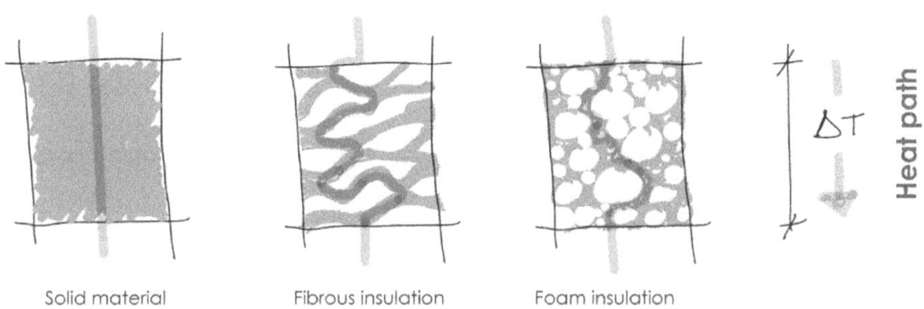

FIGURE 3.11 Designing cross-sectional heat transfer routes in diverse materials [18].

While thermal camera palettes show hotter and colder places in thermal images, that does not mean there are temperature problems or thermal features that are leaking. Several image segmentation approaches categorized building thermal pictures by temperature. The performance and execution time of temperature segmentation methods vary. For image temperature fluctuation detection, segment pixel temperature data was expressed using symbolic aggregation approximation (SAX). Image temperature disparities were detected. Thermal imaging temperature analysis is the only way to determine the damaged area of the wall [19].

3.3.1 Classification of Defects

Investigators may label the same issues "minor," "slight," "moderate," or "very severe." Crack width classifications that do not always assess damage have been abandoned. This categorization ignores the defect cause and worsening. Building research establishment defines damage according to "aesthetic," "serviceability," and "stability."

Every component's molecular structure can be modified in a two-component gel. Checking strictly site supervision has increased in recent development to improve quality. Builders and subcontractors should inspect every day for workmanship faults and rectify them immediately. Supervisory staff must also understand and oversee construction activities and craft workers.

Unclear specs and designs always lead to poor construction. Prepared blueprints and drawings simplify future work and help find and remedy faults [20].

Three faults can occur in insulated buildings.

1. *Thermal Bridge Design Flaws:* They commonly occur where building materials meet or structure composition changes.
2. *Implementation or Workmanship Defects:* Retrofitting produces them for practical reasons. Lack of worker rigor and site control generally causes them.
3. *Lifetime Defects:* Aging, settling, and humidity produce insulation failures.

Design TBs and implementation faults cause most insulated structural heat losses. These frequent effects have been mainly ignored in models, despite lowering envelope thermal resistance. These problems increase gearbox loads, reducing thermal insulation benefits in low-energy buildings.

Figure 3.12 illustrates an external wall with hollow blocks, 8 cm mineral wool insulation, and inner and outer plaster. The insulation panel is 0.3 m broad and 1 m long, thus all problems are likely width-related. Figure 3.13 shows how dimensional flaws affect insulation thermal conductivity. The insulating resistance decreased regardless of the fault, indicating a major issue.

Most problematic is MW-Sh at 2 cm (0.5% insulating volume). Thermal performance losses are 40%, compared to 36%, 28%, and 12% for openings, grooves, and crushes. Performance is 60% lost for a 4 cm fault (20% of insulating volume), 44% for a groove, 40% for a sheath, and 20% for a crush.

An incorrect heat insulation system reduces energy savings. Unexpected TBs and lower energy savings during heat insulation project preparation. Systemic project planning neglect reduces energy savings. Butt-fastened insulation without a perimeter bead can damage façades.

A considerable decrease in stickiness and stability may damage the façade. Incorrectly bonded boards might repeatedly sag, expanding plaster and damaging

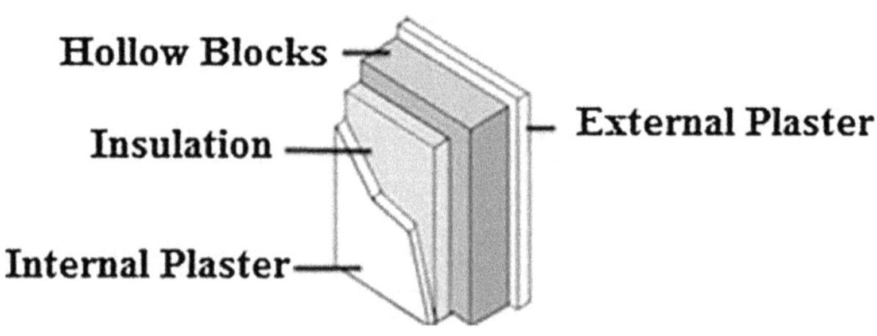

FIGURE 3.12 External structure of an insulated wall [21].

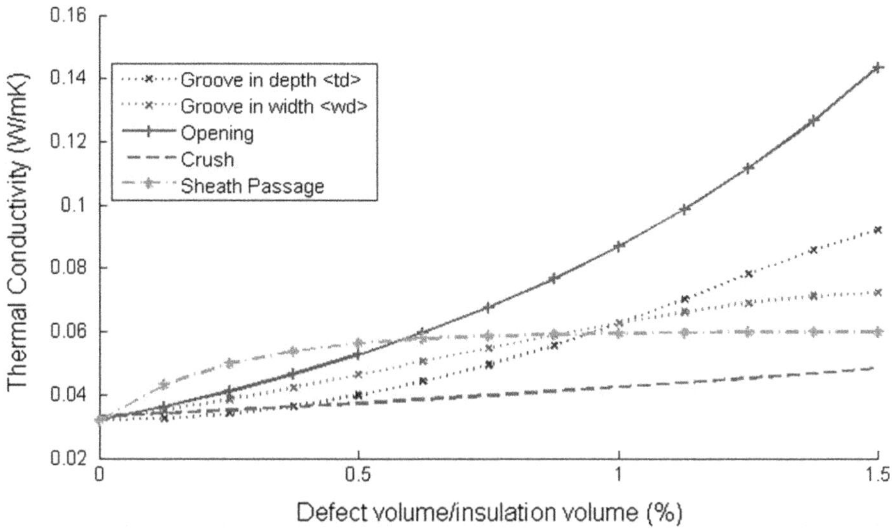

FIGURE 3.13 Type and size of defects affect heat conductivity [21].

board contact areas. Building perimeters without board gluing reduces fire safety due to vertical gas fluxes [22].

The examination concluded that failing reinforcement bar installation, element dimensioning, wall-floor slab couplings, and floor grouting cause building defects. Construction surface issues include honeycombs in concrete—unintended cavities formed by mortar failing to fill aggregate spaces—and bumps, dips, and uneven surfaces caused by missing grout and mastic sealer. Paint and plastering were missing, and rooms and moist areas had inadequate floor and wall finishing, causing post-handover issues. These issues may result from poor construction protection. Problematic building regions demonstrate that many construction faults are undetected. These defects damage pillars, the exterior, and other building sections. They were considered "generic."

Analyzing subcontracted trades helped reach more exact conclusions. When it comes to buildings, enclosures, and installations in general, subcontractors are the root of 68% of all construction issues. General issues account for only 6% of all problems after the structure is turned over. Surface flaws predominate in bedrooms, bathrooms, and kitchens. Construction flaws and post-handover defects differ due to warranties and client and contractor quality assessments.

In the first 10 years, the contractor fixes structural faults. Other living situation problems have a 2-year liability term. Minor faults must be fixed by the contractor before handover within 12 months. Even though research and best practices illustrate the benefits of quality management, the Spanish housing industry rarely incorporates it.

Contractors concentrate quality control on structural defects that could be serious during the defects liability period (DLP). During quality control and inspection, serious faults are usually reduced or removed before handover. Poor-quality subcontractors cause major issues. Once subcontractors leave the site, the primary contractor may have trouble getting them to return to remedy stains, cracks, missing fittings, or a lack of a second coat of paint. Numerous people will have started new enterprises.

Thus, post-handover finishing is required. Quality standards and client–contractor disagreements may explain the gap between construction and post-handover issues. A contractor prioritizes technical quality like foundations and structural integrity over functional quality like paintwork and aesthetics. Most consumers judge a product's practicality by its appearance, cleanliness, presentation, and feel.

Most "inappropriate installation" faults are workmanship-related, such as poorly completed sheet pile walls and poorly installed prefabricated wall planks. Some "surface" flaws are also linked to craftsmanship, such as improper grouting and mixing. Contractors subcontract most building activities to specialist trade contractors. Trade contractors perform 90% of construction work, while the principal contractor manages and coordinates. Most Spanish builders use small subcontractors, exacerbating the issue.

To clearly illustrate the key insulation techniques discussed and address potential defects and their solutions, Table 3.1 provides a concise comparative overview.

3.4 AIR LEAKAGE DEFECTS

Air leakage affects energy consumption, hygrothermal properties, interior air quality, air pressure dynamics throughout the envelope, natural ventilation, and fire safety. Building management techniques, kind of housing, age, and number of floors are the main factors determining building envelope air leakage.

An effective air barrier and its connection with other architectural elements are needed to seal the building. Implemented design solutions may affect air barrier location. On the building envelope's interior or exterior. Load-bearing structures or insulation can form the air barrier. Traditional timber-framed building envelopes use vapor barriers as air barriers in cold areas. Due to air leakage at building envelope connections and air barrier material, tightening and sealing air barrier joints is essential for building airtightness.

The effects of sealing and tightening structural floors, basement walls, chimneys, and windows on air leakage have been studied. Air leakage rates are now routinely measured before a new building is occupied [23].

Extreme and careless airtight sealing of buildings and efforts to minimize fresh air exchanges as part of a heating energy-saving program harm interior air quality and microclimate. Long-term, this can cause sick-building syndrome, which can cause health problems. Sick-building syndrome occurs when rooms and buildings worsen residents' low well-being, causing weariness or disease. This syndrome includes eye irritation, respiratory tract or skin irritation, nausea, headaches, pain, tiredness, and attention issues. These symptoms are linked to poor microenvironment, especially air quality. Research shows that up to 80% of residents of a tightly sealed structure can have health difficulties due to poor ventilation [24].

European lightweight construction uses vapor barriers to seal the building envelope. Electrical and plumbing junctions, intersections, and penetrations hinder

TABLE 3.1
Comparison of Insulation Techniques and Solutions for Defects

Insulation Technique	Description	Advantages	Limitations	Solutions for Common Defects
Thermal Insulation (General)	Application of insulating materials (e.g., fiberglass, mineral wool, foam) to walls, roofs, and floors.	Reduces heat transfer significantly, improves energy efficiency, enhances indoor comfort.	Requires precise installation, risk of moisture accumulation.	Ensure proper vapor barriers, seal joints and edges meticulously.
External Wall Insulation	Insulation fixed externally on building façades (e.g., EPS, mineral wool panels).	Effective thermal bridging reduction, external aesthetics improved.	Vulnerable to weather damage, higher initial cost.	Protective external renders, proper drainage design.
Cavity Wall Insulation	Filling cavity walls with insulating materials like foam or blown cellulose.	Excellent retrofit solution, effective thermal barrier.	Risk of moisture penetration if not properly sealed.	Regular checks for moisture ingress, effective sealing of wall cavities.
Internal Wall Insulation	Insulating internal surfaces with boards or panels.	Cost-effective, easy installation.	Reduction in internal space, potential condensation issues.	Install proper vapor control layers, ventilate adequately.
Reflective Insulation	Foil-based insulation reflecting radiant heat.	Highly effective in hot climates, space-efficient.	Limited effectiveness in cold climates, dependent on proper installation.	Proper sealing to avoid gaps, use complementary insulation methods.
Green Roof Systems	Vegetative layer over insulation and waterproofing.	Excellent thermal regulation, environmental benefits.	High initial cost, weight constraints.	Robust waterproofing, proper drainage layers.

airtightness. To stop natural and induced convection heat loss, the cold side of the insulation should have an impermeable wind barrier [25].

Air leakage or infiltration occurs when cracks and fissures in the building envelope allow air to enter. Air leakage varies from regulated ventilation, which brings fresh air into buildings. The best way to conserve energy in the future is to reduce air leakage and manage ventilation. The building's age, wind conditions, interior–exterior

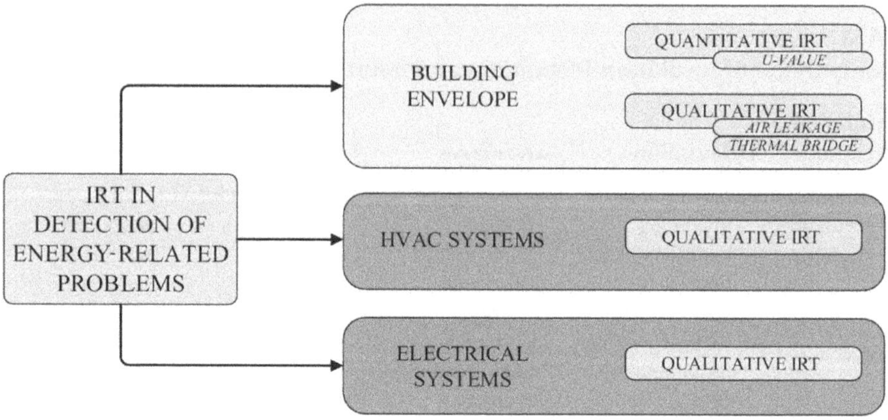

FIGURE 3.14 Building energy problem identification with IRT [26].

temperature differentials, and passive and active systems affect air infiltration. Before addressing air leakage in building envelopes, IRT should identify infiltration spots.

Figure 3.14 shows IRT energy assessments of the building envelope, HVAC, and electrical systems. According to the literature, most IRT building envelope evaluations lose energy performance [26].

3.5 ROOFING SYSTEMS

An innovative roofing system (IRS) generates a lot of cool natural light from the roof while preventing heat gain in tropical residential structures. Residents and low-rise structures gain almost 70% of heat from the roof. Up to 1 kW/m^2 of solar radiation affects roofs. Approximately 20–90% heat is absorbed. This framework has three parts: heat prevention/reduction, thermal moderation, and heat dissipation. Figure 3.15 categorizes and illustrates each component's many methods.

Daylighting and passive cooling are examined in this IRS development project (Figure 3.16).

Figure 3.17 compares cool and traditional roofs to extensive and intense green roofs (20 and 50 cm planting media). The multilayer canopy-root-soil model (MLcan) and forcings were adjusted to replicate non-vegetated roof conditions with extremely low soil moisture (0.01), no vegetation (LAI = 0), and no precipitation. White (cold) roofs had 0.95 soil reflection (albedo) and black (traditional) 0.05. Comparison: 0.2 green roof soil reflectivity. Green roofs reflect and split energy because of wetness and vegetation, unlike cool roofs. Infiltration and water storage are optimized by high-porosity, high-conductivity planting mediums and water retention panels on most green roofs [28].

3.5.1 Roof Garden

Roof gardens should be both safe and profitable due to their intended use, environmental benefits, and beautification. It is recommended to construct a garden-art forest on the platform, balcony, windowsill, parapet, and metope in addition to the main

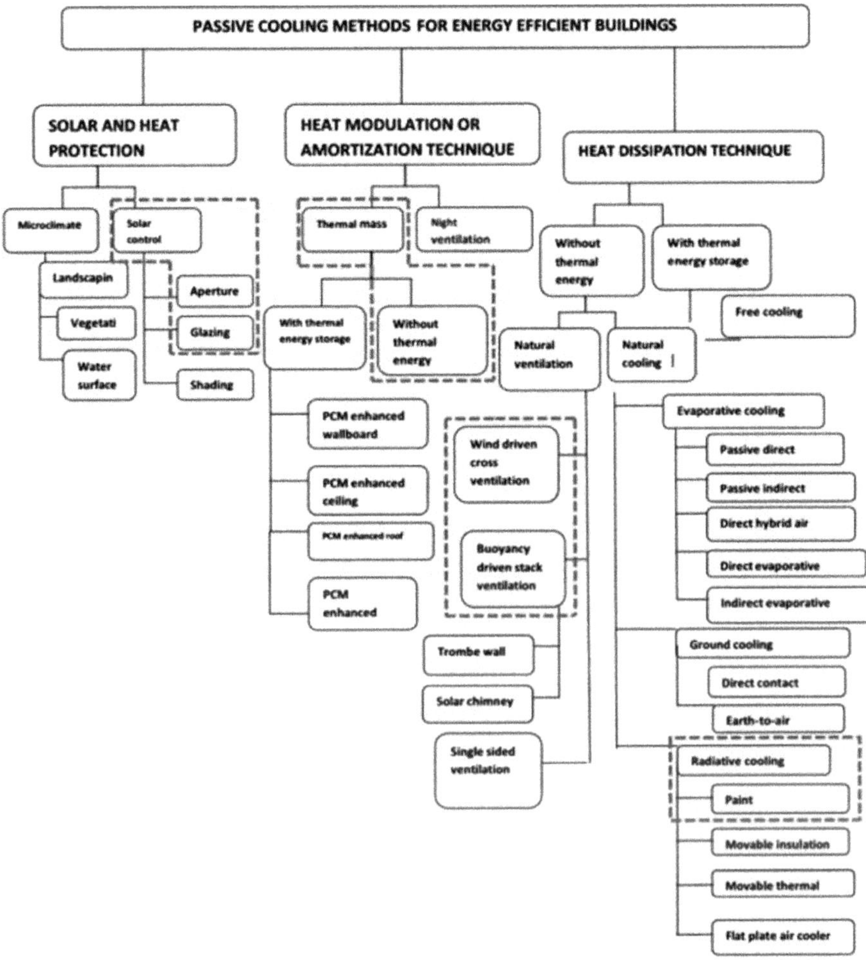

FIGURE 3.15 Passive cooling methods in energy-efficient buildings classified by important criteria [27].

body building. To save water, preserve the ecosystem and beauty, and decrease trash, clean and use dead leaves locally. Research discovered green roofs reduce summer heat transfer by 80%. In July, green roofs saved 2.2–16.7% of energy. Slope roof gardens for habit. Slopes or steps lead to the roof, turning the "wasteland in the air" into a green playground. New planting roof technology allows modular planting roofs. This promotes green towns and lowers construction expenses.

As fresh water is scarce, green buildings incorporate water-saving devices. Rainwater-collecting and water-saving appliances are popular. Popularization requires water conservation knowledge and government subsidies. Housing needs must be balanced with environmental needs. Vegetation reduces heat absorption, greenhouse impact, and ventilation difficulties. Designing the greenery system should include waterproofing, plant types, and other factors [29].

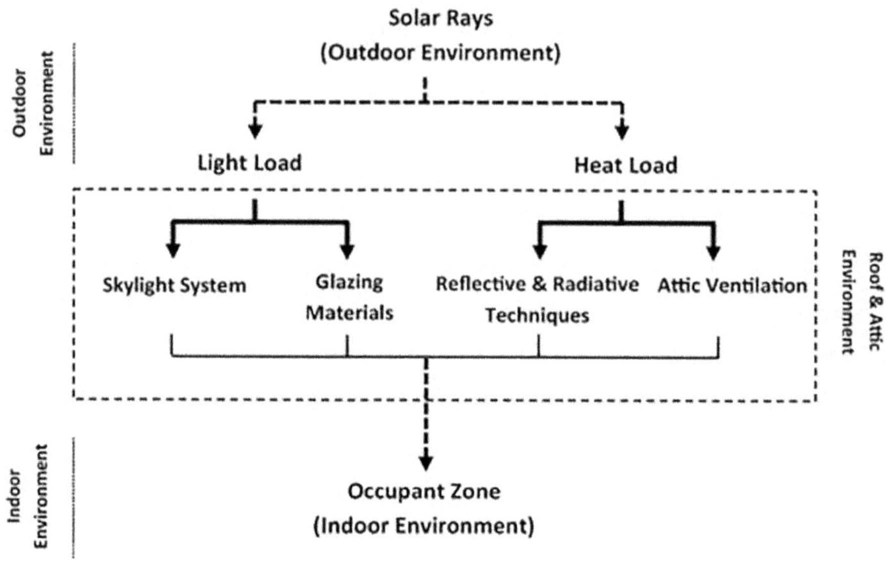

FIGURE 3.16 IRS solar design methodology [27].

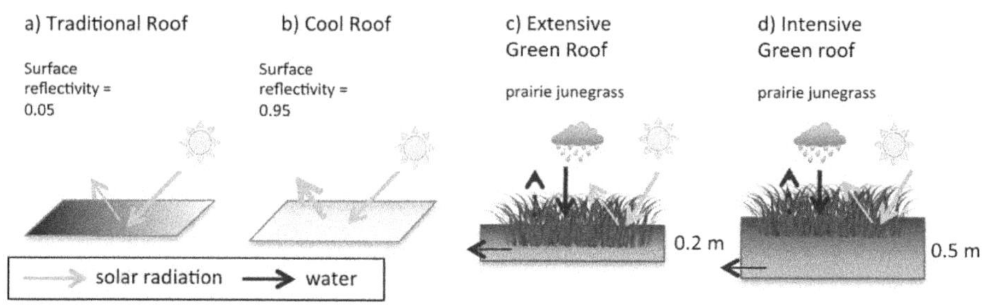

FIGURE 3.17 Four roofing techniques were simulated using MLCan: (a) standard black roof, (b) cool white roof, (c) extended green roof with 0.2 m soil layer, and (d) intensive green roof with 0.5 m soil layer [28].

REFERENCES

1. Sadineni, S. B., Madala, S., & Boehm, R. F. (2011). Passive building energy savings: A review of building envelope components. *Renewable and Sustainable Energy Reviews*, *15*(8), 3617–3631.
2. Papadopoulos, A. M. (2016). Forty years of regulations on the thermal performance of the building envelope in Europe: Achievements, perspectives and challenges. *Energy and Buildings*, *127*, 942–952.
3. Zhang, T., Tan, Y., Yang, H., & Zhang, X. (2016). The application of air layers in building envelopes: A review. *Applied Energy*, *165*, 707–734.
4. Al-Homoud, M. S. (2005). Performance characteristics and practical applications of common building thermal insulation materials. *Building and Environment*, *40*(3), 353–366.

5. Kumar, D., Alam, M., Zou, P. X., Sanjayan, J. G., & Memon, R. A. (2020). Comparative analysis of building insulation material properties and performance. *Renewable and Sustainable Energy Reviews*, *131*, 110038.
6. Alencastro, J., Fuertes, A., & de Wilde, P. (2018). The relationship between quality defects and the thermal performance of buildings. *Renewable and Sustainable Energy Reviews*, *81*, 883–894.
7. Kirimtat, A., & Krejcar, O. (2018). A review of infrared thermography for the investigation of building envelopes: Advances and prospects. *Energy and Buildings*, *176*, 390–406.
8. Rakha, T., El Masri, Y., Chen, K., Panagoulia, E., & De Wilde, P. (2022). Building envelope anomaly characterization and simulation using drone time-lapse thermography. *Energy and Buildings*, *259*, 111754.
9. Piggot-Navarrete, J., Blanchet, P., Cabral, M. R., & Perez, C. (2023). Investigating the impact of construction workmanship defects on the hygrothermal performance and airtightness of lightweight-structure wooden envelope systems. *Building and Environment*, *243*, 110705.
10. Ostańska, A. (2018). Thermal imaging for detection of defects in envelopes of buildings in use: qualitative and quantitative analysis of building energy performance. *Periodica Polytechnica Civil Engineering*, *62*(4), 939–946.
11. Chen, G., Hou, Y., Ge, H., Zhang, S., Liu, X., Guo, X., & Xie, D. (2024). Effect of thermal bridges on the energy performance of Chinese residential buildings. *Energy and Built Environment*, *6*(3).
12. Garrido, I., Lagüela, S., Arias, P., & Balado, J. (2018). Thermal-based analysis for the automatic detection and characterization of thermal bridges in buildings. *Energy and Buildings*, *158*, 1358–1367.
13. Kassim, M., Aslani, A., & Zahedi, R. (2024). Energy performance analysis of thermal insulating plaster in the different climate zones. *Thermal Science and Engineering Progress*, *47*, 102294.
14. Viot, H., Sempey, A., Pauly, M., & Mora, L. (2015). Comparison of different methods for calculating thermal bridges: Application to wood-frame buildings. *Building and Environment*, *93*, 339–348.
15. Theodosiou, T., Tsikaloudaki, K., & Bikas, D. (2017). Analysis of the thermal bridging effect on ventilated facades. *Procedia Environmental Sciences*, *38*, 397–404.
16. Gomes, A. P., de Souza, H. A., & Tribess, A. (2013). Impact of thermal bridging on the performance of buildings using Light Steel Framing in Brazil. *Applied Thermal Engineering*, *52*(1), 84–89.
17. Nardi, I., Perilli, S., de Rubeis, T., Sfarra, S., & Ambrosini, D. (2019). Influence of insulation defects on the thermal performance of walls. An experimental and numerical investigation. *Journal of Building Engineering*, *21*, 355–365.
18. O'Hegarty, R., Amedeo, G., & Kinnane, O. (2024). The impact of compromised insulation on building energy performance. *Energy and Buildings*, *316*, 114337.
19. Khan, N., Pathak, N., & Roy, N. (2019, June). Detecting common insulation problems in built environments using thermal images. In Antonio Puliafito & Aryya Gangopadhyay (Eds.), *(2019 IEEE International Conference on Smart Computing (SMARTCOMP)* (pp. 454–458). IEEE.
20. Hasan, M. I. M., Abd Razak, N. N., Endut, I. R., Samah, S. A. A., Ridzuan, A. R. M., & Saaidin, S. (2016). Minimizing defects in building construction project. *Jurnal Teknologi*, *78*(5–2).
21. Aïssani, A., Chateauneuf, A., Fontaine, J. P., & Audebert, P. (2016). Quantification of workmanship insulation defects and their impact on the thermal performance of building facades. *Applied Energy*, *165*, 272–284.

22. Vlček, P., Kubečka, K., Kubečková, D., & Vaculíková, H. (2013). Defects of insulation systems and their negative effect on the accumulation and energy saving. *Advanced Materials Research*, *649*, 143–146.
23. Kalamees, T., Alev, Ü., & Pärnalaas, M. (2017). Air leakage levels in timber frame building envelope joints. *Building and Environment*, *116*, 121–129.
24. Lis, P. (2020). The optimal amount of air for the natural ventilation of classrooms as a function of the rate of outdoor air leakage from windows. *Zeszyty Naukowe Politechniki Częstochowskiej. Budownictwo*, 108–113.
25. Gullbrekken, L., Bunkholt, N. S., Geving, S., & Rüther, P. (2020). Air leakage paths in buildings: Typical locations and implications for the air change rate. In J. Kurnitski & T. Kalamees (Eds.), *E3S Web of Conferences* (Vol. 172, p. 05010). EDP Sciences.
26. Kirimtat, A., & Krejcar, O. (2018). A review of infrared thermography for the investigation of building envelopes: Advances and prospects. *Energy and Buildings*, *176*, 390–406.
27. Al-Obaidi, K. M., Ismail, M., & Rahman, A. M. A. (2014). Design and performance of a novel innovative roofing system for tropical landed houses. *Energy Conversion and Management*, *85*, 488–504.
28. William, R., Goodwell, A., Richardson, M., Le, P. V., Kumar, P., & Stillwell, A. S. (2016). An environmental cost-benefit analysis of alternative green roofing strategies. *Ecological Engineering*, *95*, 1–9.
29. Lai, F., Zhou, J., Lu, L., Hasanuzzaman, M., & Yuan, Y. (2023). Green building technologies in Southeast Asia: A review. *Sustainable Energy Technologies and Assessments*, *55*, 102946.

4 Natural Daylight and Efficient Lighting

4.1 INTRODUCTION

Residential energy consumption constitutes 25% of total energy use in Europe. Tertiary structures and similar types contribute to 40% of energy consumption and 36% of CO_2 emissions. Enhancing energy efficiency in buildings leads to a decrease in energy consumption and CO_2 emissions. Natural light plays a crucial role in regulating physiological and psychological rhythms, influencing building energy consumption and human activity, thereby enhancing comfort and health. Natural light, owing to its variable characteristics, requires regulation and direction to enhance or substitute artificial lighting. Unregulated natural light can negatively impact the environment, as excessive solar gains elevate cooling energy consumption.

Most natural light control systems reduce the negative impacts of natural light but overlook its positive effects. To reduce a building's external heat load from solar radiation, reliance on natural light may be inadequate, resulting in increased energy consumption for electrical lighting. Windows allow natural light to enter a building; however, as the distance from the windows increases, the impact of natural light diminishes, necessitating the use of artificial lighting.

According to the International Energy Agency (IEA), artificial lighting constitutes 14% of energy consumption in the European Union (EU) and 19% globally. Neglecting daylight incurs high costs due to the electric load in buildings, which is associated with high primary energy expenses. One kilowatt of power utilized on-site requires 3 or 4 kW of primary energy, with the remainder lost as chimney heat at the power plant. Conventional coal or oil-fired power plants achieve an efficiency of 35–40% in converting primary energy into electricity, with an additional 6% loss occurring during transmission. Power plants are responsible for 50% of CO_2 emissions in affluent countries such as the United States, Japan, and Germany. Over 40% of energy consumption in developed nations is allocated to building heating, cooling, air conditioning, lighting, and other requirements [1].

Daylighting contributes to visual comfort, enhances energy efficiency, and supports the advancement of green building practices. Daylight serves as the optimal light source for color rendering and closely aligns with human visual perception. Daylight enhances mood and promotes well-being. Contemporary architecture employs natural lighting to enhance the visual environment. Daylighting is more suitable for

DOI: 10.1201/9781003752837-4

human vision than artificial illumination. Daylighting enhances the indoor atmosphere as individuals seek natural sunlight in their homes and offices. Natural daylight can enhance occupant productivity and contribute to a healthier work environment. There is an increased awareness of the interactions between building energy and the environment. The environmental impact of energy use in buildings is an emerging concern for architects and building designers. Optimizing power usage yields advantages for both environmental sustainability and economic performance. Office buildings exhibit significant electricity consumption due to artificial lighting. *A digital platform.* Limited buildings effectively combine natural sunlight with artificial lighting, and there is a lack of operational performance data. Surveys indicate that daylighting measures are infrequently utilized in commercial buildings. Approximately 50% of commercial buildings in the USA are equipped with energy-efficient lamps and ballasts, while only about 10% incorporate daylighting schemes [2].

4.2 SUNLIGHT

Green architecture conserves resources, safeguards the environment, minimizes pollutants across its life cycle, offers healthy, comfortable, and efficient spatial use, and fosters harmonious synergy with nature. Green buildings optimize multiple objectives:

1. Passive design aims to reduce energy consumption and enhance indoor comfort while also minimizing energy use during operation.
2. When standard measures fail to ensure adequate interior comfort, suitable active interventions are implemented to enhance and decrease energy consumption in buildings [3].

Assess daylight using the daylight factor. It consists of diffused skylight (SC), internally reflected light (IRC), and externally reflected light. Modify these three elements to regulate the quality and quantity of daylight. Daylighting refers to the incorporation of sunlight into building design. Natural light enhances well-being, awareness, and energy efficiency by decreasing reliance on artificial lighting. Proper utilization of windows, SCs, clerestories, and other openings effectively captures daylight.

The location and dimensions of windows influence daylighting effectiveness.

1. For each façade, assess window dimensions, elevation, and glazing specifications individually.
2. Enhance southern and northern exposure.
3. North-facing windows provide indirect light while minimizing heat gain.
4. Minimize eastern and western exposure during periods of low sunlight to prevent glare and overheating.
5. Sunlight at the horizon complicates shading.
6. The height of window heads significantly influences daylight penetration, with sufficient daylight typically reaching up to 1.5 times the head height.

7. Encourage significant indoor daylight penetration. High-wall windows, roof monitors, and clerestories enhance natural light penetration while reducing glare.

For large windows, implement advanced daylight harvesting systems such as exterior light shelves, light tubes, and increased ceiling heights to enhance the penetration of sunlight into the building.

In areas with restricted window access, utilize SCs and roof monitors. Light-colored interior surfaces enhance coverage and reduce luminance contrast [4].

Heat management is essential as sunlight penetrates the building. Proper aperture placement allows for natural assistance. Integrate daylighting with artificial lighting to enhance energy efficiency, while acknowledging the potential rise in air-conditioning demands due to increased heat gain. Designers and decision-makers must balance daylight harvesting with solar heat gains. Incorporating daylighting into building design from the outset is essential for achieving net-zero energy building (NZEB) design objectives and minimizing energy consumption.

The daylight factor is a widely utilized metric for evaluating a window's potential illuminance in a room under adverse conditions, specifically during overcast weather with reduced daylight availability. Currently, daylight factors represent the predominant metric for assessing daylighting. The Commission on Illumination (CIE) considers this definition as a fundamental lighting metric. The calculation remains unaffected by room location, as daylight factors are assessed under cloudy conditions, rendering the sun's position irrelevant. Therefore, the measurement of the sunshine factor is influenced solely by model geometry, excluding factors such as time, window orientation, or room placement.

The daylight factor is a widely used method for evaluating a window's potential illuminance in a room under adverse conditions, such as overcast skies with reduced daylight. Currently, daylight factors represent the predominant metric for assessing daylighting. The CIE regards this definition as a fundamental lighting statistic. The calculation remains unaffected by room location, as daylight factors are assessed under cloudy conditions, rendering the sun's position irrelevant. Therefore, the measurement of the sunshine factor is influenced solely by model geometry, excluding factors such as time, window orientation, or room placement [5].

4.2.1 Solar Radiation and Lighting Analyses

Building information modeling (BIM) software evaluates the effects of lighting on both the exterior and interior of buildings. BIM software incorporates an advanced solar radiation analysis feature that assists designers and engineers in optimizing building sun exposure. BIM software can illustrate the sun's position and trajectory in relation to a building model at any given time and location, enabling designers and engineers to optimize the building's placement and orientation from the outset. BIM software assesses solar gain on building surfaces, temperature variations, and radiant energy exchange. Visualize and report assessment results for any specified time interval in building thermal analysis. BIM software such as Virtual Environment (VE) can simulate both internal and external solar shading effects and evaluate them against the design. Designers may utilize the comparison to select shading systems.

BIM software utilizes advanced analysis of lighting conditions to enhance daylighting and visual comfort in buildings. Designers and engineers can assess their lighting solutions visually through the building lighting overview provided by BIM software. BIM software offers a detailed simulation of both natural and artificial light on a point-by-point basis [6].

Weather variables have a significant effect on the proportions of blue (B) and red (R) light at sun elevation angles below 20°, while the proportion of green (G) light remains relatively constant. The red-to-far-red (R:FR) ratio at low solar elevation varies from 0.8 in overcast conditions to 1.3 in sunny conditions. A study of previous tests on tree seedlings indicates that the R:FR ratio is correlated with the light quality responses of plants across various latitudes, as lower solar angles persist longer with increasing latitude. Midday measurements of sunlight spectral composition at varying solar elevation angles indicated a reduced proportion of blue (B, 400–500 nm) and red (R, 600–700 nm) light, alongside an increased proportion of green (G, 500–600 nm) light compared to sunset. The study indicated that clouds and dust cover exert a minimal influence on light spectra, primarily impacting B and R light, akin to the alterations in solar light as it passes through a plant canopy. While extensive knowledge exists regarding the effects of static light quality on gas exchange and other physiological processes in plants, there is limited understanding of the implications of dynamic light quality changes on plant biology. The impact of ultraviolet (UV) light on plants, including shoot elongation and the synthesis of UV-protecting secondary metabolites, as well as the roles of UVB receptors (UVR8) and UVA-blue light receptors, has been extensively studied. The signaling effects of UV light on plants diminish with heightened radiation, indicating that twilight is the most critical period for UV exposure. In higher latitudes characterized by prolonged darkness, variations in morning and evening light quality may serve as critical signals for plants. Multiple studies have shown that various ecotypes of tree species respond differently to red or far-red light treatments as a means of day extension, likely due to adaptations to the end-of-day light quality characteristic of their native environments. Elevated temperatures restrict the growth-enhancing effects of far-red light, indicating that light quality can interact with other environmental factors such as temperature. Comprehending the variation in plant light quality is essential for forecasting the impacts of increasing temperatures, particularly in marginal areas such as those adjacent to tree latitudinal range boundaries [7].

4.2.2 Utilization of Sunlight

Solar energy represents a clean and non-polluting renewable resource. This design incorporates an integrated solar energy building system with a conventional brick wall structure to reduce costs. The wall utilizes insulation techniques characteristic of the straw ecological composite wall. Façade windows and SCs effectively resolve indoor illumination challenges. Due to the substantial temperature difference between the interior and exterior of the ground, a ground source heat pump is employed, while the service center and its auxiliary rooms utilize solar photovoltaic (PV) panels.

Optimize solar energy utilization through applications including solar building integration, PV panels, supplementary sunlight booths, solar water heaters, and additional initiatives.

Maximize the use of geothermal energy, given the region's extreme cold, which creates a notable temperature differential between the soil surface and subsurface. This technology can be employed for thermal conversion in residential heating, thereby conserving non-renewable energy resources.

Maximize the use of natural light by incorporating insulated façade windows to meet requirements for daylighting and ventilation.

Comprehensive utilization of renewable energy and waste materials is evident; an abundance of masonry and other resources is available for ground paving or wall construction. Concurrently, various straws and other substances can function as wall insulation materials. Human and animal waste can be utilized to generate biogas for cooking, lighting, heating, and various energy requirements.

Evaluating various passive solutions and utilizing renewable energy sources, such as solar energy, significantly reduces reliance on municipal facilities, thereby substantially decreasing energy consumption during construction [8].

Although advancements have been made in forecasting sunshine availability in urban environments with intricate structures, the emphasis has primarily been on the installation of solar panels on rooftops and façades, as well as the impact of solar radiation on heating and cooling requirements. PV panels were engineered to extend beyond building façades and rooftops. Various situations, architectural typologies, and urban forms can yield distinct effects. This analysis provides guidance for decisions regarding early urban morphology indicators and building typologies.

Figure 4.1 presents the average ground and façade illuminance values for all point, slab, and contemporary block typologies. All façades, irrespective of orientation or plot density, exhibit a limited sky view factor of 60.5, in contrast to less obstructed horizontal surfaces, which have a factor of 61.0. Consequently, the reduction in illuminance on façade surfaces—15%, 10%, and 15% for point, slab, and contemporary blocks—is less pronounced than on the ground, where reductions are 45%, 33%, and 44%, respectively. Plant-related factors (PR) significantly influence the availability of sunlight in contemporary block-type structures. The PR in contemporary block typology increases from 1.3 (C1-1) to 5.0 (C4-4), while in point block typology it rises from 0.8 (P1-1) to 5.8 (P5-5), despite both experiencing a similar reduction in illuminance. The presence of PR minimally influences the sunlight availability for slab blocks. An increase in PR from 1.3 (S1-1) to 5.0 (S4-4) results in a reduction of façade and ground illuminance by 10% and 33%, respectively.

Utilizing sunlight on exposed urban surfaces is more advantageous than employing decorative gardens and building façade components in place of artificial interior systems. Increased installation of PV panels on building envelopes, coupled with advancements in PV efficiency, can enhance energy self-sufficiency.

Numerous existing buildings fail to incorporate solar energy resources into their structures and designs, thereby limiting the maximization of solar energy utilization. Building-integrated PV thermal systems can fulfil a building's domestic electrical and thermal energy requirements; however, the variability of solar energy supply complicates the integration of air collectors into the building structure. Additionally,

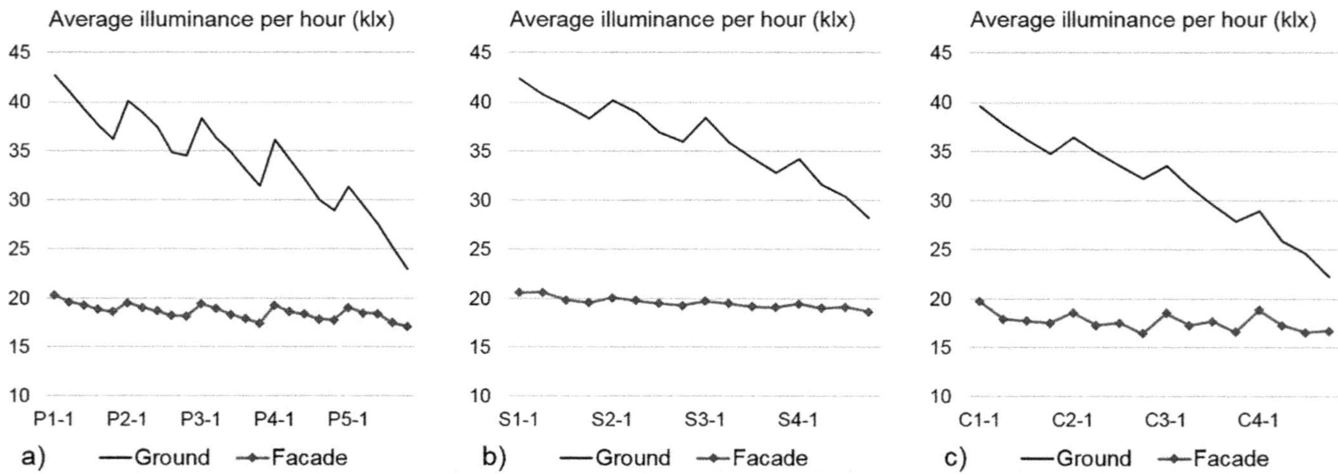

FIGURE 4.1 Average illuminance levels per hour for the ground and façade points, (a) point block cases, (b) slab block cases, and (c) contemporary block cases [9].

the design of the system is significantly affected by the building's structural load-bearing capacity. Building-integrated PVs contribute to the design and development of zero-energy buildings; however, economic and technical challenges need to be addressed. Designers emphasize the maintenance and replacement of PV system components and auxiliary equipment. The economic costs influence policy decisions and the support of the owner [10].

An NZEB generates sufficient renewable and carbon-free energy on-site to offset all energy consumption from the grid, including electrical, thermal, and natural gas, over the course of a year. Energy performance in all-electric buildings will counterbalance emissions from power grid utilization, resulting in net-zero emissions.

Due to the limited on-site renewable energy generation, NZEBs are required to maintain minimal energy demands and exhibit high energy efficiency in their form, envelope, and mechanical systems. Numerous technologies and approaches are available for utilizing solar energy in buildings, particularly in NZEBs. The primary categories of these technologies include PV, thermal, and photovoltaic+thermal (PVT), which can be further divided into two significant subcategories: non-integrated (conventional) and building integrated. Unifacial and bifacial inclined roof-mounted PV systems harness solar radiation to generate electricity from one or both surfaces. Depending on project requirements, one or multiple collectors may be utilized [11].

Energy efficiency in sustainable architecture is achieved through a combination of passive and active design strategies. Passive design solutions involve optimizing a building's orientation, thermal mass, and natural ventilation to minimize energy consumption for heating, cooling, and lighting. Strategic placement of windows and shading devices optimizes natural daylighting and reduces dependence on artificial lighting. Additionally, designing buildings to utilize prevailing winds and thermal masses contributes to temperature regulation indoors and reduces reliance on mechanical heating and cooling systems. Active design solutions improve passive measures through the integration of technologies that dynamically manage energy consumption.

The strategies include high-efficiency HVAC systems, programmable thermostats, and energy-efficient appliances. The integration of advanced technologies with passive design elements enables sustainable architecture to achieve significant energy savings and reduce overall energy consumption. The integration of renewable energy sources is crucial for achieving energy-efficient, sustainable design. Solar panels, wind turbines, and geothermal systems are commonly utilized for the on-site generation of clean, renewable energy. Solar PV panels transform sunlight into electrical energy, whereas solar thermal systems provide hot water for residential applications [12].

Figure 4.2 depicts the mechanism of spectral energy transfer in the reversible $CuSO_4$ liquid-filled energy-efficient window. The management of energy-efficient windows encompasses convective heat transfer and transmitted solar radiation, with seasonal mode switching achieved by rotating the windows to introduce liquid into the outer and inner air gaps. The energy transfer process is outlined below.

1. *Solar Radiation Transfer:* $CuSO_4$ solution's near-infrared absorption allows liquid-filled energy-efficient windows to let only visible light in, regardless of season. The $CuSO_4$ solution would absorb UV and NIR solar radiation, making it portable indoors or outdoors.

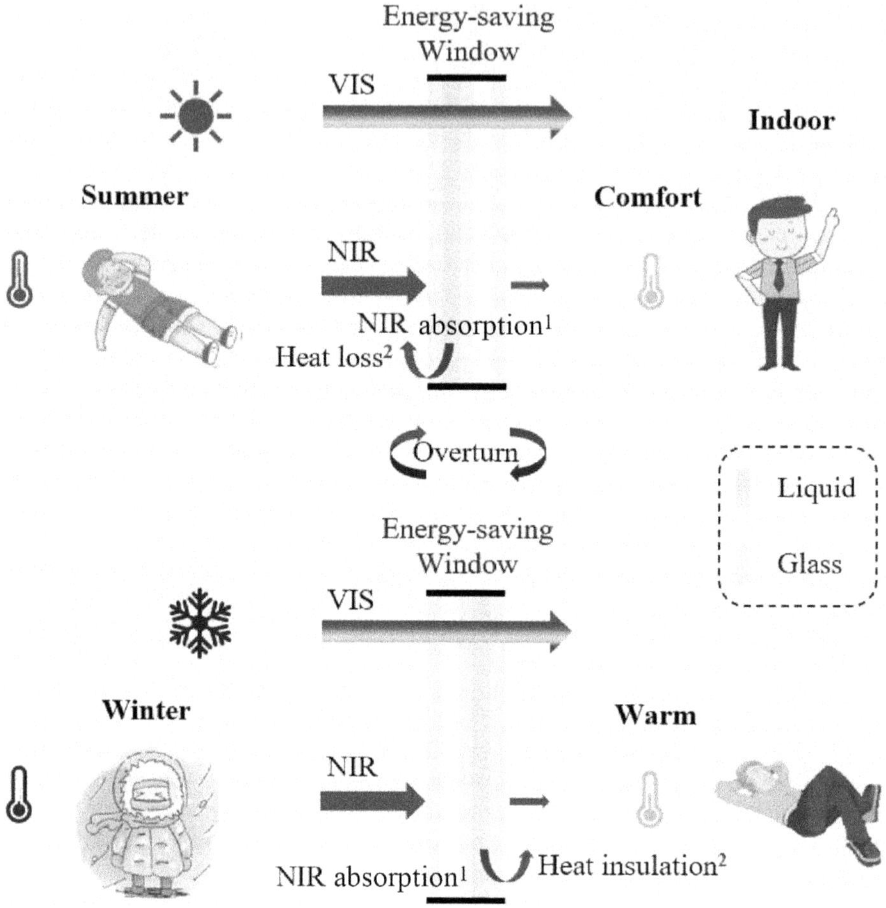

FIGURE 4.2 Process of spectral energy transfer in liquid-filled energy-efficient windows [13].

2. *Convective Heat Transfer:* The liquid-filled layer is in the external air gap in summer mode. Air gap insulation helps the $CuSO_4$ solution absorb solar energy and transport it outside via convective heat transfer. The liquid-filled layer in the inner air gap directs solar energy absorbed by the $CuSO_4$ solution inwards in winter mode, lowering heating demand.

In summer mode, mass fraction affects indoor and outdoor air temperatures, as shown in Figure 4.3. As external solar radiation intensifies in the morning, the $CuSO_4$ solution in a window's air interlayer absorbs solar radiation, while the water-filled interlayer absorbs less, reducing the temperature hike. Near noon, the interior air temperature difference between groups grows steadily. As $CuSO_4$ solution absorbs solar energy, it significantly reduces indoor solar radiation heat gain, improving summer cooling load efficiency. Like thermochromic materials, $CuSO_4$ solutions absorb NIR.

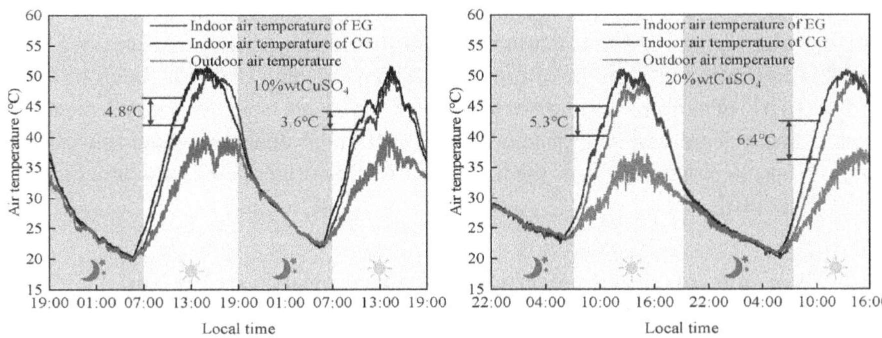

FIGURE 4.3 Indoor/outdoor air temperature of two groups [13].

The inner surface temperature of $CuSO_4$ liquid-filled energy-saving windows is lower than that of water-filled windows. The temperature difference between the two groups is positively correlated with the mass fraction of the $CuSO_4$ solution.

4.3 VITAMINS RECEIVED FROM SUNLIGHT

During the 1920s, manufacturers created glass that permitted a greater percentage of UV radiation transfer mission relative to conventional glass. During this period, the use of solar radiation became significant, rendering the improvement of sunlight entry into buildings a logical progression. Direct sunlight exposure enhances well-being and boosts infection resistance, even when transmitted through glass. Bright-light therapy is utilized in the management of several psychiatric disorders, such as seasonal affective disorder and major depression. Sunlight has a beneficial impact on length of stay, mortality rates, perceived stress, and pain levels. Bright light provides benefits that extend beyond the reduction of depression symptoms; increased light exposure may slow cognitive decline in individuals with dementia. Without suitable temporal signals from the sun or other sources, essential biological rhythms may become irregular, which could result in numerous health problems, including depression, diabetes, obesity, and cancers of the breast and prostate. Regulating biological rhythms in patients may enhance immunological activity; however, it is uncertain whether habitual exposure to sunlight through glass would assist patients in resisting infection.

Vitamin D helps the immune system, but the immunological consequences of window-filtered sunlight are unknown. Hospitals have long had sunrooms. The benefits of sunlight for patients are becoming recognized, but solar gains must be minimized to avoid overheating. Hospitals and other facilities may reduce air-conditioning needs, but the absence of sunlight increases infection, depression, and other health risks. Enhanced sunlight exposure may speed patient healing [14, 15].

Research demonstrates considerable health advantages associated with solar UV exposure, extending beyond the production of vitamin D. UVA exposure releases nitric oxide (NO) from intradermal nitrogen molecules, mostly nitrate. NO reduces cardiovascular disease risk and fights viruses, among other functions. The data indicate a correlation between seasonality and health outcomes, with peak rates occurring in winter and minimal rates in summer. Exposure to solar UVA has been linked to a reduced risk of autoimmune diseases, lower blood pressure, enhanced metabolic function, and a decreased incidence of COVID-19. Individuals living at high latitudes and travelling to sunny locations during winter demonstrate increased concentrations of 25(OH)D [16].

4.3.1 Solar Radiation Controlled the Development of Vitamin

Upon exposure to sunlight, previtamin D_3 absorbs solar UVB radiation and isomerizes into two primary photoproducts: lumisterol and tachysterol. Both photoproducts demonstrate no effect on calcium metabolism. Upon exposure to sunlight, the skin converts roughly 15% of 7-dehydrocholesterol into previtamin D_3. Increased exposure results in a photoequilibrium where previtamin D_3 is converted into lumisterol and tachysterol while also reverting to 7-dehydrocholesterol. Additionally, vitamin D_3 is synthesized from previtamin D_3 in the skin upon exposure to solar UVB radiation, where it absorbs UVB radiation and is converted into various suprasterols and 5,6-trans-vitamin D_3. Additionally, previtamin D_3 can be transformed into several toxisterols. Therefore, irrespective of the level of sun exposure, vitamin D intoxication will not occur, as any excess previtamin D_3 and vitamin D_3 are subject to photodegradation into byproducts that do not exhibit calcemic activity.

4.3.2 Factor Influencing the Vitamin Production

1. *Zenith Angle:* About 1% of solar UVB radiation penetrates the Earth's surface, even at noon during the summer. The stratospheric ozone layer effectively absorbs all UVC (200–280 nm) and UVB radiation up to approximately 290 nm. The ozone layer absorbs approximately 99% of UVB radiation in the wavelength range of 291–320 nm. Extending the path length of solar UVB through the ozone layer results in a reduction of UVB photons that reach the Earth's surface. This explanation examines the restricted synthesis of vitamin D_3 in the skin due to sun exposure in winter for individuals living above and below roughly 33° latitude.
2. *Skin Pigment:* Human evolution took place at the equator. Prolonged sunlight exposure results in melanin production, which functions as a natural sunscreen with an absorption spectrum ranging from 290 to 700 nm, facilitating the effective absorption of solar UVB radiation. Although Africans possess heavily pigmented skin, a small quantity of UVB radiation can still penetrate the epidermis, enabling the synthesis of vitamin D_3. This was demonstrated when adult individuals with skin types 2 and 5 were exposed to equivalent levels of UVB radiation in a tanning bed. White adults demonstrated a more than 30-fold increase in blood levels of vitamin D_3, whereas Black adults exhibited no significant change in their blood levels of vitamin D_3. Exposure of Black adults to five times the UVB radiation resulted in an approximate 15-fold increase in their blood levels.

Observational and epidemiological studies suggest that increased latitude correlates with a higher risk of cancers, indicating a potential role of sun-induced vitamin D_3 synthesis as a beneficial factor in these findings. Sunlight exposure induces several physiological effects on the skin, such as alterations in the immune system and enhanced synthesis of β-endorphin and nitric oxide. Numerous studies, encompassing both interventional and observational research, indicate that enhancements in vitamin D levels are associated with a decreased risk of various lethal cancers.

Vitamin D may have an indirect role in the etiology of schizophrenia. Research indicates that prenatal exposure to influenza may elevate the risk of developing schizophrenia. Vitamin D supplementation decreases the risk of influenza infection. Multiple studies associate vitamin D insufficiency with depression, Alzheimer's disease, epilepsy, and neurocognitive decline. The brain contains vitamin D receptors (VDR) and 1-alpha-hydroxylase (1-OHase). This may elucidate the elevated prevalence of vitamin D insufficiency alongside neurocognitive decline, dementia, depression, and Alzheimer's disease. Adults diagnosed with depression demonstrated significantly reduced levels of 25(OH)D within the community [17–20].

4.3.3 Impact of Vitamin D

The medical research community remains divided regarding the health effects of sunlight. From a dermatologist's perspective, sun exposure correlates with heightened risks of photoaging, immunosuppression, and skin malignancies attributable to its UV radiation. Further studies have associated sun exposure with a heightened risk of age-related macular degeneration, suicidal ideation, and behaviors, as well as episodes of bipolar disorder. Epidemiological research indicates several inverse relationships between sun exposure and various chronic diseases. Because observational studies have shown conflicting results, with some showing possible benefits and others suggesting likely harm, there is confusion about the overall effects of sun exposure. Because correlations might be the consequence of confounding variables or reverse causality, adding to the uncertainty, observational studies do not prove causation.

Conversely, multiple reports indicate an inverse relationship between sun exposure and illness, prompting speculation that sunlight may enhance health via unidentified biological mechanisms. Recent discussions have focused on the potential health benefits of vitamin D, synthesized by the skin upon exposure to UVB light, a crucial element of sunlight. The vitamin D hypothesis has been thoroughly examined due to its inverse epidemiological correlations with various chronic diseases. There is increasing evidence indicating that vitamin D lacks positive effects, as demonstrated by various randomised clinical trials and Mendelian randomization studies [21].

Sun exposure and vitamin D have distinct effects on the nervous system. Sun exposure, as indicated by magnetic resonance imaging (MRI) data not associated with vitamin D supplementation effects, directly impacts the development of multiple sclerosis, with both serving as independent risk factors for central nervous system demyelination. While white, Hispanic, and Black groups exhibit a decreased risk of multiple sclerosis associated with elevated serum 25(OH)D concentrations, this reduced risk is specifically linked to higher lifetime UV radiation exposure in white populations. The impact of UV exposure on the risk of multiple sclerosis remains evident in Black and Hispanic populations, even after accounting for various nucleotide polymorphisms associated with vitamin D metabolism. Clinical trials are currently underway to determine whether a limited wavelength of UV light can reduce multiple sclerosis in humans.

Skin exposure to sunlight provides benefits that surpass mere vitamin D production. UV radiation induces NO production in the skin and promotes the release of NO from dermal storage into the circulatory system. UVA and UVB radiation stimulate the expression of inducible nitric oxide synthase (iNOS) and endothelial nitric oxide synthase (eNOS) in mouse and human keratinocytes. UV radiation enhances blood flow independently of nitric oxide synthase by releasing cutaneous nitric oxide stores in human skin, resulting in a sustained reduction in blood pressure in healthy adults. This phenomenon may partially elucidate seasonal variations in hypertension and cardiovascular health. UV radiation is acknowledged for its potential to reduce hypertension; however, evidence regarding the effectiveness of vitamin D supplements in this context remains limited. This suggests that the mechanisms through which UV radiation exerts its effects may function largely independently of vitamin D. iNOS expression alleviates symptoms in a murine model of ulcerative colitis.

UV radiation increases the synthesis of heme oxygenase (HO-1), and the pharmacological induction of HO-1 exhibits therapeutic effects across multiple human diseases while also providing protective advantages in experimental disease models. The reaction products of HO-1, carbon monoxide and biliverdin, demonstrate anti-inflammatory effects by reducing IL-6 and tumor necrosis factor (TNF) production, as well as decreasing bilirubin levels, respectively. Bilirubin influences cholesterol metabolism and levels of peroxisome proliferator-activated receptor gamma (PPAR-γ), resulting in improved insulin sensitivity and glucose tolerance in a mouse model of diet-induced obesity. The induction of HO-1 by UV radiation serves various functions in metabolic support and immune system regulation [22].

Individuals with advanced chronic kidney disease and end-stage renal disease are prone to deficiencies in both 25-hydroxyvitamin D and 1,25-dihydroxyvitamin D. With the decline of renal function, serum concentrations of 1,25(OH)2D diminish, occurring before any changes in serum calcium or phosphorus levels. The optimal serum levels of 25(OH)D are not definitively established, but they may need to surpass 30 ng/ml in individuals with chronic kidney disease to enhance extra-renal synthesis of 1,25(OH)2D and to regulate parathyroid hormone secretion. Patients with chronic kidney disease may experience vitamin D deficiency due to multiple factors. In individuals with chronic kidney disease (CKD), a decrease in renal mass and glomerular filtration rate (GFR) leads to a reduced influx of 25(OH)D into the renal tubules, resulting in lower uptake of 25(OH)D. Impaired kidneys show decreased megalin expression, resulting in reduced reabsorption of 25(OH)D. Furthermore, albuminuria activates the megalin pathway, leading to decreased receptor availability for the reabsorption of 25(OH)D-DBP. Proteinuria causes damage to proximal tubular cells, leading to a reduction in megalin expression. CKD reduces the resorption of 25-hydroxyvitamin D (25(OH)D), resulting in decreased formation of 1,25-dihydroxyvitamin D (1,25(OH)2D) in a substrate-dependent manner (Figure 4.4).

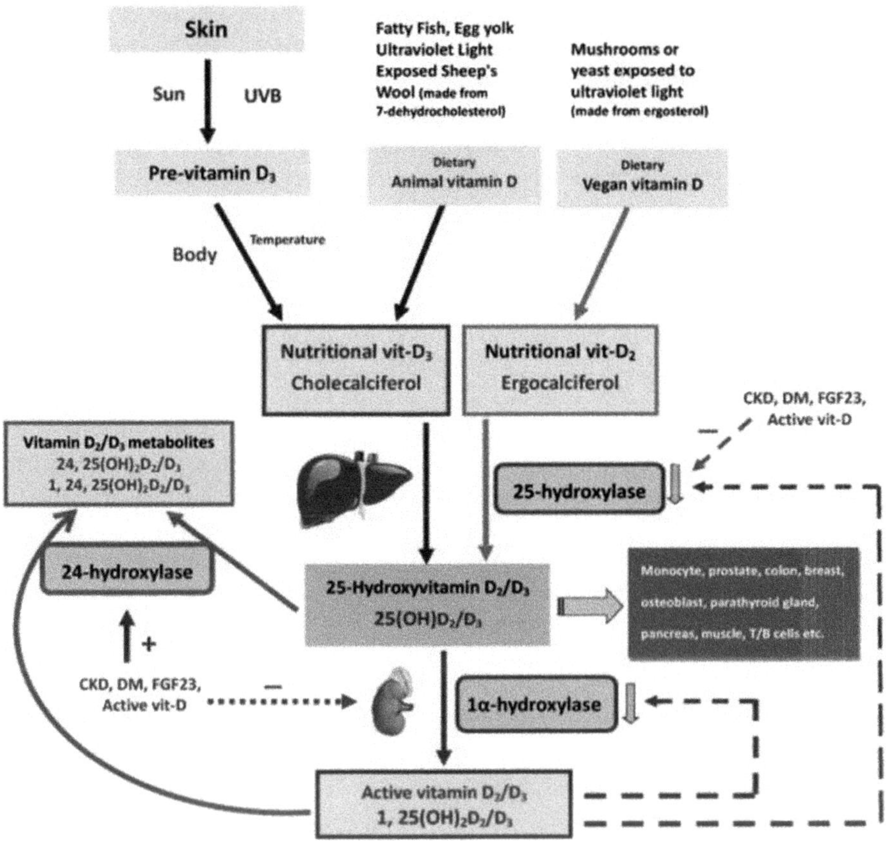

FIGURE 4.4 Metabolism of vitamin D and its regulatory mechanisms [23].

4.3.4 Natural Lighting and Daylighting

Daylighting is frequently endorsed by sustainability rating systems due to its potential to lower energy consumption from artificial lighting and possibly mitigate heat loss, depending on climatic conditions. Additionally, the health effects associated with daylighting integration include improved occupant productivity and mental well-being, as well as decreased disruption of circadian rhythms.

Fitwel, WELL, LEED, and DNGB emphasize the significance of daylight access for its circadian benefits and its positive impact on human mental and physical health. The psychological well-being section examines the impact of daylighting on mental health, noting that these credits primarily enhance visual comfort [24].

4.4 WINDOWS AND STRESS

The connection between nature and its positive effects on human health, behavior, and overall well-being is well documented. Architectural designs utilize visual windows to provide natural views and daylight in interior spaces, which may affect the occupants' experience. Structures consume roughly 40% of energy to maintain thermal comfort and sufficient daylighting in their interiors. The energy consumption of heating, cooling, and visual comfort in buildings in India accounts for 33% of the total, according to Green Rating for Integrated Habitat Assessment (GRIHA) data. This includes both commercial and residential constructions. Numerous modern buildings incorporate diverse glass materials that cover a substantial area of their surfaces. Clear glass is frequently employed in modern architecture to enhance the influx of natural light into buildings. Solar radiation penetrates glass windows more quickly than walls, roofs, and floors, resulting in a higher transmission of radiation due to the transparent properties of glass. The primary characteristics of glass are transmittance, reflectance, and absorbance. The physical and chemical properties of window glasses vary across different characteristics. The placement of windows critically affects the quantity of solar radiation entering buildings.

The use of glass enclosures, especially clear glass, results in increased heat gain and improved daylighting in buildings, potentially leading to uncomfortable conditions. The integration of metal oxides into molten glass offers an effective method for tinting clear glasses. The variations in hue can be achieved by modifying the types of metal oxides employed. The presence of iron, manganese, chromium, and vanadium oxides contributes to glass coloration. The use of colored tinted glasses can significantly reduce cooling loads by blocking a portion of solar energy and decreasing energy transmission. The application of metal coatings to the surface of glass, alongside the incorporation of metals during the glass-making process, can enhance light reflection. These glasses are designated as reflective glasses. The application of thin metallic films, including gold, silver, copper, and aluminum, to glass surfaces demonstrates significant effects on the optical properties of window glass. Prior research examined different types of window glass, such as clear and bronze variants, employing double and triple glazing with a 10-mm air gap to assess the minimum solar radiation transfer [25].

The office experience is influenced by various architectural and interior design elements, such as the proximity of a window to an individual's desk, the characteristics of the view outside the window, and the social density defined by the number of occupants in the office space. Despite well-defined preferences for windows and the acknowledged restorative benefits of natural views, there is a noticeable gap in research that investigates a comprehensive model assessing the influence of these factors on immediate comfort and discomfort in office environments, along with their subsequent effects on various well-being measures after work. The analysis utilized Hedge's conceptual framework for Sick Building Syndrome (SBS), integrating individual differences and architectural characteristics to delineate SBS symptoms as the basis for our model. The model suggests that particular individual and architectural factors may directly and indirectly affect physical and psychological discomfort in an office setting, especially via perceived environmental conditions (Figure 4.5).

The preference for natural light rather than electric light is a crucial factor in the significance of windows for building occupants. Windows function as conduits for the transfer of air, light, and sound, thereby influencing the indoor environment considerably. The preferences for windows are explicitly delineated. A window view provides insights into temporal and meteorological conditions, mitigates sensations

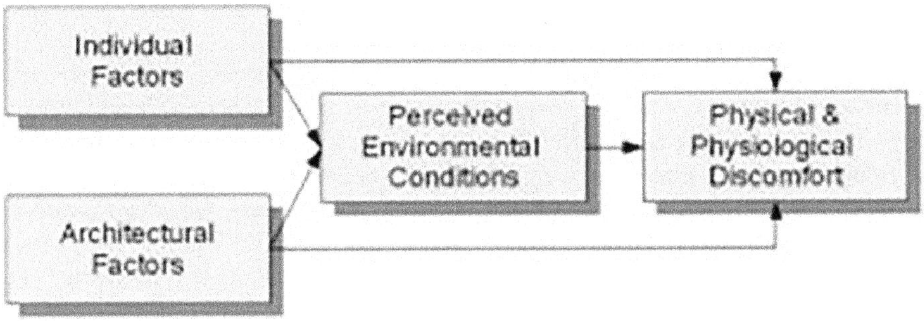

FIGURE 4.5 Basic model of individual and architectural factors consideration [26].

of confinement, and may enhance ocular health by offering a distant horizon for visual engagement. Windows present a dual function: they facilitate the entry of natural light and provide external views, while also potentially causing problems such as glare and thermal discomfort. An optimal perspective includes both the foreground and the skyline; however, effective management of the sky's intense brightness is essential. Further examination revealed that the relationship was more complex. Satisfaction levels concerning ventilation, especially regarding thermal comfort, were lowest for individuals situated next to the window.

The desire for daylight, or direct sunlight, may be abruptly diminished when it results in visual discomfort. Individuals often close their blinds and maintain them in that position, even after the source of their discomfort has been resolved. The hypothesis posited that individuals positioned farther from the window would encounter fewer problems associated with glare and heat gain. The view is a particularly captivating attribute of windows. The results showed that occupants situated farther from a window reported decreased satisfaction with the view and exhibited a greater preference for seating nearer to the window. Research indicates a distinct preference for natural landscapes over urban or constructed environments among individuals with window views. Natural environments contribute positively to human health by aiding in the recovery from mental fatigue. The hypothesis posited that urban views would result in heightened discomfort and adversely impact office impressions, whereas views of suburban and rural landscapes, characterized by abundant greenery, would mitigate discomfort and improve office impressions [26].

4.4.1 Window Views

Window views represent a crucial component of the urban built environment that affects human health; nonetheless, the differences in impacts and health benefits linked to various window view elements are still inadequately examined. Historically, window views have been categorized exclusively as either natural or unnatural. We propose a classification that includes three elements: sky, buildings, and green space. Individuals frequently spend extended durations indoors, with windows acting as the main link to the external environment. Windows play a vital role in regulating physiological and psychological states, promoting restoration, and influencing environmental perceptions. Work-related stress, traffic noise, and urban air pollution present considerable health hazards for urban residents.

Natural window views with plant landscaping can improve patient recovery, enhance employee mood, and reduce noise disturbances for urban residents. Window views that promote human health serve as a significant indicator of a healthy urban environment and societal well-being. The health benefits of window views are well documented; however, the mechanisms through which these views influence the human body, particularly concerning "window view elements," are not sufficiently investigated. Window views comprise diverse outdoor elements that individuals can observe through a building's windows.

Window views provide significant benefits for human health, especially in terms of physical and mental well-being. Researchers have looked at the health benefits of window views in the past using either large-scale data analysis or real-world proof. Studies show that patients who stay in rooms with windows heal faster and stay in the hospital less than those who stay in rooms without windows or with views that are blocked. Window views for residents in home isolation may reduce anxiety and depression, subsequently improving mental health. Enhanced window views for office staff are associated with improved health outcomes and greater overall well-being. Window views can improve students' emotional well-being, decrease stress, foster positive emotions, and diminish negative emotions. The main hypotheses include attention restoration theory, stress reduction theory, the biopsychosocial model, and the biophilia hypothesis within the framework of ecological psychology.

People have a clear aesthetic preference for parks and other green areas in cities. In addition, people's physical and emotional well-being are improved by proximity to green areas. Increased happiness and lessened perceptions of environmental noise are two benefits of having access to green areas. Adding sky features to one's range of view changes how one perceives the natural world. Most of the time, the sky appears light blue because the atmosphere scatters the sunlight. Concentration, temperature, and drowsiness are all influenced by blue light. In addition, people's health is affected by how thermal comfort is affected by the availability of sky and green space.

A variety of sky factors, including meteorological conditions and atmospheric states, have an impact on human health. Elements of a building have an inherent relationship to people's health and well-being and have a substantial impact on many aspects of people's health. The majority of research has found that natural features in window views had more positive effects on health, including less depression, anxiety, and dissatisfaction, compared to artificial elements. According to the latest study, looking out of city windows at cityscapes for a long time may be bad for the brain's activity and wave patterns, especially the parasympathetic and sympathetic responses.

Theories of stress reduction and attention restoration are consistent with this occurrence. According to other studies, it may be good for architectural features to match human preferences and adaptations in terms of how things look, how they are organized, and how big they are. These three things have a big impact on how good window views are for your health. The question of which arrangement of window view elements is best for health is still open. To learn how different combinations of sky, green space, and buildings affect people's physical and emotional health, more studies are needed.

The idea is to use a three-part system to describe window views instead of a binary system. The three parts would be sky, buildings, and green space. The sun, clouds, and atmosphere are the parts that make up the sky. Everything man-made, including buildings, roads, and public spaces, are considered part of the building elements. Among the many components of green space are hilly topography, aquatic systems, and flora. Taking into account the locations, materials, textures, and dynamic properties of items in window views, this categorization is based on their similar features (Figure 4.6).

Natural items make up the majority of sky elements, whereas man-made non-natural objects predominate in building elements, and a mix of natural and man-made objects make up green space elements. Parts of the sky usually take up the top half of the window view, while the bottom half and the middle half are occupied by the other two parts. The organic components that make up most of the green space elements have a rough, less reflective surface, a softer texture, and an uneven look. Inorganic materials with flat, reflective surfaces, geometric shapes that give the impression of strength and stability, and hard surfaces make up the bulk of the building's components. The weather has a big impact on the sky elements, which might have different textures depending on whether it is a clear or overcast day [27].

4.4.2 Window Area and Location

One of the most important factors affecting the effectiveness of natural ventilation in tall structures is the arrangement of window openings. Three factors that may affect natural ventilation are

1. the placement and orientation of windows and
2. orientation of windows
3. doors

Researchers looked at the three factors together and found that changing two of them at once had a greater impact than changing all three at once because of the mutually reinforcing effects. Ventilation and daylight efficiency are strongly affected by the window-to-wall and window-to-ground area ratios [28].

Respondents' homes were a tier down in the conceptual models' hierarchical organization (Figure 4.7). A key setting for living, leisure, and entertainment for the elderly in China is residential communities that display gated spatial arrangements surrounded by walls. Because they show how green the neighborhood is as a whole, the community-level vegetation coverage and green volume ratio were set. At the individual level, the green-window ratio was devised to quantify the degree to which visual greenness is emitted from residential windows. The hypothesis proposed that different measures of residential greenspace exposure may have some relationship with the quality of sleep experienced by older persons when assessed on an individual basis.

A healthy lifestyle must include sleep due to the restorative effects it has on one's mental and physical well-being. For the elderly, sleep disturbances can lead to a host

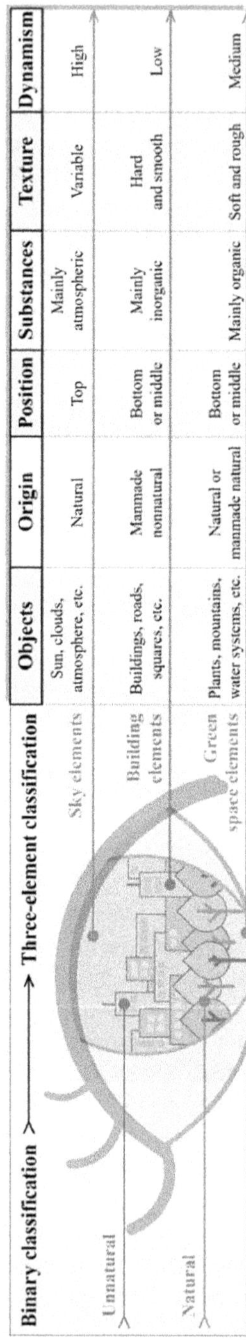

FIGURE 4.6 Three elements of window views [27].

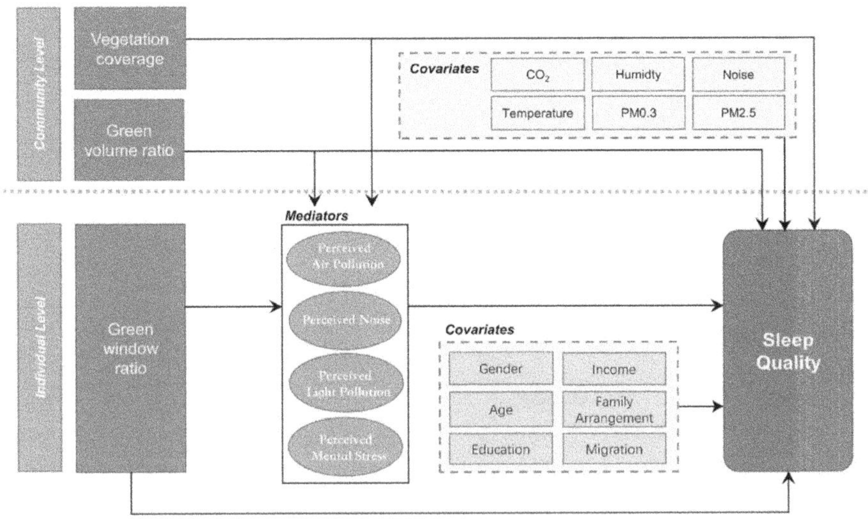

FIGURE 4.7 Conceptual models [29].

of problems, including diminished ability to function throughout the day, accelerated cognitive loss, and emotional distress (such as depression and anxiety). The quality of life for the elderly declines with time as a result of this. Encouraging older adults to have a better night's sleep should be a top priority. The quality of one's sleep is greatly affected by one's living environment, one's lifestyle choices (such as one's diet), and one's demographic traits (such as one's age and gender).

Open green areas found in urban environments, or green spaces, have a profound effect on city people' well-being and pleasure. Physical and mental health, subjective well-being, and total health have been increasingly associated with residential green spaces' accessibility, availability, and nature-related experiences [29].

4.4.3 PV Windows

With today's technology, windows can generate power in addition to their main function. The process of glazing turns the produced solar cells on the transparent glass into real solar cells. The building sector is optimistic about the many potential applications of these alternative window types. Making the most of natural light and energy production is one of the challenges of solar cell windows. The experts thought about PV ventilated glazing. When used in a working setting, the a-semi-transparent Si glazing works better than the non-transparent c-Si solar cells.

The research made use of a small workplace in Hong Kong with varying window types to analyses the integrated performance. Power usage dropped dramatically as the solar cells' transmittance ranged from 0.45 to 0.55. Natural ventilation with a PV glazing system resulted in a 6.7% savings rate, while double PV glazing resulted in a 4.8% savings rate [30].

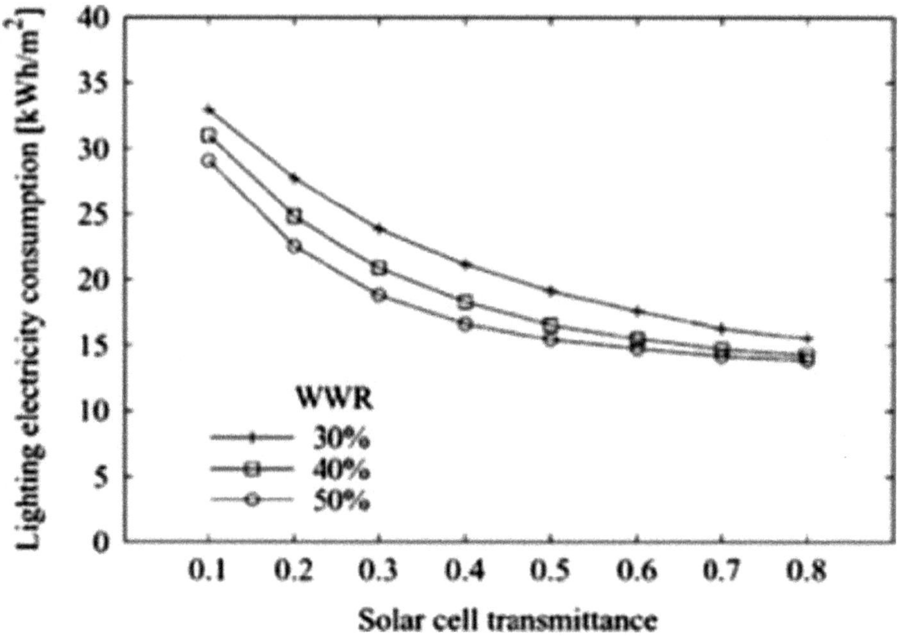

FIGURE 4.8 An analysis of the building's lighting energy usage is conducted each year by plotting the solar cell transmittance for different window sizes [31].

The efficiency of the solar cell glazing was evaluated in comparison to the conventional mode, which takes into consideration the building's initial thermal properties. Figure 4.8 shows the inconsistent illumination energy usage of PV glazing. Considering the building's total energy use is necessary for optimizing the size of the windows and the transmittance of the solar cells in light of the aforementioned discrepancies. The correlation between overall power consumption and the transmittances of solar cells is shown in Figure 4.9. The most efficient use of energy was achieved with a water-weight resistance of half and a solar cell transmittance of 40%. Furthermore, this combination enhances the building's energy performance by 54% when compared to the default option.

4.5 SIDE LIGHTING AND TOP LIGHTING SYSTEMS

Planning and constructing so-called "green buildings," which drastically reduce power consumption, is an additional, more preventative measure for protecting our limited electricity supply. A "green building" is an eco-friendly, humane, and economically beneficial alternative to conventional building practices. About 1.5 times the standard deviation is the most minor shift in how light affects people's feelings. Given the wide range of possible outcomes when comparing identically designed interiors operated under varied environmental conditions, it is prudent to provide a range of illuminations for each intended interior use. For each range, the suggested illuminance scale consists of three consecutive steps [32].

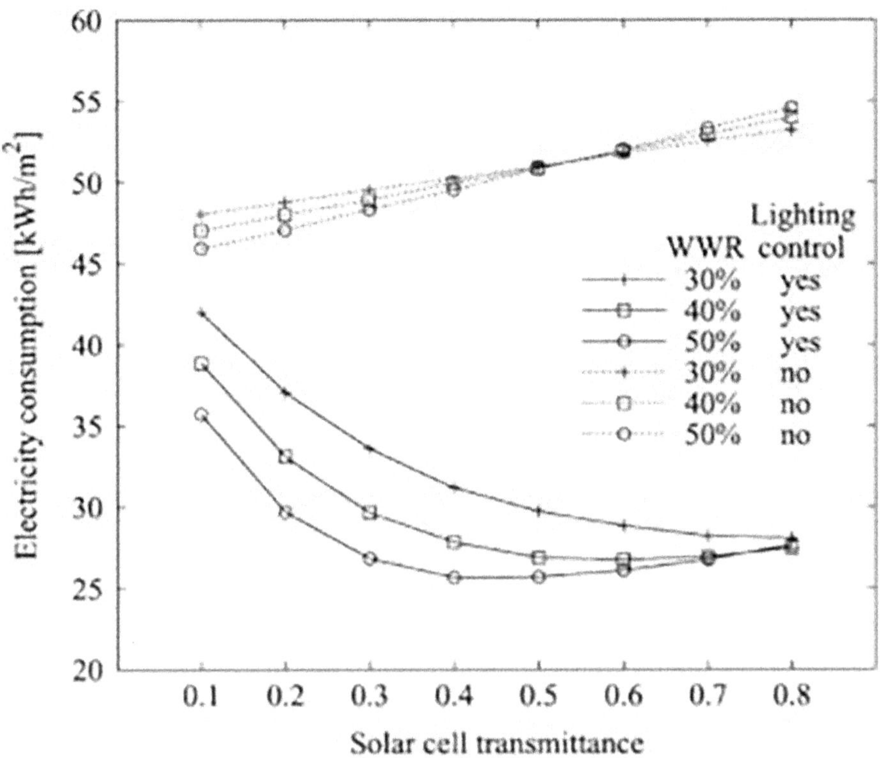

FIGURE 4.9 Annual power usage of the building's HVAC, lighting, and cooling systems in relation to the transmittance of solar cells [31].

4.5.1 Light Level Acceptability of Occupants

The suggested expansion of the building's automatic sunshine monitoring system to regulate the blind curtains at façade windows with high sunlight reflectance will help reduce sun glare. On top of that, nearby employees should be able to manually adjust the blind curtains while they are on the clock. Incorporating outdoor blinds and external louvres or reworking the building's glass system to eliminate glare are two ways to improve visual acuity in areas where it has been observed that there is an excess of light. Solar film coatings applied in conjunction with daylight-linked lighting control allow for the management of illuminance and solar heat gain in high-illuminance building regions, leading to favorable outcomes.

Radiance software simulations showed that office buildings can improve daylighting performance and cut energy costs with a well-designed light shelf placed at the right angle, height, and placement. It is necessary to investigate how renters perceive the comfort level affected by solar heat gain through fenestration. Although these results are encouraging, it is important to think about how much it will cost to maintain and replace the glazing system before making any changes or additions.

More study is needed to look into how cost-effective, user-friendly, aesthetically pleasing, and important these ideas are for the overall operation of buildings if they are actually used. It is important to thoroughly examine how solar heat gain affects the comfort of occupants, not just their visual comfort. To measure the amount of heat gain from glazing during working hours, a pyranometer and software simulations may be used [33].

It brings people closer to nature and offers directional, intensity, and color variations. Because it allows the body to make vitamin D and hormones that are related to light, sunlight is an important passive energy source for human health. It is essential for metabolism, controlling the biological clock, and keeping circadian rhythms in check. Furthermore, during business hours, there is a great deal of natural light streaming into office buildings, making it an important factor in this setting. Due to the absence of natural light in these structures, electric lighting plays a major role in the overall energy consumption. The requirement for electric lighting has been significantly reduced because of recent developments in intelligent lighting systems, but there are still opportunities to use sunshine.

To measure the amount of daylight, one can use dynamic daylight metrics like Useful Daylight Illuminance (UDI) or Daylight Autonomy (DA). UDI is the percentage of occupied hours that fall within a range that occupiers view as "useful" in terms of daylight levels. The range limitations can be adjusted depending on the application, although they usually prevent very low (<100 l×) and very high (>2500 l×) amounts of light from entering the system. The top limit is set to detect when there is too much sunlight, which can cause discomfort in the eyes and body. Furthermore, UDI and DA can be combined into one parameter by configuring an autonomous unit (500–2500 l×). When there is sufficient natural light entering the work plane, people may get things done without using artificial illumination, a feature known as DA. DA is a yearly metric that measures the percentage of workable daylight hours that are illuminated to a minimum of 300 l× or 500 l× [34].

The illuminance meters could measure anything from 0 to 20,000 light-emitting diodes. For a more precise evaluation of control system performance and their capacity to incorporate daylight in reaction to transient variations, high-frequency data acquired by a datalogger at 1-minute intervals was crucial. The west roller blind was fully closed for data related to the south orientation, while the south roller blind was totally closed for data related to the west orientation. There were a total of 24 days' worth of data collected, with 6 days allocated to each orientation and season. To describe the environment, observed data were used to determine DA. When compared to alternative calculating methods that assume constant daylight availability, this metric takes into consideration changes in daylight length.

With a maintenance factor of 0.8, 284 l× are equivalent. Starting with the signal from the photosensor, the light output was determined as a function of time δ(t) for each DLCS that was taken into consideration (Figure 4.10).

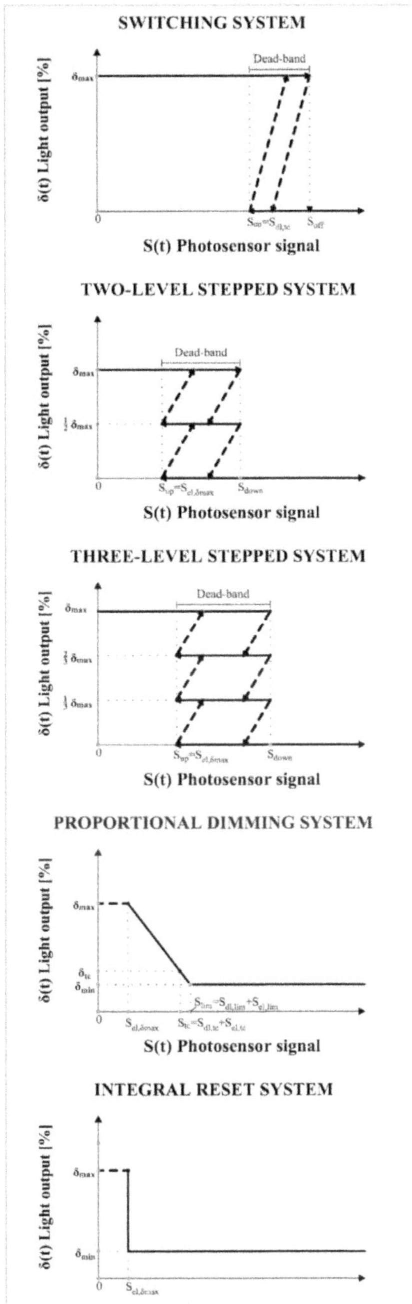

FIGURE 4.10 Chosen control algorithms [35].

REFERENCES

1. Gago, E. J., Muneer, T., Knez, M., & Köster, H. (2015). Natural light controls and guides in buildings. Energy saving for electrical lighting, reduction of cooling load. *Renewable and Sustainable Energy Reviews*, *41*, 1–13.
2. Alrubaih, M. S., Zain, M. F. M., Alghoul, M. A., Ibrahim, N. L. N., Shameri, M. A., & Elayeb, O. (2013). Research and development on aspects of daylighting fundamentals. *Renewable and Sustainable Energy Reviews*, *21*, 494–505.
3. Zhang, Y., Wang, W., Wang, Z., Gao, M., Zhu, L., & Song, J. (2021). Green building design based on solar energy utilization: Take a kindergarten competition design as an example. *Energy Reports*, *7*, 1297–1307.
4. Nearly zero energy building standard. https://nzeb.in/knowledge-centre/passive-design/daylighting/
5. Acosta, I., Munoz, C., Campano, M. A., & Navarro, J. (2015). Analysis of daylight factors and energy saving allowed by windows under overcast sky conditions. *Renewable Energy*, *77*, 194–207.
6. Lu, Y., Wu, Z., Chang, R., & Li, Y. (2017). Building Information Modeling (BIM) for green buildings: A critical review and future directions. *Automation in Construction*, *83*, 134–148.
7. Chiang, C., Olsen, J. E., Basler, D., Bånkestad, D., & Hoch, G. (2019). Latitude and weather influences on sun light quality and the relationship to tree growth. *Forests*, *10*(8), 610.
8. Zhang, Y., Wang, W., Wang, Z., Gao, M., Zhu, L., & Song, J. (2021). Green building design based on solar energy utilization: Take a kindergarten competition design as an example. *Energy Reports*, *7*, 1297–1307.
9. Tablada, A., & Zhao, X. (2016). Sunlight availability and potential food and energy self-sufficiency in tropical generic residential districts. *Solar Energy*, *139*, 757–769.
10. Chen, L., Hu, Y., Wang, R., Li, X., Chen, Z., Hua, J., ... & Yap, P. S. (2024). Green building practices to integrate renewable energy in the construction sector: a review. *Environmental Chemistry Letters*, *22*(2), 751–784.
11. Shirinbakhsh, M., & Harvey, L. D. (2024). Feasibility of achieving net-zero energy performance in high-rise buildings using solar energy. *Energy and Built Environment*, *5*(6), 946–956.
12. Olajiga, O. K., Ani, E. C., Sikhakane, Z. Q., & Olatunde, T. M. (2024). A comprehensive review of energy-efficient lighting technologies and trends. *Engineering Science & Technology Journal*, *5*(3), 1097–1111.
13. Zhang, C., Li, D., Wang, L., Yang, Q., Guo, Y., Zhang, W., ... & Pu, J. (2024). Experimental investigation of indoor lighting/thermal environment of liquid-filled energy-saving windows. *Renewable Energy*, *220*, 119704.
14. Hobday, R. A., & Dancer, S. J. (2013). Roles of sunlight and natural ventilation for controlling infection: historical and current perspectives. *Journal of Hospital Infection*, *84*(4), 271–282.
15. Qian, H., Li, Y., Seto, W. H., Ching, P., Ching, W. H., & Sun, H. Q. (2010). Natural ventilation for reducing airborne infection in hospitals. *ISME Journal*, *45*, 559e565.
16. Grant, W. B., Bhattoa, H. P., & Pludowski, P. (2024). Determinants of vitamin D levels from sun exposure: a global perspective. In *Feldman and Pike's Vitamin D* (pp. 97–113). Academic Press.
17. Clemens, T. L., Adams, J. S., Henderson, S. L., & Holick, M. F. (1982). Increased skin pigment reduces the capacity of skin to synthesise vitamin D3. *Lancet*, *1*, 74–6; PMID: 6119494. http://dx.doi.org/10.1016/S0140-6736(82)90214-8

18. Hoogendijk, W. J., Lips, P., Dik, M. G., Deeg, D. J., Beekman, A. T., & Penninx, B. W. (2008). Depression is associated with decreased 25-hydroxyvitamin D and increased parathyroid hormone levels in older adults. *Archives Genetic Psychiatry*, *65*, 508–512; PMID:18458202. http://dx.doi.org/10.1001/archpsyc.65.5.508
19. Holick, M. F., Binkley, N. C., Bischoff-Ferrari, H. A., Gordon, C. M., Hanley, D. A., Heaney, R. P., et al. (2012). Guidelines for preventing and treating vitamin D deficiency and insufficiency revisited. *Journal of Clinical Endocrinology Metabolism*, *97*, 1153–1158; PMID:22442274. http://doi.org/10.1210/jc.2011-2601
20. Wacker, M., & Holick, M. F. (2013). Sunlight and Vitamin D: A global perspective for health. *Dermato-Endocrinology*, *5*(1), 51–108.
21. Heiskanen, V., Pfiffner, M., & Partonen, T. (2020). Sunlight and health: Shifting the focus from vitamin D3 to photobiomodulation by red and near-infrared light. *Ageing Research Reviews*, *61*, 101089.
22. Erem, A. S., & Razzaque, M. S. (2021). Vitamin D-independent benefits of safe sunlight exposure. *The Journal of Steroid Biochemistry and Molecular Biology*, *213*, 105957.
23. Liu, W. C., Zheng, C. M., Lu, C. L., Lin, Y. F., Shyu, J. F., Wu, C. C., & Lu, K. C. (2015). Vitamin D and immune function in chronic kidney disease. *Clinica Chimica Acta*, *450*, 135–144.
24. McArthur, J. J., & Powell, C. (2020). Health and wellness in commercial buildings: Systematic review of sustainable building rating systems and alignment with contemporary research. *Building and Environment*, *171*, 106635.
25. Gorantla, K. K., Shaik, S., & Setty, A. B. (2018). Day lighting and thermal analysis using various double reflective window glasses for green energy buildings. *International Journal of Heat & Technology*, *36*(3), 1121–1129.
26. Aries, M. B., Veitch, J. A., & Newsham, G. R. (2010). Windows, view, and office characteristics predict physical and psychological discomfort. *Journal of Environmental Psychology*, *30*(4), 533–541.
27. Yao, T., Lin, W., Bao, Z., & Zeng, C. (2024). Natural or balanced? The physiological and psychological benefits of window views with different proportions of sky, green space, and buildings. *Sustainable Cities and Society*, *104*, 105293.
28. Chen, X., Yang, H., & Lu, L. (2015). A comprehensive review on passive design approaches in green building rating tools. *Renewable and Sustainable Energy Reviews*, *50*, 1425–1436.
29. Zhang, J., Zhou, S., Xia, T., Yin, Y., Wang, X., Cheng, Y., ... & Zhao, B. (2024). Residential greenspace exposure, particularly green window-views, is associated with improved sleep quality among older adults: Evidence from a high-density city. *Building and Environment*, *253*, 111315.
30. Li, P., Guo, W.-W., & Qiu, Z.-Z. (2011). Energy performance of PV windows used in different cities of China. In *ASME 2011 Power Conference Collocated with JSME ICOPE 2011* (pp. 575–579). American Society of Mechanical Engineers.
31. Hee, W. J., Alghoul, M. A., Bakhtyar, B., Elayeb, O., Shameri, M. A., Alrubaih, M. S., & Sopian, K. (2015). The role of window glazing on daylighting and energy saving in buildings. *Renewable and Sustainable Energy Reviews*, *42*, 323–343.
32. Sandanasamy, D., Govindarajane, S., & Sundararajan, T. (2014). Energy efficient lighting systems in green buildings: An overview. *International Journal of Advanced Engineering Research and Science*, *1*(2), 44–50.
33. Kwong, Q. J. (2020). Light level, visual comfort and lighting energy savings potential in a green-certified high-rise building. *Journal of Building Engineering*, *29*, 101198.

34. Gabriel, E., de Campos Grigoletti, G., Meller, G., & Zambonato, B. (2024). Daylight and energy performance of side lighting systems in an office room in a subtropical climate. *PARC Pesquisa em Arquitetura e Construção*, *15*, e024003.
35. Bellia, L., & Fragliasso, F. (2019). Automated daylight-linked control systems performance with illuminance sensors for side-lit offices in the Mediterranean area. *Automation in Construction*, *100*, 145–162.

5 Renewable Energies

The global rise in renewable energy (RE) utilization over recent decades signifies a purposeful change toward cleaner and more dependable power generation. Data from the International Renewable Energy Agency (IRENA) indicates that RE capacity has doubled since 2010, demonstrating its increasing viability and acceptance. This significant expansion transcends the four primary categories of RE: solar, wind, hydro, and biofuels. These areas have benefited from the growing appeal of RE due to advancements in technology, operations, and cost competitiveness. The rapid advancements in RE, particularly in solar and wind power, are noteworthy. Innovations in technology and economies of scale have significantly reduced the price of solar photovoltaic (PV) energy [1].

Despite notable advancements in RE, a significant portion of global energy consumption in 2022 continued to derive from fossil fuels, including coal, oil, and natural gas. This results from several factors include the following:

1. existing infrastructure,
2. the regulatory and legislative environment,
3. challenges in energy storage, and
4. grid integration due to the intermittent nature of specific RE sources.

Many countries' "green recovery" initiatives seek to enhance economic growth by investing in RE and energy efficiency following the COVID-19 pandemic. These concepts should be monitored in the coming years to evaluate their long-term implications.

The contemporary global energy landscape is marked by disparities in energy accessibility, resulting in significant energy inequality. The lack of access to reliable and affordable energy solutions significantly impacts a substantial portion of the population, particularly in low- and medium-income countries. The absence of energy access constitutes a major barrier to development, as energy is crucial for economic growth and the betterment of living conditions. There exists a dual challenge of addressing energy poverty while ensuring that proposed solutions align with the global change toward greener and more sustainable energy sources [1].

DOI: 10.1201/9781003752837-5

The emergence of a few Green Building Rating Systems (GBRS) allows businesses to create environmentally sustainable structures customized to the specific climate and landscape of each region. Due to the efforts of its parent organization, the United States Green Building Council, in collaboration with international council members under the World Green Building Council, to modify their database and guidelines for various contexts, Leadership in Energy and Environmental Design (LEED) functions on a global level, positioning it as one of the few GBRS. Local GBRS may provide enhanced service to clients in areas where global GBRS lacks the necessary database or adaptation to operate effectively, particularly in regions with limited data availability. Limited research has investigated the life cycle relationship of GBRS with decarbonization in buildings. Modifications to the façade aimed at enhancing thermal load were evaluated as part of the renovation techniques tested against GBRS. However, due to the exclusion of certain criteria in the study, it remains challenging to identify which factors significantly reduce emissions and which may need adjustments for enhanced effectiveness [2].

Energy is crucial, prompting numerous investors and operators to explore new resources and technologies to reduce reliance on external influences. End customers seek sustainable building concepts characterized by minimal energy consumption and reduced operating costs, prompting a shift in the real estate market toward this innovative approach. An integrated and forward-looking strategy facilitates the emergence of green buildings.

Green buildings, when compared to traditional structures, enhance indoor environmental quality and occupant health while achieving energy usage reductions of up to 30%. Buildings significantly contribute to climate change and other energy-related environmental issues due to their excessive energy consumption. Buildings, encompassing residential, commercial, and institutional sectors, account for 30% of Canada's secondary energy consumption and are responsible for approximately 29% of the nation's CO_2 equivalent emissions. In the USA, buildings account for 39% of primary energy consumption and 70% of electrical usage, indicating a comparable trend internationally. Energy usage in buildings contributes an additional 38% of CO_2 emissions, 52% of SO_2 emissions, and 20% of NO_2 emissions in the USA, exacerbating the existing environmental burden [3].

5.1 ROLE OF RE IN GREEN BUILDINGS

Environmental factors are responsible for more than 25% of all deaths and illnesses in the world. This shows how poor environmental quality negatively affects people's health. Environmental quality is essential as it attracts tourists, subsequently leading to increased income, job creation, and overall economic development in the region. Humans consistently pursue economic expansion to fulfil their production and survival requirements. Frequently, individuals overlook the interrelation among physical, chemical, and biological environments. Globally, individuals are mobilizing to transition from fossil fuels to RE sources. A potential approach to promote the adoption of RE sources involves reducing the obstacles to investment in sustainable initiatives. By adopting these measures, companies can reduce their reliance on fossil fuels and other outdated energy sources. Green finance serves as a mechanism to support companies or projects that positively impact the environment and the economy [4].

The utilization of RE sources is essential for the sustainability of the planet. The primary driver of climate change is the release of harmful greenhouse gases into the atmosphere due to the combustion of fossil fuels, which underpin the current energy framework. Solar, wind, hydro, and geothermal power represent RE sources that contribute to environmental sustainability. Solar power and wind power harness energy from the solar and kinetic force from the wind to generate electricity without pollution. Geothermal energy utilizes the Earth's intrinsic heat, while hydropower exploits the kinetic energy of flowing water to produce electricity. These sources mitigate air pollution and environmental degradation due to their significantly reduced environmental impact. Widespread adoption of RE sources is essential to mitigate climate change and safeguard the environment for future generations.

Obstacles will arise on the path to an RE-driven green economy. Modernization of the grid and energy storage technologies requires significant investment to incorporate RE sources into existing infrastructure. RE sources such as solar and wind power are intermittent; thus, they need efficient storage to provide continuous electricity. Additionally, RE may encounter opposition from the entrenched interests of the fossil fuel industry. Technological advancements are consistently improving the effectiveness and cost-effectiveness of RE sources. Advancements in wind turbine technology are enhancing the efficiency of wind energy capture, while solar panels are increasingly effective in converting solar radiation into electrical power.

Battery storage systems are currently experiencing significant innovation. These methods appear effective for maintaining grid stability and storing surplus RE. Research and development initiatives, along with policy frameworks that promote investment in RE, can enable a smoother transition. To garner support for this change, it is crucial to enhance public awareness of the environmental and economic benefits of RE. Educational programs can promote a more sustainable culture and foster responsible energy usage habits. A collective initiative across all levels is necessary to transition to an RE-based green economy.

Governments must intervene to equalize opportunities for RE companies, promote research and development, and establish supportive policy frameworks. Private companies must invest in new technology and enhance existing infrastructure to increase the accessibility of RE sources on the grid. International cooperation is essential to ensure a fair and equitable transition for all nations, facilitate the sharing of best practices, and promote technological advancements. Finally, individuals can contribute by being conscious of their energy consumption and by supporting companies and initiatives that align with their sustainability values. Collaboration among governments, corporations, and citizens is essential to overcoming obstacles and harnessing the potential of RE, thereby fostering a sustainable, equitable, and environmentally friendly economy [5].

5.1.1 Use of Solar Systems

The sun's light and heat can be utilized perpetually. The fundamental concept underlying solar energy building designs is presented below:

1. Energy derived from the sun's conduction, convection, and radiation is utilized.
2. A well-designed structure can regulate internal temperature by leveraging natural processes.
3. The materials of a building either reflect, transmit, or absorb solar radiation upon contact with the surface.
4. The sun's rays generate consistent airflow patterns within the designated areas.

Due to the fundamental impact of solar heat, architects and engineers must consider factors such as (1) building material thickness, (2) density, (3) specific heat, (4) surface absorption and reflection coefficients, (5) degree of smoothness or roughness, (6) fullness or cavity, and (7) heat conduction coefficients when designing heating and cooling systems for a structure. Strategic architectural planning enables the harnessing of both active and passive solar energy [6].

5.1.2 Wind Energy

Wind power is increasingly recognized as a prominent strategy for energy transformation due to its low-carbon and renewable benefits. A loan or investment that supports the growth of the social economy while simultaneously reducing carbon emissions exemplifies green financing. Unlike conventional financing, green financing can drive energy-intensive sectors toward a low-carbon transition and support low-carbon sectors in obtaining sufficient investment funding. The global energy transition now incorporates green financing as a strategic approach. Significant capital investment in green energy projects has been facilitated by green financing, accelerating the energy transition and enhancing the wind power sector. The relationships among carbon intensity, green funding, and RE utilization were analyzed through a vector error correction model. The findings indicate that green financing contributed to a reduction in carbon intensity by promoting the expansion of RE generation.

The financial market can influence the expansion of wind power in multiple ways [7] in the following ways:

1. *Innovations and Patents:* Green technology can reduce production costs, accelerate the development of the new energy sector, and address technological constraints, as indicated by a panel vector autoregressive model. Green financing played a crucial role in this process by supplying significant capital for the research and development of new technologies. Green technology in Southeast Asian countries has lagged behind current advancements. International technology transfer can facilitate rapid advancements in new energy technologies and the expansion of wind power.

2. *Information Disclosure:* The findings indicate that green bonds serve as a feasible mechanism for financing RE projects. Initially, it is essential to establish a robust information disclosure system.
3. *Market for Capital:* A robust capital market may facilitate the effectiveness of green bonds in advancing the RE sector. This research identified a unidirectional causal relationship between green bonds and wind energy through the application of the autoregressive distributed lag model. Deepening capital market reforms could enhance the wind energy industry's ability to attract both local and foreign investment, thereby accelerating its expansion.
4. *Utilization Efficiency:* The effects of green bonds on the efficiency of RE utilization were positive in both the short term and long term. The wind power sector rapidly scaled production due to investments in energy efficiency from green banks, encompassing both public and private capital.
5. Issuing green bonds has been identified as an effective method of green finance for promoting RE development in six Southeast Asian countries. The extent of trade openness significantly influenced the growth of wind power utilization. The wind power sector is driven by the objective of enhancing green finance development to attract increased foreign investment.
6. *Spillover Effect:* The development of wind power was influenced by spatial factors resulting from significant variations in regional wind energy resource availability, technological advancements, and industrial policies [7].

The built environment can produce RE on-site by incorporating wind energy devices into structures. The integration of wind turbines into building façades and openings represents an innovative approach to onsite energy generation. In urban environments, aerodynamic façade design enhances energy capture by 22% by channeling wind flow to the wind energy system, resulting in increased wind velocity and a reduction in turbulence by approximately 30%. The mean wind flow in a building can be enhanced by 25% through the implementation of recessed and curved walls for openings, and by 33% using converging and diverging open corridors. Adjusting fillet radius and diameter permits building openings to generate 650% more wind power than free-stream conditions.

The integration of turbines into building openings or the deployment of turbines between two buildings presents a compelling potential application of micro wind energy. Architects and engineers have employed computational fluid dynamics (CFD) to illustrate wind energy systems in tall structures from an aesthetic and design perspective. The wind speed at the building's peak is significantly elevated due to shear flow and predominantly undisturbed free flow conditions. These regions can optimize wind energy utilization through the installation of turbines. Figure 5.1 illustrates installation of tall wind turbines alongside two prospective locations for smaller wind turbines.

The validity of a building-integrated wind energy system is significantly contingent upon its economic viability. The design of wind energy systems that can compete with other RE sources, while accommodating architectural integration, aesthetics, functional requirements, and environmental conditions, presents a significant challenge for research institutions and designers, particularly in terms of maximizing

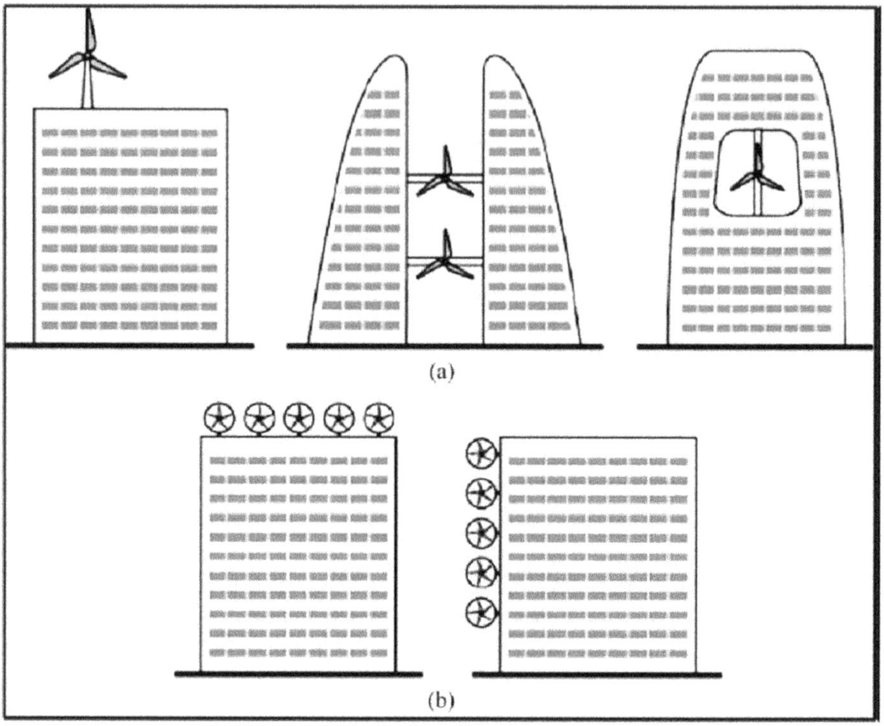

FIGURE 5.1 Wind turbines on tall buildings can be integrated in two ways: (a) on top of the building or (b) on the side of the building, depending on their size [8].

return on investment (ROI). By 2027, the small wind turbine sector is projected to generate $309 million in revenue. The efficient harnessing of high winds reveals the financial advantages of integrating wind turbines or systems onto tall structures. The system's costs can be recovered over its lifespan. The investment is projected to yield returns within 15 years, aiming for a performance of 0.2 kWh/€ at an average annual wind speed of 6.5 m/s.

Due to the novelty of many building-integrated and urban wind energy systems, it is difficult to verify manufacturers' longevity promises. Government incentives enhance the affordability of RE projects, thereby facilitating the utilization of these resources and mitigating the impacts of global warming through the promotion of environmentally sustainable energy sources. Private investors may lack the financial capacity to invest in building-integrated wind turbines that do not generate profit over their operational lifespan.

Recent studies have employed machine learning techniques to predict wind pressure and flow patterns associated with buildings. An unsupervised machine learning approach, specifically a clustering algorithm, was employed to classify the pressure patterns by evaluating the pressure coefficient on two buildings with differing gap widths (Figure 5.2). Wind tunnel data were analyzed using three distinct clustering algorithms: K-means, K-medoids, and density-based spatial clustering of

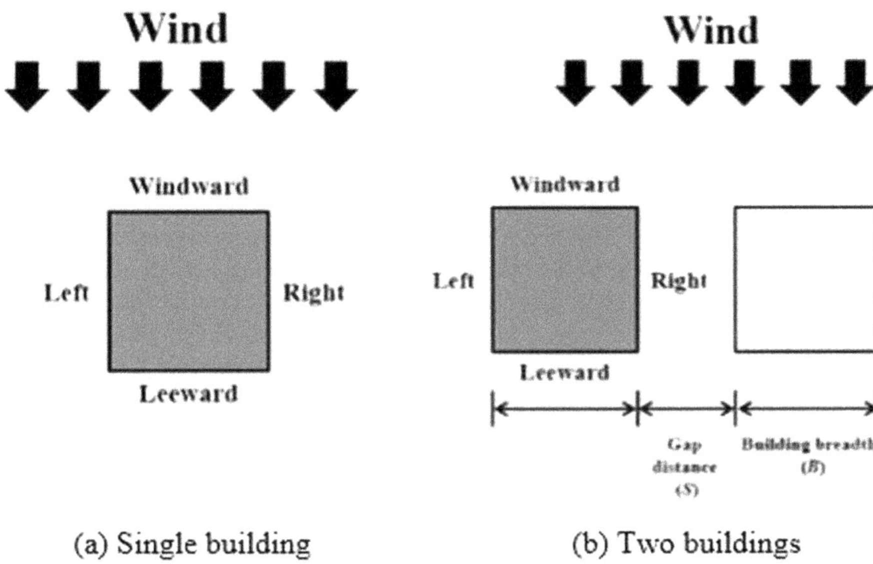

FIGURE 5.2 Wind direction and model setting [8].

applications with noise (DBSCAN). The results indicate that clustering methods yield more informative pressure patterns and local wind pressure coefficients around a square-shaped single building and two adjacent building models.

5.1.3 Geothermal Energy

This study on geothermal energy examines two main types of ground source heat pump (GSHP) installations: water-to-water (W-W) systems and brine-to-water (B-W) vertical borehole heat exchangers (BHE). Open systems, also referred to as water-to-water systems, harness the heat energy of subterranean water sources. The entrance temperature range for these systems typically remains consistent, between 8°C and 15°C. The primary advantage of W-W systems is the utilization of the consistent and mild temperature of groundwater for heating or cooling structures, resulting in significant energy reduction. Consequently, they effectively condition space. The versatility and adaptability of closed systems, including brine-to-water systems, are improved. Various environments, frequently characterized by spatial constraints, utilize B-W systems. Horizontal B-W heat exchangers are commonly employed for this purpose. Vertical BHE utilize a heat transfer fluid (brine) to effectively transport geothermal energy to the heating, ventilation, and air conditioning (HVAC) systems of buildings in closed-loop configurations. Closed systems, including B-W heat exchangers, are favored in various contexts due to their adaptability [9].

Proximity in the installation of GSHPs is critical for optimizing the utilization of shallow geothermal energy. The feasibility and performance of surrounding GSHP systems are significantly influenced by their ability to mitigate interference among them. Different criteria are employed to reduce thermal interference in open-loop and closed-loop systems, as the former extracts groundwater while the latter circulates heat transfer fluids. Factors such as temperature variations and modeling scenarios determine the minimum distances required for GSHP installations to coexist without adversely affecting their efficiency. Minimal distances must be meticulously evaluated to optimize the exploitability of geothermal resources while ensuring the reliability and sustainability of GSHP [10].

Figure 5.3 illustrates three conventional methods by which geothermal borefields operate in mild climates. The initial option entails utilizing the borefield alongside a GSHP to deliver heating to the building. Both active and passive cooling regimes are effective methods for cooling buildings. Passive cooling, also referred to as direct ground cooling, utilizes the borefield as a heat exchanger, facilitating the transfer of heat from the building into the ground without requiring additional GSHP. A temperature differential between the borefield fluid and the structure is necessary for adequate heat transfer; additionally, a maximum temperature for the fluid must be established. Active cooling enables the utilization of a GSHP, facilitating elevated borefield temperatures. A reversible GSHP can function as the heating and cooling system for the building.

Figure 5.4 illustrates that each of the four load profiles was initially allocated to a separate quadrant. In Q4, a heating-dominated thermal load is observed; as the simulation advances, the temperature profile decreases due to the borefield extracting more heat than it supplies, with the final year of operation characterized by a temperature-limited load. The classification aids in the comprehension of active cooling in geothermal borefields [11].

Designers either propose or require micro-generation technologies that exhibit low or zero carbon emissions. The broad adoption of micro-generation is currently impeded by several limitations, including design and technological issues.

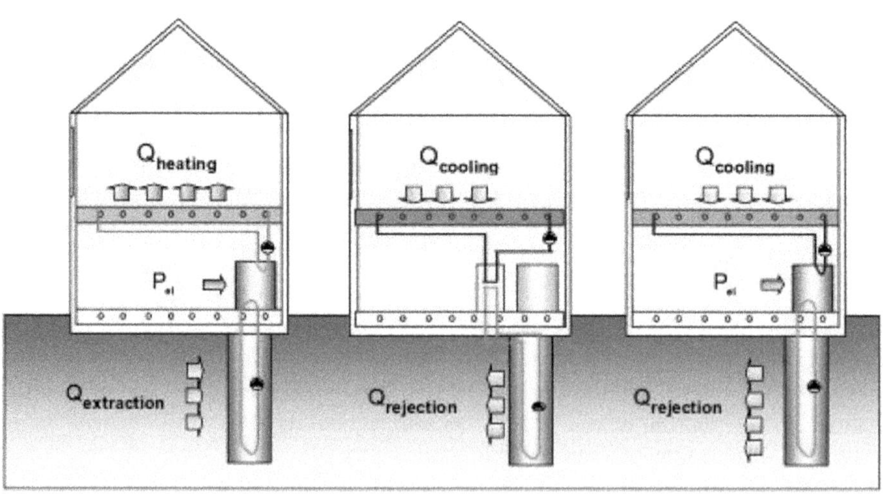

FIGURE 5.3 Geothermal borefields can either actively chill or passively heat water. When actively cooling, a GSHP is necessary, however when passively cooling, a heat exchanger is sufficient [11].

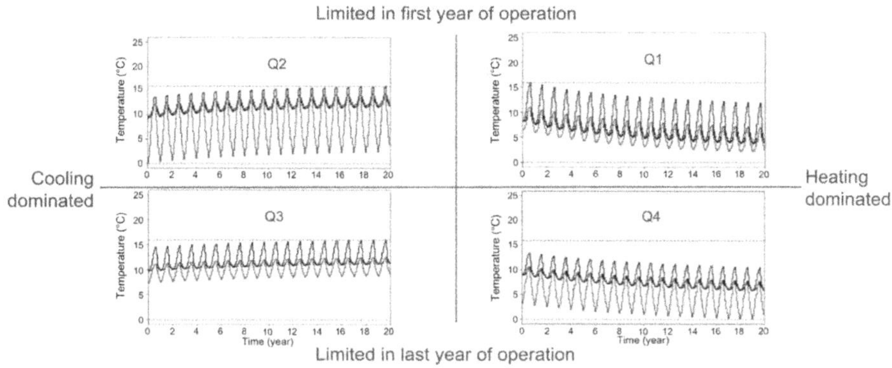

FIGURE 5.4 Illustrates of the four borefield quadrants previously discussed. The load profiles characterized by heat dominance are Q1 and Q4, while those characterized by cool dominance are Q2 and Q3. The temperature profile resulting from the heating load is represented by the red curve, while the cooling load is depicted by the blue curve. The temperature profiles are presented over the course of the 20-year simulation [11].

The micro-generation plant must operate continuously to achieve maximum profitability; otherwise, it will be inefficient. The design of micro-cogeneration technology is complicated by these challenges. Micro-cogeneration, commonly referred to as combined heat and power (CHP), is a method for simultaneously producing electricity and heat.

To achieve maximum CO_2 reduction in a development, building designers encounter several challenges, including:

1. identifying the optimal scale of a combined heat and power plant;
2. The integration of absorption chillers in large CHP plants contributes to a reduction in CO_2 emissions;
3. The incorporation of supplementary RE sources into CHP or combined cooling, heating, and power (CCHP) plants.

District heating and cooling plants demonstrate superior efficiency, a broader range of fuel options, and enhanced pollution management compared to localized systems. Cogeneration plants, also referred to as heat-only boiler stations or CHP systems, serve as the fundamental components of district heating systems. Both typically depend on the combustion of primary energy carriers. A primary distinction between the two systems is that heat-only boiler stations generate heat exclusively, whereas cogeneration plants simultaneously produce both heat and electricity. Engineers must priorities appropriate design considerations for the deployment of CHP and CCHP systems in mixed-use developments and other applications [12].

The CCHP achieves a maximum CO_2 reduction of 25.1%; however, the absorption chiller's parasitic power consumption and inferior coefficient of performance (COP) relative to contemporary vapor compression machines restrict the absorption cooling to a total emissions reduction of 1.2%. A biomass boiler, fulfilling 80% of annual heating requirements, will reduce CO_2 emissions from the project by 27.8%.

Wood boilers do not possess the quick heat output modulation capabilities inherent in gas-fueled boilers.

Thus, managing size to meet both base load and demand peaks requires the use of more responsive facilities, such as gas-fired boilers. Conversely, CHP plants are constrained in fulfilling electrical demand, as they are required to dissipate surplus heat into the environment upon reaching their heating capacity. CCHP systems will fulfil 42% of the thermal energy demand at the crossroads, while biomass boilers will account for 38%. Both systems will equally contribute to a reduction in CO_2 emissions, each achieving a 13% decrease. Overall CO_2 emissions have decreased by 26%. Currently, absorption cooling fails to achieve a reduction in CO_2 emissions of even 0.5%.

5.2 ENERGY SYSTEMS USED IN GREEN BUILDINGS

Architectural designs and constructions are increasingly influenced by green building ideas aimed at environmental protection. The integration of RE sources in building design is essential for environmental sustainability. This encompasses systems for the collection, conversion, and storage of energy. Green buildings utilize RE sources and energy efficiency measures, or a combination of both, to produce eco-friendly, low-carbon waste energy, in contrast to traditional energy systems that rely significantly on fossil fuels [13].

5.2.1 Sustainable Energy Systems

Currently, there is a critical necessity to address the significant issues of pollution and climate change. Researchers promote the amalgamation of solar and thermal technologies to attain optimal results, particularly in pertinent fields such as absorption and adsorption cooling. In areas with abundant sunlight, hybrid PV/thermal systems employing absorption cooling techniques are crucial for significantly decreasing energy usage and carbon dioxide emissions. Research indicates that solar-powered cooling systems are both practical and efficient, suggesting their potential to replace traditional compression cooling systems while reducing greenhouse gas emissions.

Research on built environments has increasingly emphasized sustainable design and green architecture. It is essential to summarize the current theoretical perspectives, trends, applications, and limitations related to the development of environmentally sustainable green buildings to assess their environmental impact and energy efficiency. Previous research has proposed a theory indicating a strong correlation between green building performance and environmental assessment levels. Consequently, comprehensive studies emphasize the necessity of identifying and incorporating sustainable energy performance indicators in environmental evaluations and green implementations. Evaluating performance necessitates consideration of essential components of sustainable energy performance indicators, including building energy efficiency, thermal performance, and material efficiency.

The promotion of sustainable cities has become a significant issue for many developing nations, driven by the notion that 21st-century urban areas must priorities environmental awareness and intelligence. Sustainability encompasses studies related to people, planet, and community, representing a significant global issue.

Defining sustainability enhances the understanding of the value of sustainable cities. "Green infrastructure" is central to the sustainability movement, aiming to transform urban planning by integrating the relationships among people, the environment, economic factors, and technology. Sustainability comprises three primary components: environmental, socio-cultural, and economic. The components are intricately connected to the enhancement of individuals' well-being, as illustrated in Figure 5.5.

Figure 5.6 presents the findings of a comprehensive study on the environmental impacts of buildings in the USA, which are expected to be similar in other

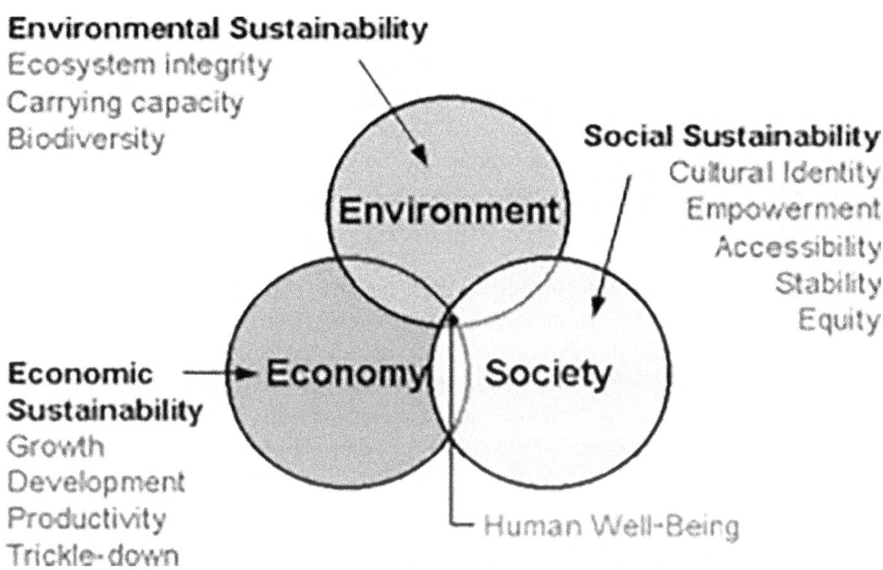

FIGURE 5.5 The basis of sustainable developments [14].

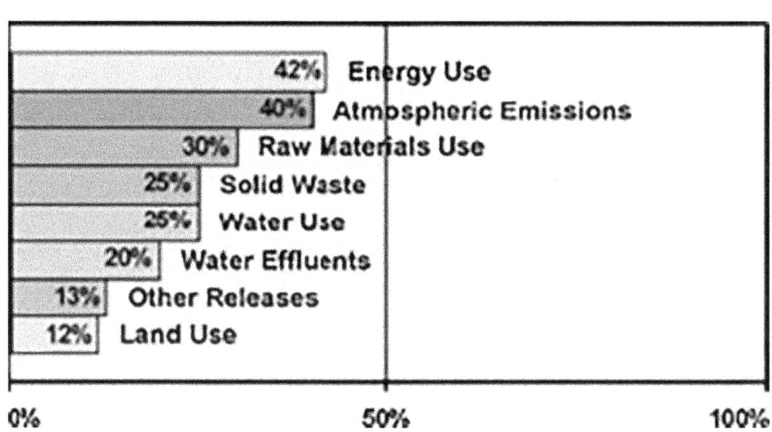

FIGURE 5.6 The environmental impacts of buildings [14].

industrialized countries. This underscores the considerable negative impacts of these structures on the environment. The primary negative consequences that require innovative solutions are energy consumption and petroleum emissions.

5.2.2 Solar Energy

Recent studies demonstrate that machine learning optimization techniques, such as reinforcement learning, genetic algorithm optimization, and particle swarm optimization, significantly enhance the management, energy efficiency, and economic optimization of solar cooling units. These developments underscore the potential significance of optimization and machine learning in enhancing the effectiveness and efficiency of solar cooling systems. Machine learning algorithms enhance system efficiency through the analysis of complex data patterns, the prediction of system behavior, and optimization of control processes. This leads to enhanced energy utilization. Ultimately, these technologies enhance the efficiency and sustainability of solar cooling systems by enabling adaptation to varying weather conditions, optimizing energy consumption, and improving operational performance.

Numerous authors have identified the various challenges and barriers that solar cooling faces in achieving acceptance within the industrial sector. Issues related to the availability and reliability of the cooling process, insufficient incentives relative to solar PV systems, and the absence of global standards exemplify these challenges. The reliance on backup heat sources is necessitated by the imbalance between cooling demand and solar irradiation. The elevated expense of solar collectors, inefficiencies in solar energy utilization, and insufficient integration within building designs present significant challenges.

Solar cooling systems provide sustainable, cost-effective, and environmentally friendly air conditioning solutions. A more robust strategy is necessary for marketing sun cooling systems, given the uncertainties surrounding solar climate interventions and the limitations of "cooling credits" as a viable solution.

Solar cooling technology has various sustainable applications, each presenting distinct advantages and disadvantages. Despite their high cost and substantial space requirements, vapor compression systems offer a high COP. Peltier thermoelectric systems, despite being compact and quiet, are costly and exhibit a low COP. Stirling cooling systems, despite their complexity and expense, offer significant efficiency advantages. Rankine and other thermo-mechanical systems exhibit complexity; however, they demonstrate high efficiency in electricity generation. The selection of technology is determined by specific application requirements, complexity, cost, and performance factors [15].

The transition from fossil fuels to RE sources is a significant challenge for governments globally, driven by the implications of climate change. Extreme weather occurrences and the continuous sequence of the warmest years on record are both attributable to climate change. Consequently, 179 nations and the European Union convened in Paris for 2 weeks in December 2015 to negotiate an agreement aimed at limiting the increase in global temperatures to well below 2°C, with an ideal target of less than 1.5°C [16].

5.2.3 Energy Harvesting, Conversion, and Storage

Combining multiple types of energy harvesters into a single unit may enhance their environmental flexibility. This would enable the unit to concurrently collect various forms of ambient energy. Consider an implanted device that integrates a piezoelectric energy generator (PENG) with a biofuel cell (BFC), utilizing a fundamental resistance–capacitance (RC) high-pass filter. This hybrid device utilizes biofluid glucose and kinetic energy from respiration to generate power. The output energy and service life of the original devices can be improved by employing either of the two energy harvesting technologies, either concurrently or independently. The integration of devices produces a synergistic impact when gadgets with different operating systems work together. This facilitates more efficient harvesting of multiple energy types, as the total output is considerably greater than the sum of its two individual components.

An alternative is to develop a system for energy conversion and storage. This approach enables the efficient utilization of electrical energy harvested from the environment, facilitating its storage in energy storage devices for future use. Various energy storage devices are applicable, including supercapacitors, lithium-sulfur batteries, and lithium-ion batteries. For instance, exposure to sunlight initiates a photo-charging process in the semiconductor that converts solar energy into electricity. This is subsequently stored in the lithium battery through an electrochemical process. This stored electrochemical energy can subsequently power the electronics [17].

Future structures are increasingly adopting passive design principles, which significantly reduce energy consumption for temperature regulation, in alignment with global efforts to minimize energy use. It is feasible to reduce energy consumption by over 50% relative to current buildings through the enhanced building envelope of a passive house, characterized by superior insulation, increased air tightness, and a design that optimally harnesses solar radiation. Enhancing insulation can reduce the air conditioning load by up to 50%, although this estimate is significantly influenced by the local environmental conditions of the building model. To significantly reduce energy consumption in buildings, improvements in insulation and air tightness are expected to continue in the future. Improved insulation not only decreases thermal loads but also results in the accumulation of additional waste heat on wall surfaces.

The temperature difference between the interior and exterior of walls increases because of enhanced insulation, which inhibits heat loss to the outside environment. This results in an increase in the surface temperature of the walls. During the night, as waste heat is released and external temperatures decrease post-sunset, the temperature differential reverses. Structures are progressively outfitted with sensors and Internet of Things (IoT) devices to satisfy the growing demand for smart homes and smart buildings. Each device ideally operates independently of the grid power, as the required electric voltage is typically less than 5 V, and the power consumption remains low. Recent proposals have emerged to integrate energy-harvesting technologies with self-powered smart sensors and devices. These systems produce electricity utilizing renewable and/or underutilised energy sources or waste heat generated on-site.

PV and wind power systems exemplify RE sources that offer potential solutions. However, these systems exhibit limitations, including intermittency, and their generally large scale significantly influences building design regarding integration. Utilizing electricity generated from the unused or waste heat of buildings to power sensors represents a feasible alternative [8]. Buildings exhibiting a temperature difference with their environment often harbor several underutilized or waste heat sources. The exterior walls constituting a building's envelope serve as a pertinent example. A temperature gradient is commonly present across a building's envelope, representing a loss of heat, as the primary function of the building envelope is to protect occupants from harsh external weather conditions.

A heat sink is a component frequently utilized in energy harvesting systems employing thermoelectric generators (TEGs). Heat enters the system due to the temperature differential in the absence of any moving components. Figure 5.3a illustrates the heat flow of the system throughout the day. The energy-harvesting block produces electricity by transferring heat from the accumulated wall heat to the hot side of the aluminum absorber plate, subsequently flowing through the TEG. The phase change material (PCM) reduces the temperature on the cold side, thereby enhancing the efficiency of heat-to-electricity energy conversion (Figure 5.7). During the completion of the transition from solid to liquid, residual heat accumulates in the PCM as latent heat. Figure 5.7b illustrates the heat flow during nighttime.

Latent heat is transformed into usable heat by the PCM as the sun sets and ambient temperatures decrease. Heat is released into the atmosphere after being transported from the PCM to the absorber plate through the TEG. This heat was formerly accessible during daylight hours. During this procedure, both the hot and cold sides are inverted. The orientation direction does not influence the power generation capability of the TEG. A reversal of the current is anticipated, however. The device generates energy through its heat dissipation, while the PCM naturally reverts to a solid state. This procedure is referred to as PCM regeneration.

For the PCM to operate effectively as a heat sink the subsequent day, regeneration is necessary. Regeneration occurs through the natural diurnal variation, necessitating no additional energy. The energy-harvesting block is capable of supplying power bi-daily, while the phase change material is naturally recharged overnight for subsequent use. Heat transfer into and out of the energy-harvesting block is directly proportional to the magnitude of the surface's diurnal temperature gradient.

The utilization of the Stacking Ensemble Surrogate Model (SESM) in production of energy is additionally corroborated by its biological non-toxicity. Consequently, human breast cancer cells (MCF-7 cells, Michigan Cancer Foundation-7) are cultivated on SESM to evaluate their biological non-toxicity. Cell viability on SESM is 96% after 2 days of culture, demonstrating its minimal toxicity to biological systems.

SESM can serve as a secure energy material within the framework to assess the efficacy of transforming solar energy into useful light and heat, as well as the related storage and utilization mechanisms. The related performances are analyzed by the creation of a model house measuring 45 mm x 45 mm x 30 mm. The exterior of the model house is covered with a 1 mm thick SESM layer (Figure 5.8a and b). The house can capture and store solar energy, which can later be employed during colder months to generate electricity and supply light in dim settings.

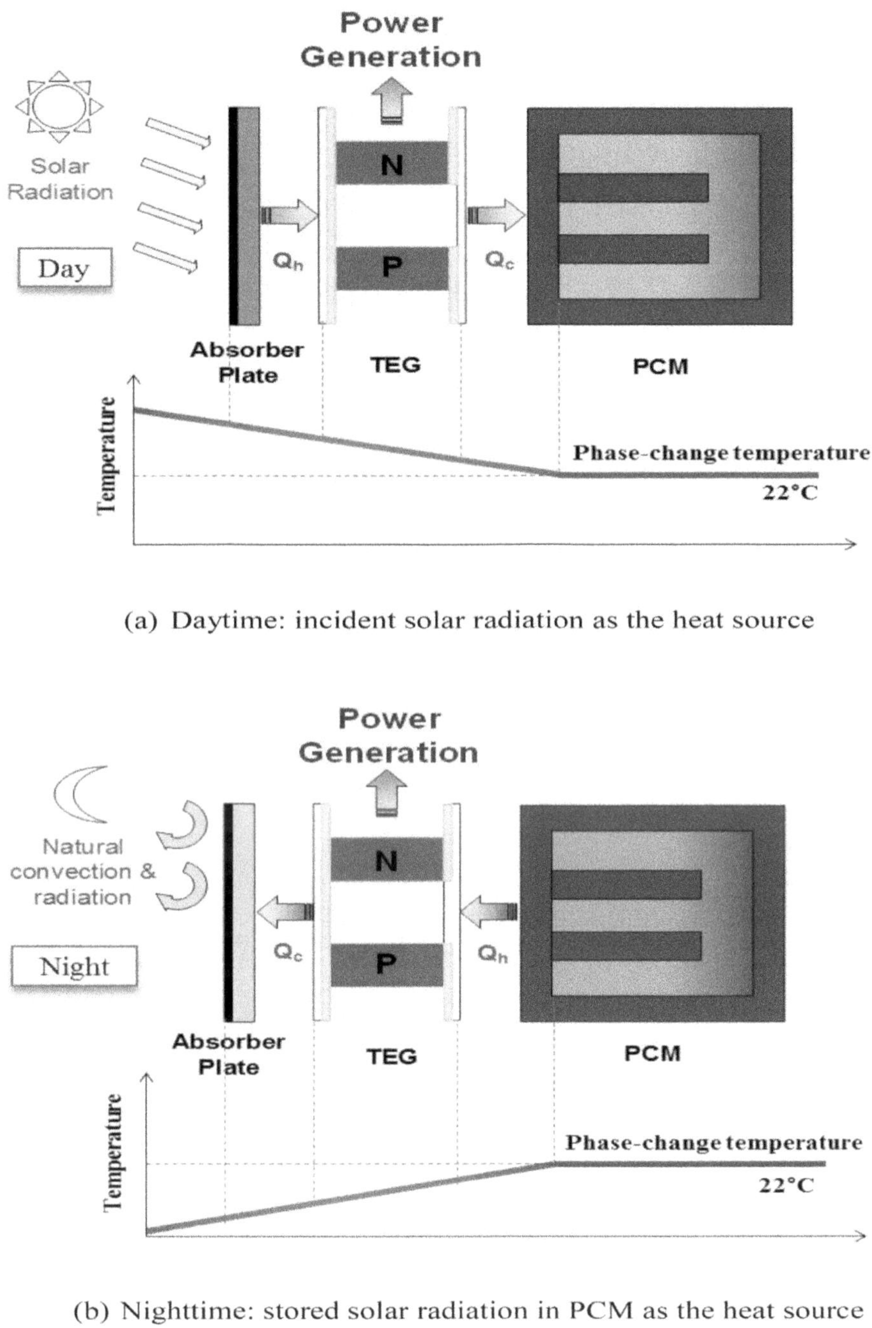

FIGURE 5.7 Concept of building integrated energy-harvesting block [18].

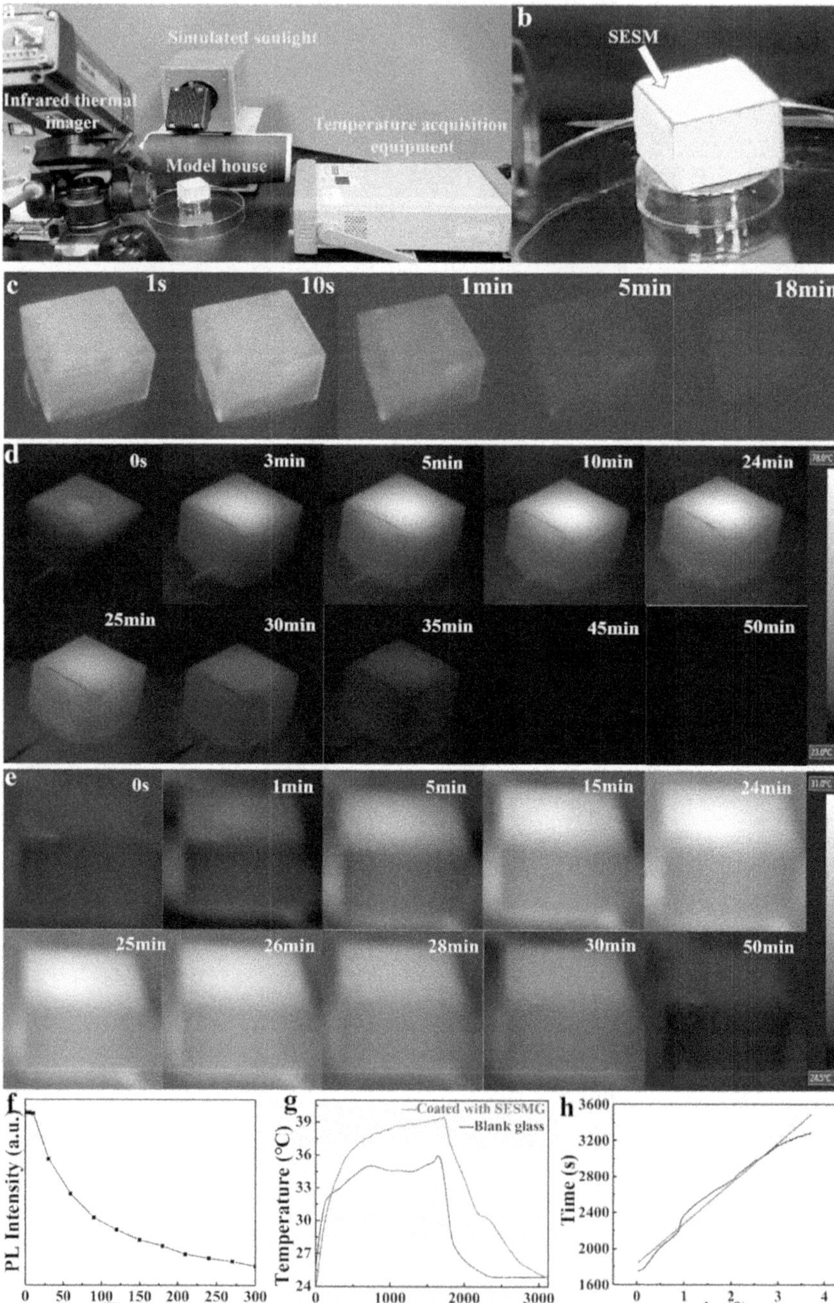

FIGURE 5.8 (a,b) The model facilitates a reversible energy storage and release process. (c) The reversible transition from ground state to excited state in PLMs, along with the reversible melting-freezing transition of TD in SESM. (d,e) The model house is positioned

10 cm from the optical lens, under a simulated sunlight source with a power output of 1000 W/m^2. (f) The model house's blue fluorescence emission from SESMs embedded with PLMs and the intensity of their fluorescence degradation. (g) Model house temperature profiles with and without SESMs after 30 minutes of simulated solar (1 kW/m^2) irradiation and 25 minutes of darkness; SESMG: Stacking Ensemble Surrogate Model Generation. (h) The slope of the irradiation time against $-\ln(\theta)$ of the model house painted with SESMs reflects its system time constant (τs) [19].

The reversible transition from ground state to excited state in Persistent Luminescence Materials (PLMs), along with the reversible melting-freezing transition of time-dependent (TD) (Trimming the UCERF3-TD Hazard Tree) in SESM, facilitates a reversible energy storage and release process. The model house is positioned 10 cm from the optical lens, under a simulated sunlight source with a power output of 1000 Wm^{-2}. The surface-mounted self-luminous SESM installed in the home is capable of sustaining a consistent blue light output, even in the absence of direct sunlight, following a charging duration of 5 minutes. The persistent blue light observed from various angles within the home indicates that the self-luminous SESMs demonstrate effective light conversion, storage, and release (Figure 5.8c and f).

In future work, new high-performance PLMs can be used to replace the commercial PLMs, further improving the efficiency of solar-to-light conversion [19]:

$$\text{Conversion efficiency}_{\text{solar to light}} = \frac{\int P_{\text{emitted light}}\,dt}{\int P_{\text{solar light}}\,dt}$$

The application of an infrared thermal imager to capture infrared images of the house's façade, alongside the use of thermocouples for measuring internal temperature variations, enables an assessment of the model house's response to solar radiation. The external surfaces treated with SESMs and those uncoated exhibit a slight increase in temperature within 24 minutes of simulated sunlight activation, followed by a gradual return to room temperature upon deactivation. The model house with SESM coatings differs from the one without them in two respects. The solar absorption of the SESM-coated model house is primarily concentrated on the roof, whereas the uncoated model house exhibits uniform solar absorption across its surface.

The difference in light transmission rates between bare glass and SESMs is the primary cause of this phenomenon. To mitigate the persistent thermal stress induced by omnidirectional light irradiation, the model house coated with SESMs can be subjected to localized heating. Furthermore, when solar irradiation is discontinued, the external temperature of the house treated with SESMs is markedly lower than that of the house without such treatment. This indicates that the house equipped with SESMs utilizes its interior heat more effectively (Figure 5.8d and e). Figure 5.8g illustrates the interior temperatures of the model house, comparing conditions with and without SESM coatings to support this assertion. The findings indicate that both the coated and uncoated models undergo a rapid temperature rise in the initial phase.

The model home coated with SESMs exhibits slightly lower interior temperatures compared to the uncoated version, attributable to enhanced light transmission from unprotected glass. The solid–liquid phase change and thermal energy storage capabilities of SESMs are evident in the model house's temperatures, which range from 36°C to 39°C under solar radiation exposure. The thermal energy storage of the SESMs in this section is calculated to be 1262 J, equivalent to 10 g multiplied by 126.2 J/g. Heat exchange with the environment likely accounts for the temperature variations between 33°C and 35°C in the model house without SESMs. Subsequently, the model house is placed in a dark environment devoid of sunlight, resulting in significantly higher interior temperatures for the SESM-coated model house compared to the uncoated version.

The effective release of thermal energy generated by the solar-thermal conversion of the nano-structured carbon-photothermal material (NSCP) in the solar-absorbing energy storage materials/modules (SESMs) accounts for this outcome. The SESMs in this section exhibit a released thermal energy of 1268.8 J, calculated as 10 g × 126.88 J/g. The maximum temperature (T_{max}) is recorded at 40°C, the surrounding temperature (T_{surr}) at 24°C, and the transmittance (τ_s) at 447. The use of the light-absorbing material (NSCP) is limited to 1%, despite the solar–thermal conversion efficiency being roughly half that of PV efficiency [27]. Installing SESMs on windows or walls enables the absorption and transformation of solar energy throughout the day, owing to their effectiveness in solar-to-light and solar-to-thermal conversion.

Input electric energy can be conserved by emitting the converted light and heat during nighttime. Modifying the thickness of SESMs enables the adjustment of thermal energy storage capacity. Adjusting the type and concentration of doped PLMs in SESMs allows for modulation of the emitted light's color and intensity. The energy cost savings from lighting and air conditioning enable self-luminous SESMs to recoup their expenses over time [19].

5.3 SOLAR ENERGY

5.3.1 Distributed Solar Heating System

Research on solar heating systems incorporating seasonal thermal energy storage (STES) has extensively focused on system design, capacity matching optimization, and operational performance. The year-round performance test results indicate that the STES system can sufficiently heat a home throughout the year, attributed to the 28.4% net stored energy currently available. An STES system at the building scale was developed, featuring a total solar collector area of 41.6 m^2 and a 36 m^3 underground water tank, as illustrated in Figure 5.9a. Figure 5.9b illustrates the energy-efficient two-story research house, which utilizes a solar thermal system for heating and cooling. The system achieved a solar fraction of 100% for space heating with a total load of 15 GJ.

5.3.2 Energy Management System

To balance power system supply and demand with intermittent renewable energy sources (RES) like solar power, an energy management system (EMS) is needed. The emergence of new communication technologies, including cloud computing, advanced

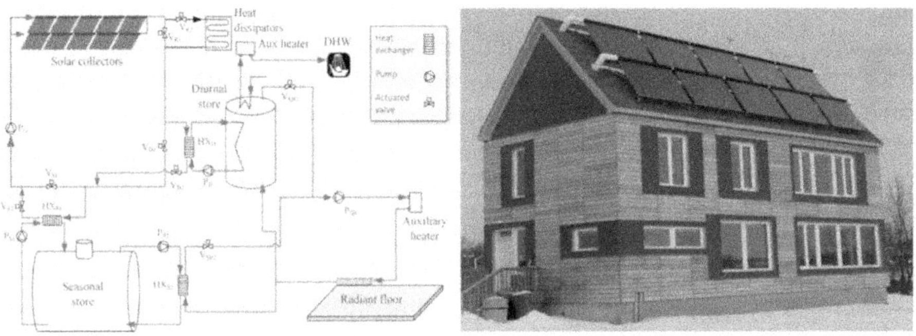

(a) (b)

FIGURE 5.9 Schematic diagram of the system and research house. (a) The solar heating system with STES. (b) The two-story research house in Ottawa [20].

metering infrastructure (AMI), the IoT, and big data, has transformed the traditional grid into a smart grid. The conventional EMS has evolved into an integrated energy management system (IEMS) to address the technological requirements of the smart grid. EMS can be categorized as either real-time or predictive; however, an IEMS integrates communication and technological advancements to optimize grid load and power supply by merging these two control types. An EMS typically focuses on a singular power source, whereas IEMS is capable of incorporating multiple energy systems and employing various control methodologies.

The majority of the research articles have focused on enhancing the system's storage, energy dispatch and supply, consumption, and equipment sizing. The objective was to emphasize the interrelationship of each parameter and the significant concepts and design considerations associated with them within the IEMS [21].

A smart microgrid consists of a network of distributed generators and load resources that are geographically proximate and can be managed effectively. It has multiple sources of alternating current power, with at least one utilizing RE such as solar or wind. Smart grids, despite their intermittent nature, facilitate enhanced control over resources, thereby enabling efficient and reliable power delivery. Smart grids, characterized by their distributed and flexible architecture, provide a viable and sustainable solution for electricity generation in developing countries. The renewable smart hybrid microgrid is ideal for EMS dispatch due to its requirement for the simultaneous intelligent communication of multiple load and supply sources. The term "smart grid" encompasses various implementations of microgrids, including renewable smart home microgrids, residential home microgrids, and RE home systems [21].

5.4 HYBRID RE

5.4.1 Biomass Energy

Biomass serves as an efficient and versatile energy source, capable of conversion into biofuels, heat, and electricity from diverse organic materials. Biomass exhibits a lower carbon footprint relative to alternative energy sources and is categorized into renewable and sustainable. Employing local individuals and utilizing regional resources, it contributes to the economic enhancement of the area. It is dependable as it does not rely on weather conditions, unlike solar or wind energy. The initial application of a net-zero energy building features a distinctive biomass-driven system for cooling, heating, and power generation. Their findings demonstrate that the system achieves an 86% reduction in carbon dioxide emissions, a 47% cost savings, and an 11% decrease in energy consumption. An efficient biomass-fired hybrid system designed to satisfy the heating, cooling, and electrical requirements of a historical structure was assessed through an energy evaluation. The analysis indicated that biomass quality significantly impacts system performance regarding irreversibility, leading to the conclusion that it requires appropriate enhancement. A biomass-fired heater designed for heating buildings and homes in Spain was developed and rigorously tested (Figure 5.10).

The proposed system decreases primary energy consumption and carbon dioxide emissions by more than 90% relative to traditional fossil fuel-based systems, indicating its potential to significantly contribute to the green transition paradigm. Research indicates that employing a double-effect absorption chiller for cooling yields the lowest cost of $190 per MWh. It indicates that biomass cost significantly influences system design, and the proposed clean system results in lower emissions relative to regional systems.

Figure 5.11 illustrates the progression of thermal output from the photovoltaic-thermal (PVT) panels and heater, emphasizing the importance of two separate renewable combinations and enabling an examination of the effects of varying solar irradiation on system efficiency. During periods of low solar availability, the system

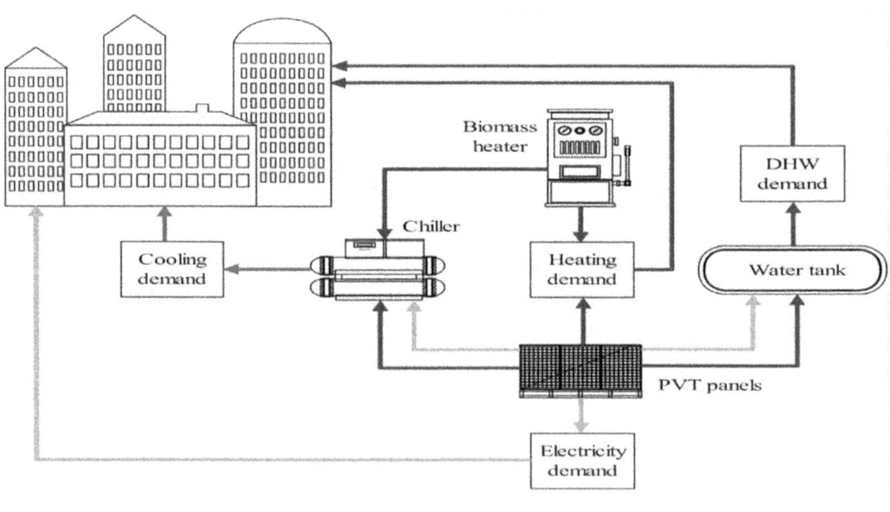

FIGURE 5.10 Visual representation of the smart building system's interaction with the grid [22].

Note: DHW = Domestic Hot Water.

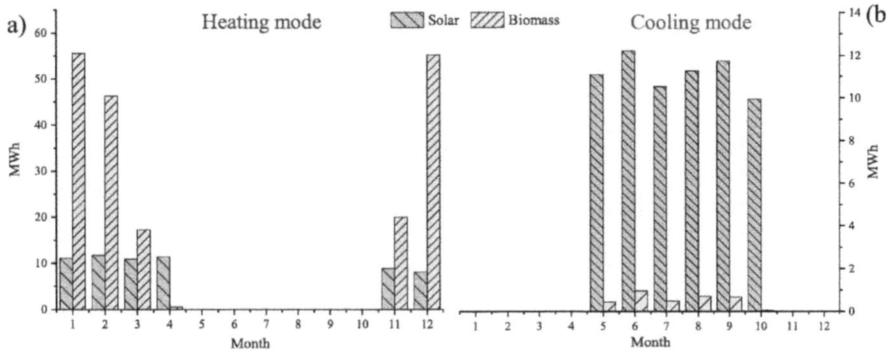

FIGURE 5.11 The monthly changes in heating from biomass and solar energy: (a) heating and (b) cooling modes [22].

predominantly utilizes biomass to power the absorption chiller and fulfil heating requirements, as illustrated in Figure 5.11. During hot weather with intense sunlight, the energy system of the smart building increasingly utilizes solar power in the summer months.

5.5 ECONOMICS OF RE

Energy significantly impacts sustainable economic growth, serving as a vital element in industrial processes. The increasing gap between global energy demands and available supplies has exacerbated pre-existing climate and environmental issues, including acid rain, haze, glacial melting, and rising sea levels. There is an urgent global necessity to address these factors, as recent research has identified them as the principal barriers to achieving objectives for reducing carbon emissions. The 2021 assessment report of the United Nations Intergovernmental Panel on Climate Change (IPCC) identifies greenhouse gas emissions from fossil fuel combustion as the primary driver of global warming.

Researchers in energy economics have extensively examined various aspects of energy transition, including its characteristics, pathways, experiences, and economic effects. Governments play a crucial role in facilitating energy transitions by establishing legally binding targets for energy development and enacting relevant policy measures.

RE is essential for promoting cleaner production and achieving environmental sustainability due to its non-polluting and sustainable nature. The growth of RE utilization is significantly endorsed by the 2016 Paris Agreement. By 2050, RE sources are projected to fulfil 90% of global energy consumption, as indicated in the International Energy Agency's report, "Net Zero by 2050: A Roadmap for the Global Energy Sector" (2021). Therefore, the utilization of RE is a crucial strategy for the global reduction of carbon emissions. Governments can significantly enhance energy efficiency and promote the expansion of the RE sector by formulating a renewable energy policy (REP).

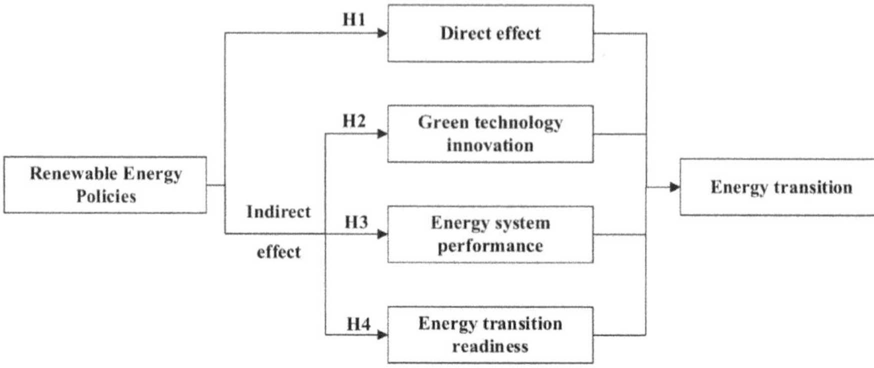

FIGURE 5.12 The energy transition and RE policy link and the mechanisms that influence it [23].

Empirical evidence regarding the mechanisms by which REP influences ET is currently inadequate. This study offers possible explanations for the relationship between REP and ET, as shown in Figure 5.12, which delineates the influence mechanisms in our analysis. REP has the potential to enhance the institutional framework of ET, thereby facilitating the progression of a clean, low-carbon, secure, and efficient energy system. Policy instruments like green electricity trading can help eliminate interprovincial barriers that hinder the transmission of renewable electricity. The enhancement of RE integration and utilization promotes an institutional environment that supports higher penetration rates of RE.

Adjusting feed-in tariff strategies decreases costs linked to renewable generation and secures revenue for producers of RE. This improves the installation and generation of RE, enhancing energy supply security, stabilizing the energy market, and facilitating the transition to low-carbon or decarbonized energy systems [23].

5.5.1 Green Finance

Green finance refers to financial mechanisms aimed at supporting environmentally sustainable initiatives that tackle climate change issues. The objective is to augment funding from banks, micro-credit institutions, insurance companies, and investments for sustainability initiatives. The financial sector is being incorporated into the shift toward low-carbon and resource-efficient markets via green finance. The focus is on improving manufacturing capacity by minimizing environmental pollution through the application of green technology and knowledge, as well as the augmentation of energy and resources.

Ecosystem neglect in conventional finance worsens environmental and socio-economic challenges. Two primary factors make green finance essential for low-carbon green growth. It links financial development, economic progress, and environmental protection. This financial model combines ecological preservation with economic profitability, effectively addressing the complexities present in both environmental and financial sectors.

Private-public investments boost green jobs and revenue. Growing the green economy and investing in green energy infrastructure reduces carbon emissions.

Maintaining biodiversity and ecosystems, they save energy and resources. Green money fosters economic growth through ecologically friendly enterprises, like traditional finance. Global Environmental Goals (GEG) relies on green finance to support and drive sustainable economic development, according to this report [4].

5.5.2 Limitations

Provinces' economic, political, and environmental characteristics may not reflect the nation's or other countries' diversity. In addition to green finance and RE, environmental policies, green technologies, and technology transfer can enhance environmental recovery. GEG and environmental performance should be examined using these characteristics in future studies. Ensuring a cause-and-effect relationship between institutional renewable energy (IRE), green money, and outcomes may be difficult. Despite our superior statistical methods, causality in social and environmental systems is still complex.

REFERENCES

1. Hassan, Q., Algburi, S., Sameen, A. Z., Tariq, J., Al-Jiboory, A. K., Salman, H. M., ... & Jaszczur, M. (2024). A comprehensive review of international renewable energy growth. *Energy and Built Environment*. www.sciencedirect.com/science/article/pii/S2666123323001186
2. Mulya, K. S., Ng, W. L., Biró, K., Ho, W. S., Wong, K. Y., & Woon, K. S. (2024). Decarbonizing the high-rise office building: A life cycle carbon assessment to green building rating systems in a tropical country. *Building and Environment*, *255*, 111437.
3. Ragab, K. M., & Orhan, M. F. (2024). Evaluating conventional and renewable energy systems for green buildings: A case study on energy efficiency and cost optimization. *Case Studies in Thermal Engineering*, 105233.
4. Qing, L., Abbas, J., Najam, H., & Ma, X. (2024). Investment in renewable energy and green financing and their role in achieving carbon-neutrality and economic sustainability: Insights from Asian region. *Renewable Energy*, *221*, 119830.
5. Shah, W. U. H., Hao, G., Yan, H., Zhu, N., Yasmeen, R., & Dincă, G. (2024). Role of renewable, non-renewable energy consumption and carbon emission in energy efficiency and productivity change: Evidence from G20 economies. *Geoscience Frontiers*, *15*(4), 101631.
6. Yüksek, İ., & Karadağ, İ. (2021). Use of renewable energy in buildings. *Renewable Energy-Technologies and Applications*, *23*, 23–162.
7. Xu, B., & Lin, B. (2024). Green finance, green technology innovation, and wind power development in China: Evidence from spatial quantile model. *Energy Economics*, *132*, 107463.
8. Kwok, K. C. S., & Hu, G. (2023). Wind energy system for buildings in an urban environment. *Journal of Wind Engineering and Industrial Aerodynamics*, *234*, 105349.
9. Ouerghi, F. H., Omri, M., Nisar, K. S., Abd El-Aziz, R. M., & Taloba, A. I. (2024). Investigating the potential of geothermal energy as a sustainable replacement for fossil fuels in commercial buildings. *Alexandria Engineering Journal*, *97*, 215–229.
10. Verhelst, C. (2012). Model predictive control of ground coupled heat pump systems in office buildings (modelgebaseerde regeling van grondgekoppelde warmtepompsystemen in kantoorgebouwen).

11. Coninx, M., De Nies, J., Hermans, L., Peere, W., Boydens, W., & Helsen, L. (2024). Cost-efficient cooling of buildings by means of geothermal borefields with active and passive cooling. *Applied Energy*, *355*, 122261.
12. Elsarrag, E., & Alhorr, Y. (2013). Green building practices: Optimisation of CCHP and biomass heating for maximum CO2 reduction in a mixed-use development. *International Journal of Sustainable Built Environment*, *2*(1), 99–108.
13. Wang, R., & Zhai, X. (2018). *Handbook of energy systems in green buildings*.
14. GhaffarianHoseini, A., Dahlan, N. D., Berardi, U., GhaffarianHoseini, A., Makaremi, N., & GhaffarianHoseini, M. (2013). Sustainable energy performances of green buildings: A review of current theories, implementations and challenges. *Renewable and Sustainable Eenergy Reviews*, *25*, 1–17.
15. Ali, F. H., Al-Amir, Q. R., Hamzah, H. K., & Alahmer, A. (2024). Unveiling the potential of solar cooling technologies for sustainable energy and environmental solutions. *Energy Conversion and Management*, *321*, 119034.
16. Lu, Y., Khan, Z. A., Alvarez-Alvarado, M. S., Zhang, Y., Huang, Z., & Imran, M. (2020). A critical review of sustainable energy policies for the promotion of renewable energy sources. *Sustainability*, *12*(12), 5078.
17. Yun, S., Zhang, Y., Xu, Q., Liu, J., & Qin, Y. (2019). Recent advance in new-generation integrated devices for energy harvesting and storage. *Nano Energy*, *60*, 600–619.
18. Byon, Y. S., & Jeong, J. W. (2020). Annual energy harvesting performance of a phase change material-integrated thermoelectric power generation block in building walls. *Energy and Buildings*, *228*, 110470.
19. Yin, Y., Chen, H., Zhao, X., Yu, W., Su, H., Chen, Y., & Lin, P. (2022). Solar-absorbing energy storage materials demonstrating superior solar-thermal conversion and solar-persistent luminescence conversion towards building thermal management and passive illumination. *Energy Conversion and Management*, *266*, 115804.
20. Wang, B., Liu, Y., Wang, D., Song, C., Fu, Z., & Zhang, C. (2024). A review of the photothermal-photovoltaic energy supply system for building in solar energy enrichment zones. *Renewable and Sustainable Energy Reviews*, *191*, 114100.
21. Falope, T., Lao, L., Hanak, D., & Huo, D. (2024). Hybrid energy system integration and management for solar energy: A review. *Energy Conversion and Management: X*, 100527.
22. Wang, D., Almojil, S. F., Ahmed, A. N., Chaturvedi, R., & Almohana, A. I. (2023). An intelligent design and environmental consideration of a green-building system utilizing biomass and solar having a bidirectional interaction with the grid to achieve a sustainable future. *Sustainable Energy Technologies and Assessments*, *57*, 103287.
23. Lee, C. C., & Wang, T. (2024). The impact of renewable energy policies on the energy transition--An empirical analysis of Chinese cities. *Energy Economics*, *138*, 107838.

6 Mechanical Energy Storage Systems for Green Buildings

Industrial processes release greenhouse gases (GHGs), which play a significant role in the global warming phenomenon. Power and thermal energy produced 25% of global GHG emissions in 2010. As the main source of electricity, fossil fuels also create the most air pollution. Because of their enormous potential to mitigate climate change, renewable energy-generating technologies are thus garnering more and more interest. Large-scale adiabatic compressed air energy storage (CAES) technologies, conventional power plants, and pumped hydropower could contribute to the goal. The Mobile Energy Storage Systems (MESS) store extra energy for grid demand when demand exceeds supply. A system that integrates renewable energy sources with power grids, known as MESSes, can reduce GHG emissions. MESS generate GHGs and consume substantial energy and resources throughout its lifecycle, from building to decommissioning.

To understand the environmental impacts of different MESSes, a life cycle assessment (LCA) is needed. Also important are the net energy ratio and life cycle GHG emissions. The former might assess MESS energy efficiency. Net energy ratio measures how efficiently a technology provides energy to an economy. Comparing total intake to possible output measures energy consumption's impact. The net energy ratio of a technology often matters most [1].

Electricity is generated and used simultaneously. As demand changes, power must be given at the proper level. When electricity is scarce, voltage and frequency quality degrade. Power production and consumption are generally far apart. Long transmission lines increase investment and energy loss. However, daily and seasonal renewable energy fluctuations make problems worse. Permanent and temporary energy storage is a popular solution to these issues. Energy storage produces multiple types of energy that can be used for a productive job.

Stakeholders in the field may find new opportunities due to the energy's portability and storage capability. The storage of electrical energy is quite challenging. To improve energy efficiency, it is necessary to enhance energy storage. One critical step towards improving energy storage is discovering new materials. Another

DOI: 10.1201/9781003752837-6

is understanding the properties of both current and new materials. A wide variety of methods exist for storing energy, such as

1. thermal,
2. electrical,
3. mechanical,
4. chemical, and
5. electrochemical systems.

The most studied and easily implemented approaches are mechanical pumped hydro storage systems (PHSS) and thermal energy storage (TES) systems using latent heat and phase change materials (PCMs).

For energy conservation, energy storage is essential, and it also improves the reliability and performance of energy systems and reduces the imbalance between supply and demand. Because the system becomes more cost-effective, less energy and capital are squandered, and premium fuels are saved. A power generation plant's performance could be improved with storage and load levelling, leading to lower generation costs and more energy conservation. Potential approaches for storing thermal energy include the use of PCMs [2].

The environmental impact, price volatility, and scarcity of fossil fuels drove scientists to find environmentally friendly renewable energy alternatives. Renewable energy, energy-efficient machinery, and waste heat recovery are among the solutions used. Potentially replacing fossil fuels in the near future, renewable energy sources are environmentally friendly and can be sustained indefinitely. The price of renewable energy has dropped significantly over the last decade, with photovoltaics (PVs) seeing a 90% drop and wind power a 50% drop. Renewable energy sources' capacity increased dramatically with the price drop, necessitating more effective storage. To maintain renewable energy output, energy storage device research is crucial. Figure 6.1 shows the primary energy storage technologies: mechanical, electrochemical, thermal, electrical, and magnetic.

Figure 6.2 shows the important factors to consider when choosing renewable energy storage systems. The figure indicates that choosing an energy storage system for renewable power sources involves numerous factors. Geothermal energy is less subject to daily and seasonal variations than solar and wind energy. Combining hybrid renewable energy sources reduces intermittency (again). Electricity output and renewable power source reliability may determine the size and type of energy storage system needed. An additional crucial factor to think about while selecting the appropriate energy storage technology is its intended use. Different energy storage systems are required for various purposes, such as transportation, households, and businesses. As an example, peak power usage is not a concern in the industrial use case due to the consistent energy needs. Energy usage varies each year in the house scenario, with a peak power and variations in line with population growth. Conversely, transportation energy use peaks at various times throughout the day, including morning, noon, and night.

The requirements of the application should be satisfied by an appropriate energy storage system. The current state of technology is a major factor in determining the

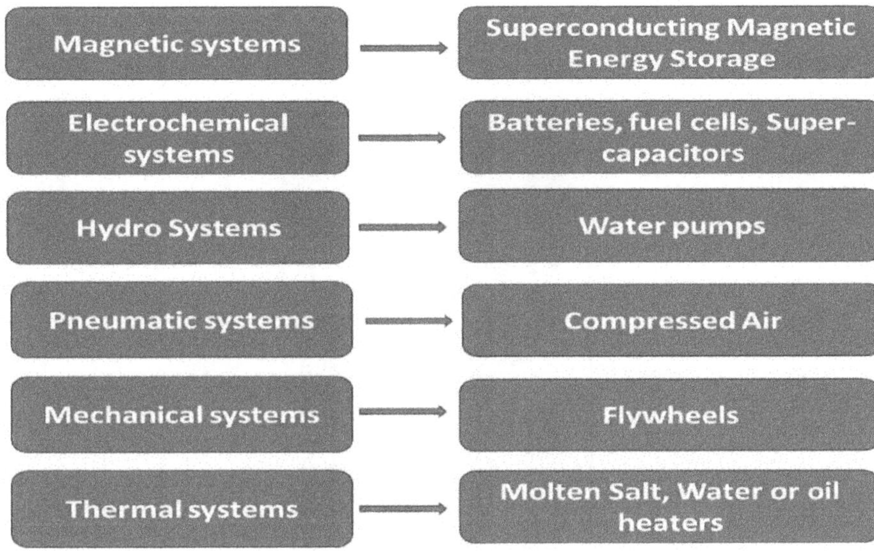

FIGURE 6.1 Main energy storage techniques [3].

optimal energy storage system. A variety of system sizes will be able to handle varied energy capacities in the mature energy storage technology, all while maintaining an acceptable cost and lifespan. Renewable energy storage systems that use the grid as their principal energy storage benefit from the grid connection. When energy demand is low or a grid interruption occurs, only an auxiliary energy storage system will be used. Although these systems have little impact on the environment, it can nevertheless be considered when deciding on the most effective renewable energy storage options. Hydro pump storage, for example, necessitates a large amount of land and produces visual pollution; disposing of batteries, which include some harmful components, can also have an adverse effect on the environment [3].

6.1 PUMPED HYDROELECTRIC STORAGE

PHES stands for pumped hydro energy storage. Using power during off-peak hours, charging pumps water from a lower reservoir to a higher-altitude reservoir for gravitational potential energy storage (PHES). Turnaround, which pumps fluid from a higher to a lower reservoir under peak demand, converts gravitational potential energy into mechanical and electrical energy. PHES round-trip efficiency is 70–80%. It is the greatest and cheapest large-scale energy storage option. In 1981, the 770 MW Nagarjuna Sagar facility became operational, marking the first pumped storage project in India. Another more recent addition is the Tehri Pumped Storage Hydroelectric Power Plant. Within the next 5–6 years, India plans to statewide develop pumped hydro storage capacity of 10 GW. Pumped hydro projects take longer to develop than other energy storage technologies due to the permitting process. Its low cost, long lifespan, and ability to supply significant amounts of power for multiple hours are its

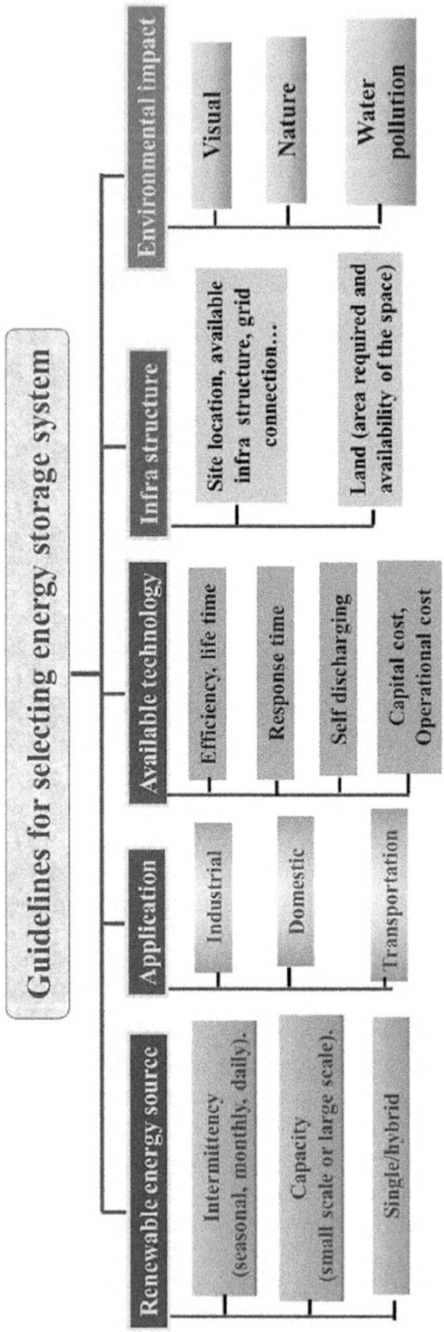

FIGURE 6.2 Energy Storage System selection recommendations [3].

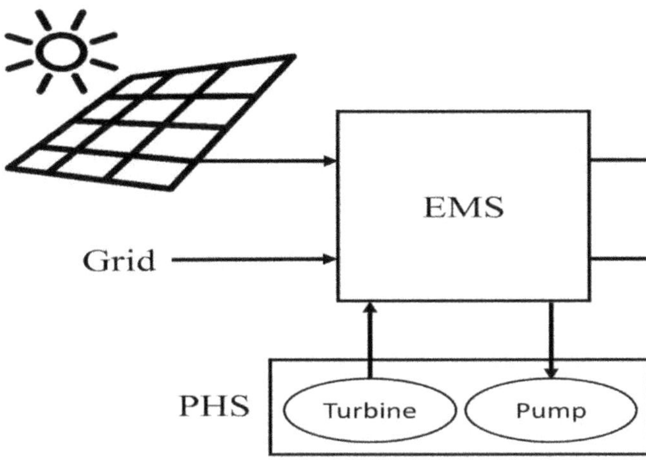

FIGURE 6.3 Schematic diagram of the microgrid [5].

primary benefits. A further benefit is the energy it produces, which is easily accessible [4].

This section discusses renewable hydrogen and pumped hydro storage initiatives and policies worldwide to help design and optimize storage for hybrid vehicles and renewable energy systems. To accelerate urban adoption of net-zero energy commercial buildings, this research optimizes pumped hydro and renewable hydrogen storage.

Figure 6.3 shows the suggested system schematic layout of the microgrid's (MG's) energy storage, PV, and grid connection. PHS construction requires only a pump and turbine. The pump stores water, while the turbine generates energy. Demand comes from the irrigation pump and farmhouse electricity. The Energy Management System (EMS) controls pump power, turbine flow, and irrigation schedules. This system integrates PHS, PV, energy demand, and irrigation [5].

Interest in smaller-scale installations, particularly in buildings, has been prompted by PHES's features. Nevertheless, there is a lack of information regarding the long-term viability of PHES in buildings, both in terms of technical aspects and financial considerations. Given the dearth of both theoretical research and data on real installations, many questions remain about the accessibility and characteristics of the essential equipment (such as water turbines or water storage tanks) for this type of installation. It is also important to consider the building's capacity for energy storage and how it impacts consumption. Consider energy storage and water delivery system interactions. Water distribution systems can supply demand and are compatible with PV installations, but their complementary use in a building is unknown.

6.1.1 PHES Modelling

Two reservoirs of varying heights are used for PHES. Pumping water from the lower reservoir to the higher reservoir stores energy. The water can then be sent through a turbine from the upper reservoir to the lower reservoir to recover that energy. The energy stored E (J) is determined by [6]

$$E = \rho Vgh \tag{6.1}$$

where

ϱ = water density (kg/m^3),
V = volume of water (m^3),
g = gravitational acceleration (m/s^2)
h = height difference (m)

Considering that the reservoirs' height differences are far more than the water level variations, the energy stored E (kW h) is roughly determined by taking into account a constant water density of 1000 kg/m^3 and a gravitational acceleration of 9.81 m/s^2.

$$\boldsymbol{E \approx Vh / 367} \tag{6.2}$$

where

V = the volume of water at the higher reservoir (m^3) and
h = the height difference between the reservoirs (m).

In terms of volumetric energy density, this gives PHES 0.056 W h/l at 20 m. Among the many energy storage technologies available, lithium-ion batteries offer a density of 600 W h/l, which is a very high level.

Electronic pump control allows partial load use of higher-capacity pumps. Costs and efficiency must be considered while choosing an electronically controlled pump. For electrically regulated pumps, "it is preferable to exploit any excess power at low efficiency than dump it as power losses." This logic ignores the cumulative cost of electronic pumps. Figure 6.4 shows the efficiency and cost curves for electronically controlled pumps over a range of values similar to those for non-electrical pumps. The data came from around 150 duty points.

6.1.2 Technical Analysis

The two primary proposed technical conditions relevant to the pipe design's bounds should be considered while choosing the PHES pipe diameter. The suggested head loss ratio (H_{ratio}), computed by dividing the H_{loss} for the specified pipe diameter by the average static head, must not exceed 15%, and the fluid velocity (V_f) within the pipe must not exceed 10 m/s. Due to the potential severity of consequences, such as cavitation, exceeding this limit, the maximum water velocity (Vf,max) has been set at 10 m/s. To find the pumping/generating mode flow rate (QP/QT) and Vf within the penstock, one needs to know the diameter of the penstock.

By iteratively applying the approach to all of the suggested diameters, we may establish an acceptable range for the pipe diameter. The procedure starts with a

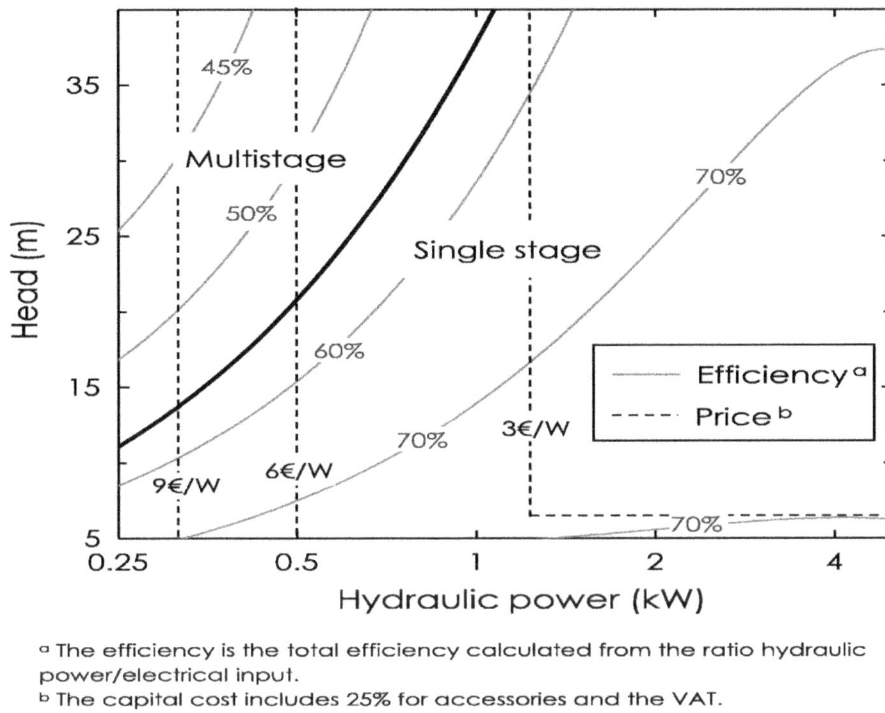

FIGURE 6.4 Centrifugal pumps with electronic control have the highest efficiency and capital costs [6].

common large diameter value and then gradually decreases it until the desired maximum value, Vf, is reached. For every suggested pipe diameter, a correlation is created between PP, Hpu, Htu, and QT using a polynomial regression method that produces high R-squared values.

A decrease in total E_{lost} may result from an increase in H_{pu}, leading to higher power consumption, which is an important consideration. On the flip side, the $E_{turbine}$ can see a drop when in generating mode. However, this depends on how the hybrid renewable energy system (HRES) is designed, particularly on the capacity of each part that makes up the system. After that, the data is combined and normalized so that a trade-off analysis can be performed among the variables H_{pu}, H_{tu}, E_{pump}, and $E_{turbine}$ [7].

6.1.3 Influence of the Building Height and Load

A building's PV-PHS system efficiency depends on several elements. Weather and installation space considerably affect PV energy output, while building height and roof capacity greatly affect PHS energy generation. The building's occupancy density affects component efficiency and energy demand profile compliance, which affects the total coverage rate. Annual dynamic simulations yielded and evaluated the following energy fluxes:

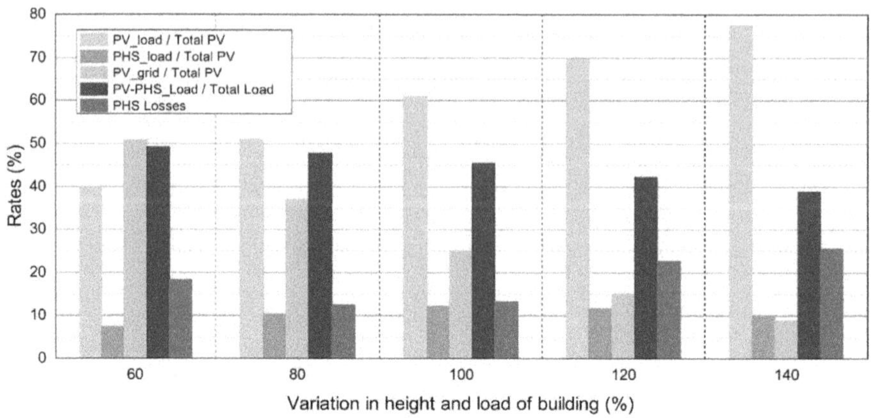

FIGURE 6.5 Load demand and building height affect normalized mean yearly energy flows [8].

(1) The PV energy output divided by the total PV energy output is known as the PV load/total PV ratio.
(2) A phosphorus storage load/total PV energy production is the ratio of the two, which measures how quickly the system can supply the load's energy demands.
(3) The PV grid/total PV ratio, which represents the ratio of PV energy input to total PV energy output.
(4) The PHS load/load ratio, which represents the relationship between the rate of system coverage and the need for load.
(5) The percentage of energy lost by the storage system in comparison to the energy input to the pump.

Figure 6.5 shows that PV energy supply to load demand rises with demand. This is due to more efficient solar panel use to meet rising energy needs. Thus, extra energy transmitted to the grid is greatly decreased. Even with increased use, the PHS provides a lot of energy for structures between 80 and 132 m, thereby increasing building height is advantageous. The PHS energy delivery rate drops as the building height climbs from 158 to 184 m and the demand increases from 120% to 140% of the starting demand. Running the pump uses more energy, and demand quickly uses up PV energy. Figure 6.5 shows that load requirement affects covering rate more than height. Thus, tall buildings can only install so much PV.

6.2 COMPRESSED AIR ENERGY STORAGE

High-pressure air stores energy in the mechanical energy storage CAES system. Commercial energy storage systems exist in various sizes worldwide. Figure 6.6 shows a CAES system's major components and operation. CAES operates through compression, storage, and expansion. In charging mode, compressors pressurize air with power off-peak. Air is cooled and kept in a reservoir after being heated and

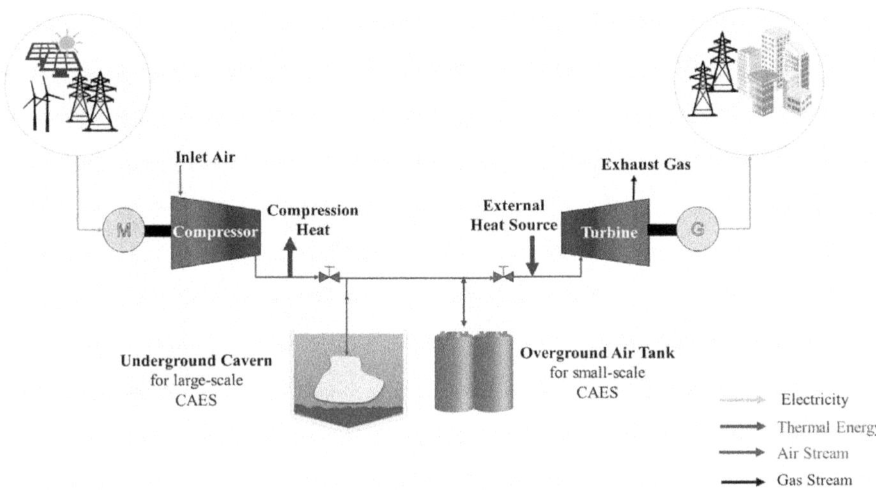

FIGURE 6.6 Schematics of the operating principle of the CAES plant [9].

compressed to high pressure. Heating, TES, and air dissipation are thermal energy uses. Discharging releases high-pressure air for power generation. The system then expands through energy-producing turbines after external heating.

Due to its high energy capacity and power rating, CAES is appropriate for stationary and large-scale energy storage applications. CAES has poor energy and power density, meaning it stores little energy and power per unit volume compared to air. To store a significant amount of energy, larger containers are needed, as seen in the underground storage of two commercial CAES plants: Huntorf in Germany (532,000 m^3) and McIntosh in the USA (270,000 m^3).

The system expands through turbines that create power once an external heat source preheats it.

In the mid-2020s, India will have the most population. India's economic growth, rural electrification, and urbanization will improve energy demand in the next few decades. India aims to generate 40% renewable electricity by 2030. To achieve this target, India is deploying more renewable electricity technology annually. Turbines generate power once an external heat source preheats the system [10].

CAES stores air in underground caves or tanks. Heat and expand compressed air powers turbines during peak demand. CAES supports long-term storage and is affordable. Energy storage and industrial operating technology benefits compressed air businesses, including manufacturing and mining. Traditional CAES devices offer 60–65% round-trip efficiency. Innovative technologies like adiabatic compressed air energy storage (A-CAES) reuse compression heat to increase efficiency [11].

The safe and convenient generation and management of compressed air make it popular in industry and services. Air compression is needed in most factories. Compressed-air generation is energy-intensive and expensive for most industries. A traditional compressed-air system's energy efficiency is unknown. As compressed air

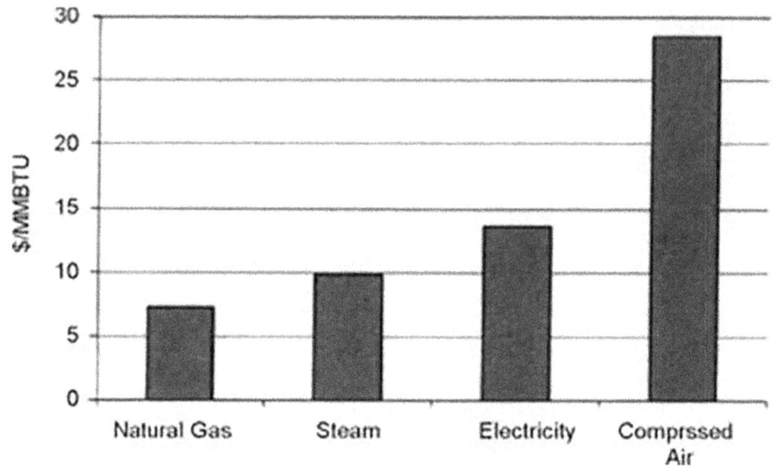

FIGURE 6.7 Cost of energy delivery modes [12].

is expensive, only utilize it when it enhances safety, production, or labor (Figure 6.7). Compressors, dryers, and associated equipment use 70–90% of electricity.

6.2.1 Working Stage

CAES stores energy for later use. This saves extra renewable energy at low demand. Large-scale CAES systems lose heat during compression and must expand to remove it. Temperature change during expansion changes the material form. Storage efficiency improves when compression heat is used during discharge.

Air energy storage systems use motorized compressors. Air entering the storage cavern is compressed. The discharge phase benefits from heat exchangers heating air with extracted heat.

Increasers, or air turbines, generate work that generators turn into power. Five phases comprise CAES' energy conversion process. Air is compressed and heat is extracted during charging. Heat exchangers direct discharged air to expanders for heating and work generation. As seen in Figure 6.8, air turbines and electrical generators generate electricity in subsequent phases.

Large-scale CAES devices are designed for load shifting on grids. Small systems are ideal for guaranteeing a continuous power supply, especially with renewable energy. For air storage and pressurization, large-scale CAES systems require an accessible and impermeable cavern. In the absence of a cavern, modular canisters can store air at precise pressures.

A modular system with three 7-l cylinders that discharge into an air turbine was examined. Maintaining the experimental system's operating pressure below 5 bar prevented heat or expansion losses and set a benchmark. Mechanical efficiency, which measures air turbine power output from compressed air, was 95%, while system efficiency was 97%. Carbon dioxide's higher density allowed the same group to reach

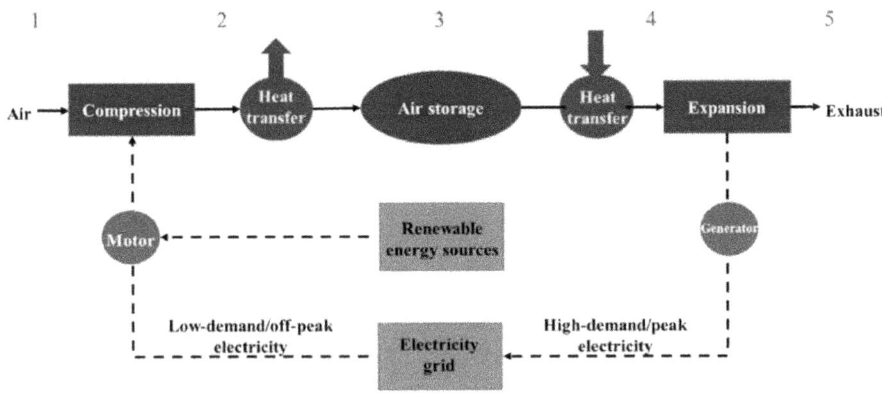

FIGURE 6.8 Operational stages of CAES [13].

79% efficiency at 3 bar operating pressures by replacing air with carbon dioxide in a closed-loop system [13].

D-CAES compressors compress air during charging, and surplus electricity powers them if energy production exceeds demand. Air is compressed and pushed into a high-pressure chamber underground. When the PV field's power output falls short, the D-CAES system's discharge mechanism expands compressed air in the turbines to generate energy. A heat recovery device heats compressed cave air with turbine waste heat to improve D-CAES round-trip efficiency.

The preheated air-fuel mixture is then burned in the combustor. Turbines generate electricity by expanding combustion byproducts at high temperatures. D-CAES reduces renewable power's intermittent nature by coordinating output with energy use, a major benefit of coupling it with PV [14].

Urbanization and energy demand are increasing the need for local renewable energy and energy storage. A-CAES may work in decentralized applications, although its usefulness is uncertain. A distributed A-CAES system for urban building infrastructure will be created. To optimize the design of a distributed A-CAES system for urban building applications, thermodynamic, techno-economic, and environmental factors are considered. The A-CAES system is sized and designed for application needs using simulation and optimization.

In Figure 6.9, solar, A-CAES, and grid energy sources power the building during various EMOSs. All facilities employ practically all power sources for electrical demands. The A-CAES system's unique role in meeting load needs across EMOSs is highlighted. Solar energy contribution to load demand is uniform across strategies A, C, and D as the renewable configuration uses the same solar PV capacity (Figure 6.9a, c, and d).

Despite peak solar intensity in May and June, March and August have the highest PV self-consumption at 211.5 MWh. This is because March and August have higher load demand than May and June. Solar energy supplies 54% and 55.7% of total electricity to meet demand in May and June, corresponding with increased

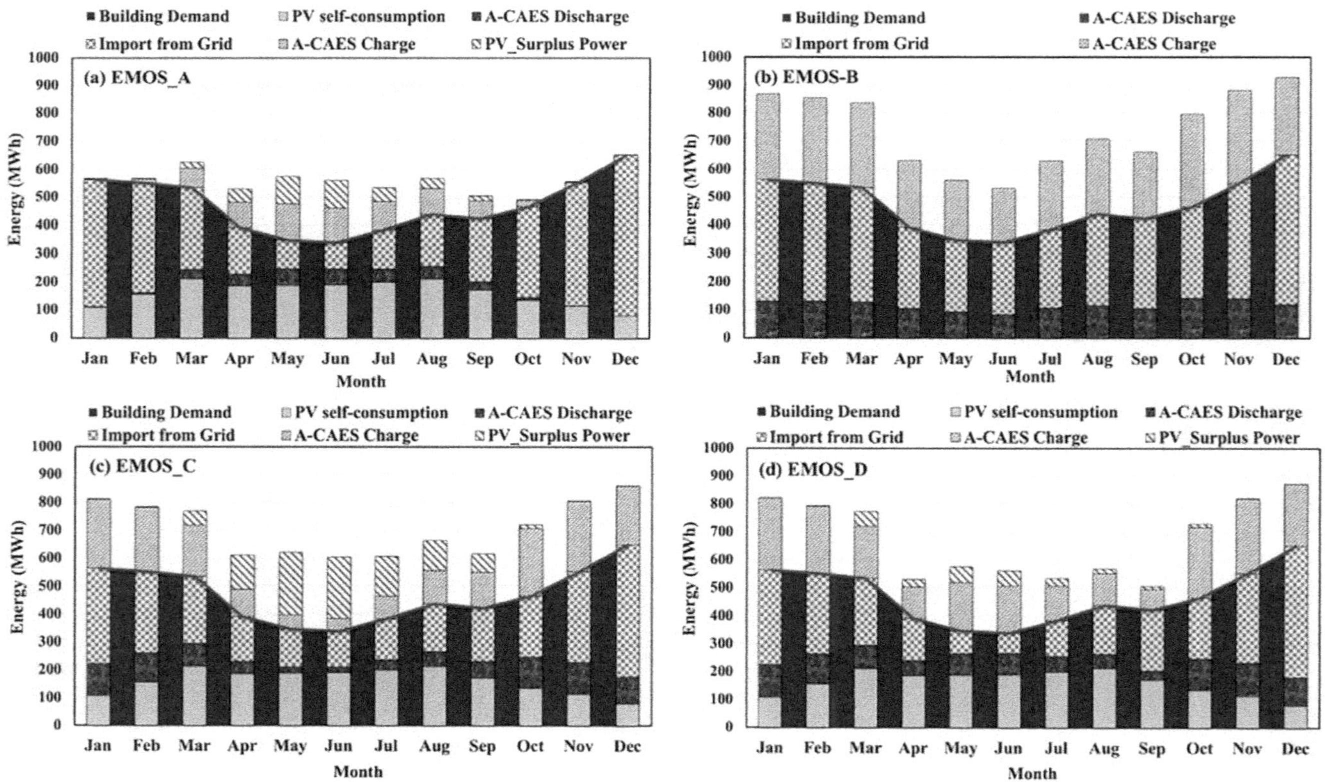

FIGURE 6.9 Multiple energy sources respond to building model electrical energy demand for different EMOSs monthly under scenario I (zero sale) [15].

solar irradiation and decreased energy needs (Figure 6.9a). During peak demand and low solar intensity, like the first two and last 3 months, A-CAES supplies 0.02–2.35% of electricity.

A-CAES supplies 6.5–18% of power in late spring and summer, lowering load demand and increasing solar intensity. For renewable integration and load shifting, the EMOS-C-designed A-CAES system (Figure 6.9c) is distributed year-round. Thus, under low solar intensity, high building energy demand, and peak-hour electricity pricing, the A-CAES system supplies 14–24% of electricity. Even with solar integration and load-shifting, A-CAES provides the least summer energy.

A-CAES stays fully charged most of the time due to grid interaction during peak hours and abundant solar availability with decreased load demand in summer, resulting in huge solar PV-surplus electricity. To take advantage of solar radiation in spring and summer, EMOS-D limits load-shifting to high-demand winter and autumn. Under EMOS-D, the A-CAES system supplies more energy than techniques A (winter) and C (summer), as shown in Figure 6.9d.

The selected EMOSs configure A-CAES based on solar energy, grid time-of-use power tariff, and load demand. The initiative aims to minimize LCOE and improve A-CAES load demand fulfilment and dependability. The techno-economic feasibility of the proposed system is assessed using alternative scenarios for PV surplus electricity management. The model's applicability across case studies is verified by post-optimization sensitivity analysis of major input parameters like energy resource amount and AST volume [15].

6.2.2 CAES Commercialization Challenges

Artificial air storage, such as pressure vessels, is too expensive for CAES in places lacking deep salt caverns or mine shafts. Liquefied air storage and modular artificial air storage can make large-scale CAES economically viable. Complex liquid air storage-CAES technology Salt Cavern Compressed Air Energy Storage (SCCAES) increases energy conversion efficiency and energy density by 5–10 times, lowering high-pressure air storage expenses. The economic benefits of SC-CAES require substantial research on system performance and key components such as

(1) cold storage,
(2) regenerative heat exchangers, and
(3) liquid expansion.

Mechanical inertia restricts the compressor and expansion unit. It takes several minutes for CAES to switch between energy storage and power-generating states, which fails to meet MG demands for real-time stabilization of variable renewable energy generation and rapid load-power balance. CAES must be used with supercapacitors, flywheel energy storage, and batteries to provide a complicated integrated design and optimal control difficulties for energy storage systems [15].

6.3 FLYWHEEL ENERGY STORAGE

Batteries, supercapacitors, flywheels, and superconducting magnetic energy storage are sophisticated ESS. Grid-connected technologies integrate renewable energy and regulate frequency and peak. Energy storage system (ESS) integrates renewable energy system (RES), smoothes power, regulates frequency, and improves energy quality. Growing popularity of battery energy storage system (BESS) installations. Because BESS operates often and harshly, contemporary power solutions prioritize battery durability.

Power grid applications that minimize frequency changes benefit from FESS's long lifespan, efficiency, and fast response time. EV PV charging stations can equal flywheel hybrid power smoothing efficiency. More than standard ESS, flywheel energy storage system arrays (FESAs) can improve power system efficiency, stability, and security [16].

Kinetic or flywheel energy storage helps machines work smoothly and gives a lot of power and energy density. Flywheels use motors or generators to transfer kinetic energy into and out of them. Flywheels are often made with permanent magnet machines due to their efficiency, low rotor losses, and high power densities.

Generating kinetic energy by charging a flywheel. To prevent frictional losses and safety hazards during long-term storage, vacuum-operated magnetic bearings sustain the rotating mass. The rotor bearing system can maximize performance using mechanical, magnetic, or combined components. Although cryogenic cooling solutions need energy, superconducting magnetic bearings (SMBs) are suited for high-speed applications. Flywheels should have a high power density and rotational velocity as energy stores expand in proportion to the square of velocity, unlike mass, which grows linearly. Systems engineering and materials research provide the necessary technology. They used to be made of steel, titanium, or aluminum, but now the flywheel rotor and housing are composites. Researchers have improved rotor materials and designs to allow 10,000 rpm speeds.

High power and energy density, longer working life, good round-trip efficiency, low environmental impact, and tens of thousands of cycles are all features of flywheel energy storage Batteries and supercapacitors have higher power density, cost, noise, maintenance, and safety than flywheels. A hybrid system with flywheel supercapacitors boosts power density. This allows pulsed power exchange and large energy storage [17].

6.3.1 Working Principle

Charge is stored as kinetic energy in FESS. Flywheel friction is reduced by vacuuming the system. External drives—electrical or mechanical—power the flywheel. Equation 6.1 indicates that the system's potential energy is directly proportional to the flywheel's mass multiplied by its angular velocity squared [18]:

$$E_k = \frac{1}{2} I \omega^2 \tag{6.3}$$

where

E_k = flywheel energy,

I = moment of inertia, and

ω = flywheel angular velocity.

Flywheel mass and diameter determine the moment of inertia, as illustrated in Equation 6.4 [18]:

$$I = \frac{1}{2} m r^2 \tag{6.4}$$

where

m = mass of the flywheel, therefore,

As stated in Equations 6.3 and 6.4, increasing the flywheel's mass, size, and rotational speed increases energy storage. The material of the flywheel determines its maximum speed. Shape can determine a flywheel's moment of inertia. Equation 6.3 calculates flywheel rotor inertia for steel solid cylinders:

$$I = \frac{1}{4} r^4 \pi \alpha P \tag{6.5}$$

where

α = length of the cylinder, and

p = density of the cylinder material.

Maximum energy density per volume and mass is shown in Equations 6.6 and 6.7:

$$e_v = K_\sigma \tag{6.6}$$

$$e_m = \frac{K_\sigma}{p} \tag{6.7}$$

In Equations 6.6 and 6.7,

e_v, e_m = kinetic energy per unit volume.

K = flywheel shape factor,

σ = maximum stress, and

p = density.

Flywheel shapes include discs with constant tension, conical discs, discs with constant thickness (pierced and unpierced), discs with rims, and thin rims. Shape affects flywheel form factor [18].

The suggested Residential Hybrid Micro Grid System (RHMGS) topology is presented in Figure 6.10. The flywheel stores and backups electricity from the wind and PV in this RHMGS. Through a storage unit, a variable renewable electricity system must provide continuous power to residential applications to supplement wind or PV output.

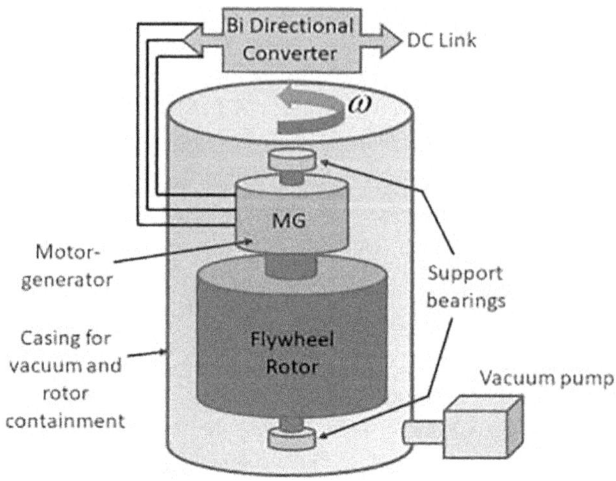

FIGURE 6.10 Configured flywheel system for electrical storage [19].

The rotor, as illustrated in Figure 6.10, needs to be a component of a larger system to enable the storage and extraction of electrical energy.

6.3.2 Flywheels in the Storage Landscape

Energy storage needs have soared because weather-dependent renewables like solar and wind have replaced massive coal-fired steam facilities. Renewable electricity is intermittent and grid inertia has diminished, lowering frequency stability. Large steam plants, like flywheels, have high mechanical inertia and react swiftly to frequency-changing supply and demand. Replacement synthetic inertia comes from sub-second reaction energy storage. Most often, lithium-ion (Li-ion) batteries are utilized as they can store energy for 1–2 hours and be used elsewhere, allowing revenue stacking. Manufacturer warranties cover a certain term. A more recent and extensive study found flywheels to be the best cost-effective rapid reaction technology without deterioration (Figure 6.11).

6.3.3 Characteristics of FESS

Flywheels have high power and energy densities, fast charge and discharge rates, high round trip efficiency, quick reactivity, and minimum environmental impact. Significant longevity is above 20 years. Charge status is easy to assess as rotation speed is temperature and wear-independent. Higher self-discharge rates than typical storage mediums and potentially hazardous flywheel rotors if poorly made are downsides. Flywheel self-discharge incurs maintenance loss, unlike other storage systems.

Heating or cooling many battery systems reduces their small self-discharge. Low maintenance makes flywheels durable. Due to their short charge–discharge cycles,

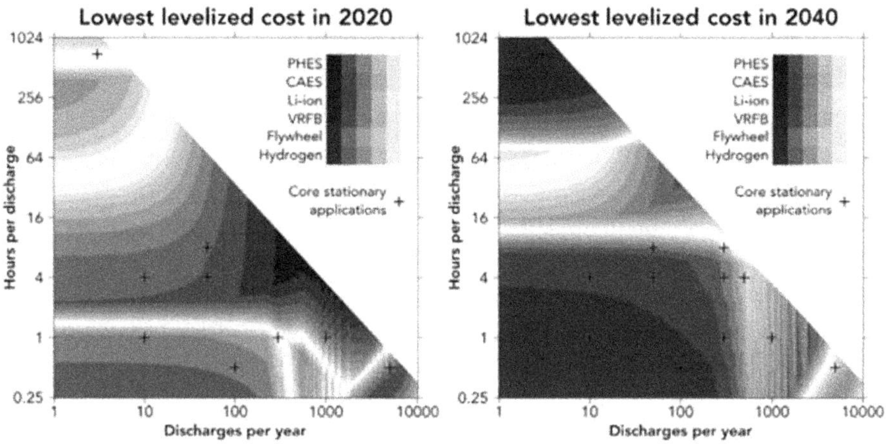

FIGURE 6.11 Lowest levelized cost of storage [19].

flywheels survive longer than other energy storage systems. Deepness of discharge (DoD) does not affect charge–discharge cycle speed; functionality and application do. According to estimations, FESS can perform normally for over 20 years after hundreds of thousands of charge–discharge cycles. With 90–95% round-trip efficiency, the device can transport huge amounts of power in seconds. FESS is also eco-friendly because its building material is non-toxic and it does not create pollutants.

Flywheels may be customized for energy storage because their power and energy ratings are distinct.

MG and power electronics dimensions affect a flywheel's power rating, whereas rotor speed and size affect energy rating.

Flywheels are often used for short power bursts.

Dimensions distinguish FESS units from other energy storage methods. Battery power densities may be 5–10 times lower than those of flywheels. Flywheel systems can replace batteries in transportation and spacecraft due to their longer lifespan and reduced volume [20].

6.3.4 FESS with Wind Generation Units

Field-oriented control (FOC) is studied and applied to the machine side converter for wind turbine control, including DFIG and IM-based FESS. Wind is intermittent, thus FESS and DSTATCOM are used. A fuzzy logic controller dampens the wind turbine's active power output to regulate FESS active power. Distributed control strategies for FESS and wind-producing units propose a local knowledge-based dispatch problem solution.

Fuzzy logic controllers for the FESS provide active power smoothing of wind fluctuations. This method has been shown to work for operating and managing DFIG-based wind turbines with matrix converters and IM-based FESS. FESS's surge power absorption improves transient and steady-state behavior in wind farms and other systems [21].

6.3.5 FESS with Solar Generation Units

PV panels convert sunlight into electricity. The sun's radiation changes daily and annually. This causes fluctuations in electrical energy production. Electrical energy storage systems like FESS can absorb PV electricity or compensate for weather-related power interruptions.

FESSes are integrated into PV power plants to reduce solar energy's instability. In combination with an appropriate control strategy, a micro-FESS can reduce PV generation changes in a PV panel system. The FESS is used with PV in Figure 6.12. FESS is used with solar units to reduce power fluctuation and uncertainty [22].

To maintain MG stability and provide electric power when RES (wind or solar) power is unavailable or insufficient, a backup energy source or energy storage system, usually battery storage, must be established. A single PV system cannot create enough power for a day, yet solar power is abundant. When solar or wind power is absent, a diesel generator (DGen) can provide power. It may work at partial loads when efficiency is low, but environmental contamination, high maintenance, and operating costs make it unsuitable.

To stabilize the MG, rapid-response ESS absorb or provide electricity. When choosing an ESS, consider capacity reserve, reaction time, energy density, storage costs, security, environmental consequences, storage length limits, and conversion response time. Battery storage in various capacities is ideal. The storage is practical, trustworthy, easy to use, and efficient. Batteries produce toxic waste, GHGs, and other environmental hazards. Battery technology is expensive, low-energy, requires key chemicals, and has a short lifespan in some storage applications.

Remote load sites are best electrified by HRES. HRES includes hydrogen generators, ESS, and wind or PV. HRES can meet energy needs at a reduced cost in areas where upgrading the grid is impossible or too expensive. Due to their high cost, limited charge/discharge cycles, and slow response time, they may not be suitable for small-scale applications.

FESS have a high power density, long service life, many charge/discharge cycles, and fast response time. Production and disposal of batteries have no environmental impact. A high charge-discharge cycle count makes the FESS excellent for power

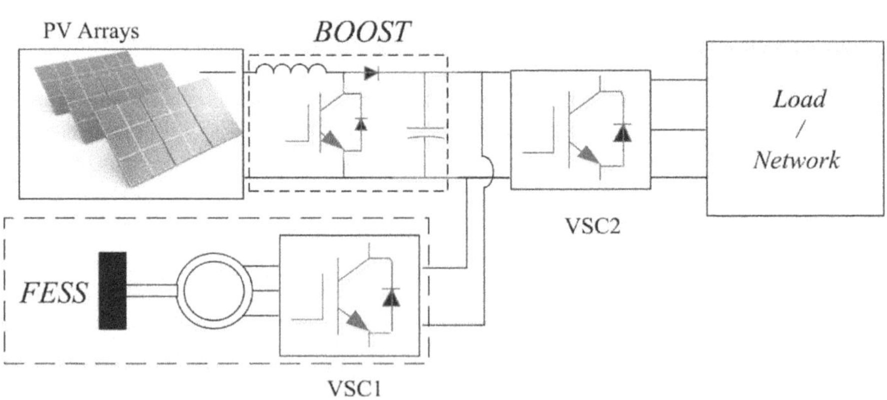

FIGURE 6.12 FESS application alongside PV [23].

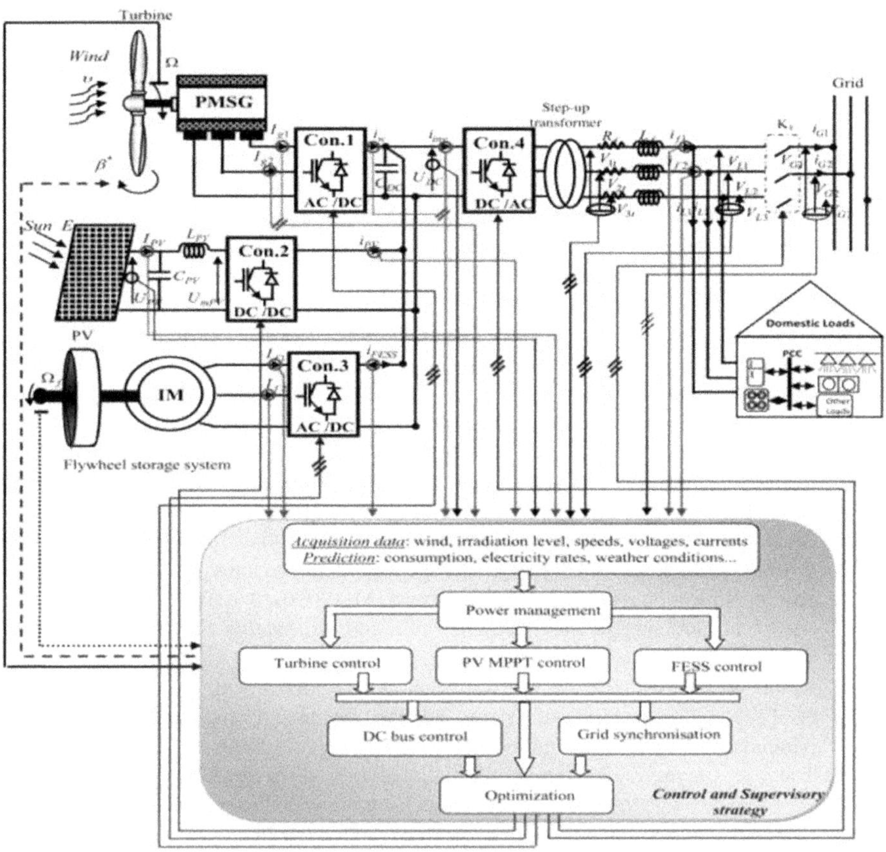

FIGURE 6.13 Turbine PMSG architecture for the microgrid power generating system Connections 1 and 2 AC/DC DC/DC [25].

smoothing, power quality enhancement, renewable energy integration, and system stability [24]. A permanent magnet synchronous generator (PMSG) architecture for the MG power generating system with a turbine is shown in Figure 6.13 [25].

REFERENCES

1. Kapila, S., Oni, A. O., Gemechu, E. D., & Kumar, A. (2019). Development of net energy ratios and life cycle greenhouse gas emissions of large-scale mechanical energy storage systems. *Energy*, *170*, 592–603.
2. Sharma, A., Tyagi, V. V., Chen, C. R., & Buddhi, D. (2009). Review on thermal energy storage with phase change materials and applications. *Renewable and Sustainable Energy Reviews*, *13*(2), 318–345.
3. Olabi, A. G., & Abdelkareem, M. A. (2021). Energy storage systems towards 2050. *Energy*, *219*, 119634.
4. Rohit, A. K., Devi, K. P., & Rangnekar, S. (2017). An overview of energy storage and its importance in Indian renewable energy sector: Part I–Technologies and Comparison. *Journal of Energy Storage*, *13*, 10–23.

5. Mousavi, N., Kothapalli, G., Habibi, D., Das, C. K., & Baniasadi, A. (2020). A novel photovoltaic-pumped hydro storage microgrid applicable to rural areas. *Applied Energy*, *262*, 114284.
6. e Silva, G. D. O., & Hendrick, P. (2016). Pumped hydro energy storage in buildings. *Applied Energy*, *179*, 1242–1250.
7. Alqahtani, B., Yang, J., & Paul, M. C. (2023). Design and performance assessment of a pumped hydro power energy storage connected to a hybrid system of photovoltaics and wind turbines. *Energy Conversion and Management*, *293*, 117444.
8. Lahmer, Y., Chaker, A., & Nedjar, A. (2024). Performance evaluation of grid-connected photovoltaic with pumped hydro storage system in high-rise building. *Energy for Sustainable Development*, *81*, 101470.
9. Bazdar, E., Sameti, M., Nasiri, F., & Haghighat, F. (2022). Compressed air energy storage in integrated energy systems: A review. *Renewable and Sustainable Energy Reviews*, *167*, 112701.
10. King, M., Jain, A., Bhakar, R., Mathur, J., & Wang, J. (2021). Overview of current compressed air energy storage projects and analysis of the potential underground storage capacity in India and the UK. *Renewable and Sustainable Energy Reviews*, *139*, 110705.
11. Thunder Said Energy. Compressed air energy storage. https://thundersaidenergy.com/downloads/compressed-air-energy-storage-costs-and-economics/.
12. Saidur, R., Rahim, N. A., & Hasanuzzaman, M. (2010). A review on compressed-air energy use and energy savings. *Renewable and Sustainable Energy Reviews*, *14*(4), 1135–1153.
13. Olabi, A. G., Wilberforce, T., Ramadan, M., Abdelkareem, M. A., & Alami, A. H. (2021). Compressed air energy storage systems: Components and operating parameters–A review. *Journal of Energy Storage*, *34*, 102000.
14. Tong, Z., Cheng, Z., & Tong, S. (2021). A review on the development of compressed air energy storage in China: Technical and economic challenges to commercialization. *Renewable and Sustainable Energy Reviews*, *135*, 110178.
15. Bazdar, E., Nasiri, F., & Haghighat, F. (2024). Optimal planning and configuration of adiabatic-compressed air energy storage for urban buildings application: techno-economic and environmental assessment. *Journal of Energy Storage*, *76*, 109720.
16. Ji, W., Hong, F., Zhao, Y., Liang, L., Du, H., Hao, J., ... & Liu, J. (2024). Applications of flywheel energy storage system on load frequency regulation combined with various power generations: A review. *Renewable Energy*, 119975.
17. Koohi-Fayegh, S., & Rosen, M. A. (2020). A review of energy storage types, applications and recent developments. *Journal of Energy Storage*, *27*, 101047.
18. Nkomo, N. Z., & Alugongo, A. A. Flywheel energy storage systems and their applications: A review. *International Journal of Engineering Trends and Technology*, *72*(4), 209–215.
19. Pullen, K. R. (2019). The status and future of flywheel energy storage. *Joule*, *3*(6), 1394–1399.
20. Amiryar, M. E., Pullen, K. R., & Nankoo, D. (2018). Development of a high-fidelity model for an electrically driven energy storage flywheel suitable for small scale residential applications. *Applied Sciences*, *8*(3), 453.
21. Cimuca, G. O., Saudemont, C., Robyns, B., & Radulescu, M. M. (2006). Control and performance evaluation of a flywheel energy-storage system associated to a variable-speed wind generator Ind Electron. *IEEE Transactions*, *53*, 1074–1085.

22. Abbezzot, C., Tran, T., Poggi, P., Serre-Combe, P., Perrin, M., & Muselli, M. (2013, March). *Using a flywheel associated to PV power plant in order to increase the integration of PV into island electrical grid*. Spain: Bilbao.
23. Arani, A. K., Karami, H., Gharehpetian, G. B., & Hejazi, M. S. A. (2017). Review of Flywheel Energy Storage Systems structures and applications in power systems and microgrids. *Renewable and Sustainable Energy Reviews*, *69*, 9–18.
24. Soomro, A., Pullen, K. R., & Amiryar, M. E. (2021). Hybrid PV system with high speed flywheel energy storage for remote residential loads. *Clean Technologies*, *3*(2), 351–376.
25. Boukettaya, G., & Krichen, L. (2014). A dynamic power management strategy of a grid connected hybrid generation system using wind, photovoltaic and Flywheel Energy Storage System in residential applications. *Energy*, *71*, 148–159.

7 Electrochemical Energy Storage Systems for Green Buildings

Energy storage systems (ESSs) increase wind and solar energy utilization by reducing planning as well as grid network operating uncertainty. ESSs are a flexible alternative to wind and solar energy, which have great grid penetration, but their economics depend on stationary applications and require careful research. Time-shifting, supplementary services, power quality, etc. are stationary uses. Pumped hydro storage (PHS) and compressed air energy storage are popular technological and economic time-shift and supply capacity solutions. Valve Regulated Lead Acid (VRLA), Lithium–ion (Li–ion), Sodium Sulfur (Na–S), Nickel–Cadmium (Ni–Cd), and flow batteries (FBs) are more adaptable and can power any ESS. Although the technology is advanced, the economics of the stationary system are still developing. Understanding how the storage section, power conversion system, and balance of plant affect capital, life cycle, and energy delivery costs is crucial.

7.1 SUSTAINABLE ELECTROCHEMICAL ENERGY STORAGE DEVICES

Figure 7.1 displays this research's techno-economic modeling system limits. Wind, solar, and the grid are excluded from system constraints.

Capital cost and battery cycle life determine levelized storage technology comparisons. The most expensive choice for bulk energy storage and transmission and distribution investment deferral, even with uncertainty, is nickel–cadmium battery cells due to their high capital cost. Vanadium Redox Flow Batteries (VRFBs) are optimal for bulk energy storage and delaying transmission and distribution investments, but their short discharge duration makes them the most expensive option for power applications, frequency regulation, and voltage support.

Batteries, fuel cells, and electrolyzers have many applications and serve many industries. Electrochemical technologies offer more than building energy storage. They stabilize the electricity grid by generating power on demand as a distributed energy source. They can optimize energy utilization during peak demand or offer backup power during outages as flexible storage devices.

DOI: 10.1201/9781003752837-7

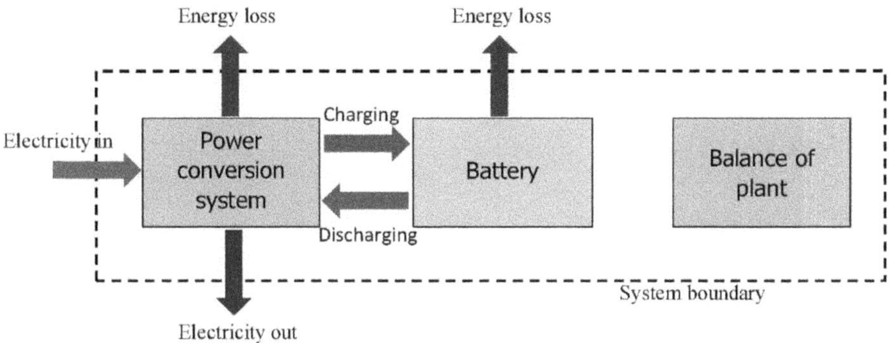

FIGURE 7.1 Energy-flow electro-chemical energy storage system components [1].

7.1.1 Importance of Energy Storage Applications in Buildings

Large electricity-generating plants often produce waste energy. Solar power is intermittent, making it difficult for the existing system to maintain a consistent load profile. Modern energy infrastructure must balance renewable energy integration with grid stability. Addressing intermittency requires energy storage. Energy storage devices store extra electricity and release it when supply is low. Mechanical, chemical, and thermal systems store electrical energy. PHS provides for 95% of energy storage capacity. Chemical methods are a modest portion, but battery storage capacity grows annually.

For cities with many small-scale energy power devices, storage options must be explored. Electrochemical systems are advantageous due to their energy density, flexibility, diversity, and efficiency. Electrochemical systems can quickly respond to electricity needs due to their fast response time. This makes them ideal for metropolitan areas that need fast power acquisition. Although electrochemical systems can be expensive, they offer more storage possibilities for structures. This range allows tailored solutions to meet built environment needs [2].

Chemical and mechanical storage systems work well for building power management. This suggests that they are built to store and release energy over minutes (short-term) to months (seasonal) rather than handle power peaks. Power-oriented technologies like flywheels and super-capacitors are rarely used for these applications, while the battery is the most well-known chemical energy storage device. The following types of batteries are suitable for application development [3]:

- Lead acid batteries
- Molten salt (sodium sulfur, sodium metal hydride) batteries
- Lithium-ion batteries (Li–ion batteries)
- Flow batteries.

Due to a 30% roundtrip efficiency and safety concerns, hydrogen-based ESSs are not currently appropriate for building-scale use.

Electrochemical energy storage systems (ECESS) are a major energy storage competitor because of their minimal maintenance needs, 70–80% efficiency, 10 Wh/kg to 13 kW/kg electrical energy storage, and easy fabrication. Maintenance is expensive, lifespan depends on depth of discharge (DoD), and cycling times are minimal for large facilities.

7.1.2 Lead Acid Batteries

Most people use rechargeable lead–acid batteries. PbO_2 positive and negative electrodes are surrounded by an electrolyte-mitigated sulfuric acid (H_2SO_4) solution. Lead–acid batteries have low response time, daily self-discharge, efficiency, capital cost, and large-scale stabilization.

7.1.3 Lithium–Ion Battery

The battery's low weight, good performance, fast response time, and great cycle economy make it ideal for computers and phones. Nanoscale Li–Ion battery electrode and electrolyte materials are being developed to increase power and specific energy.

7.1.4 Nickel–Cadmium Battery

Ni–Cd batteries are easy to maintain and work from –20°C to –40°C. Because cadmium and nickel are poisonous, Ni–Cd batteries cost 10 times more than Li–ion batteries and hurt the environment. Renewable energy is abundant in Ni–Cd batteries. Efficiency depends on Ni–Cd battery technology.

7.1.5 Sodium Sulfur Battery

Beta alumina solid electrolyte with molten sodium and sulfur electrodes make up the NaS battery. Other batteries are smaller than this 244.8 MWh. It focuses on high-power energy storage. Low self-discharge, high energy densities (150–300 Wh/L), energy efficiency (89–92%), low maintenance and material costs, non-toxic components, and recyclability characterize this storage system. In grid-connected applications, NaS batteries increase power quality and peak shaving. A high-temperature (574–624 K) chemical reaction and 80 $/kW/year liquid electrode costs are downsides. Only large power plants utilize expensive auxiliary systems.

7.1.6 Flow Batteries

With two soluble redox couplings, two liquid electrolytes in separate external tanks maximize flow battery energy storage. Tanks supply electrolytes to the cell stack, where a tiny membrane lets ions pass. The reduction–oxidation of electrolytes generates electricity. Schematic of flow batteries is shown in Figure 7.2. Unlike fuel cells, FBs perform reversible chemical processes. The materials in rechargeable batteries do not degrade over many cycles because they use reversible electrochemical reactions to ferry ions between electrodes. Electrode size and stack count determine

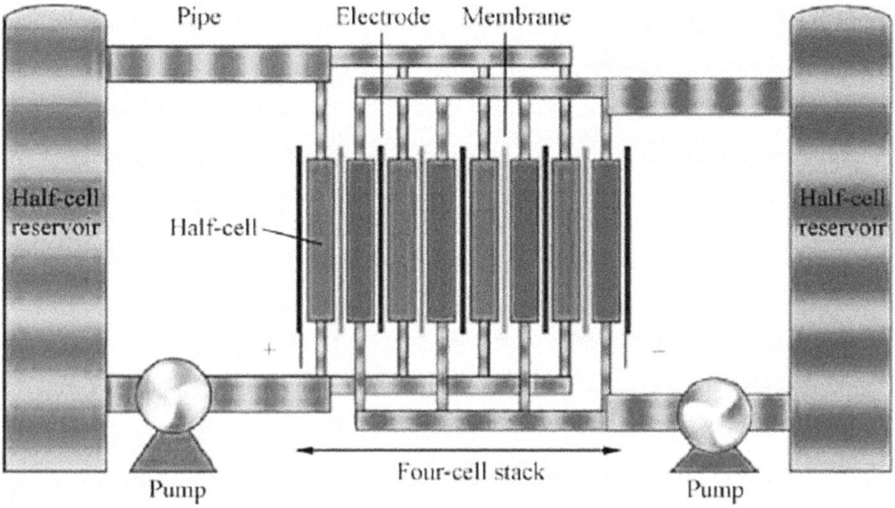

FIGURE 7.2 Schematic diagram of flow battery [4].

fuel cell power. Fast-discharging FBs last 10,000 cycles. FBs are fast, efficient, and ambient-temperature.

7.2 SUSTAINABLE MATERIALS

Due to renewable energy growth and the global objective of net-zero emissions, electrochemical energy storage (EES) technologies have improved. Electrode manufacture for batteries and supercapacitors is complex, expensive, and hazardous, requiring innovative sustainable energy storage solutions. Nature provides abundant energy storage materials for flexible or non-flexible batteries and supercapacitors to address growing energy storage needs. Building batteries and supercapacitors from natural fibers is essential for sustainable electronics. Dangerous compounds in regular batteries should be discussed. Biocompatible energy storage devices promote sustainable energy and reduce environmental effect. Many UN sustainable development goals are affected by research on green energy materials [5].

Electronics development that uses natural resources efficiently reduces harmful and poisonous electrical waste. Due to green electronics advancements, new electronics technology has advanced environmentally. Green electronics promote natural fibers. The reasons are cost, mechanical quality, natural availability, and biological decay. Transistors, energy generators, sensors, and ESSs use natural fibers.

Most bast fiber EES devices use metal oxide-modified activated carbon (AC), carbon nanotubes, and conducting polymers. Current collector, supercapacitor active electrode, and battery anode are 3D structured carbon activated chemically or mechanically. Supercapacitor electrodes made of surface-coated jute fibres (JF) modified with conductive polymers (CPs) are eco-friendly, flexible, and inexpensive. Supercapacitors' energy storage will improve using pseudocapacitive materials like

MnO_2, RuO_2, and WO_3. A conductive polymer covering can make bast fiber an electrode. A metal oxide coating, electrodeposited on the electrode, could boost its energy storage. Carbon or graphene surfaces boost energy storage.

Supercapacitors and batteries depend on the separator and active electrodes. High tensile strength and cellulose content make bast fibers good separators following composite modification. To maximize the device's operating window and life cycle, novel biocompatible electrolytes like gel or aqueous solutions must be investigated. Its packaging will also affect the outcome, as soft sustainable materials like insulative polymers require protective films instead of rigid packaging [5].

7.2.1 Green Material-Based Battery

Electric vehicles, grid-scale storage, and batteries are becoming more significant due to electronic device use. These chemicals conserve the environment and mitigate global warming, proving that biomass is crucial for future technology. Biomass types are shown in Figure 7.3. With biomass-based batteries, sustainable energy storage has improved.

Environmentally friendly, safe, and high-performance battery separators, polymer electrolytes, and carbon electrodes are revolutionising energy storage. The next-generation supercapacitors need these components for rapid charge-discharge, high cycle life, and increasingly eco-friendly, non-toxic, and adaptable designs. Supercapacitors are popular for their high power, fast charging, and cycle stability. Two types of supercapacitors store charge:

(1) Electric double-layer capacitors (EDLCs) and
(2) Pseudocapacitors.

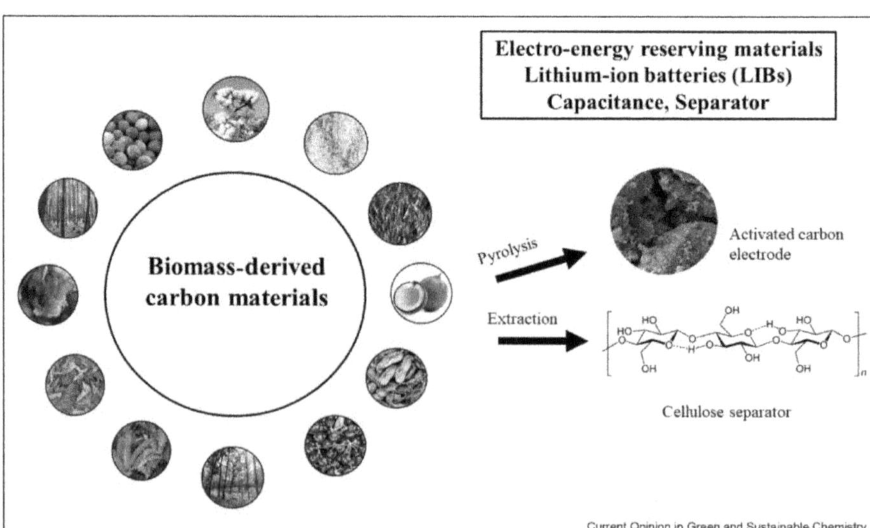

FIGURE 7.3 Biomass for battery components [6].

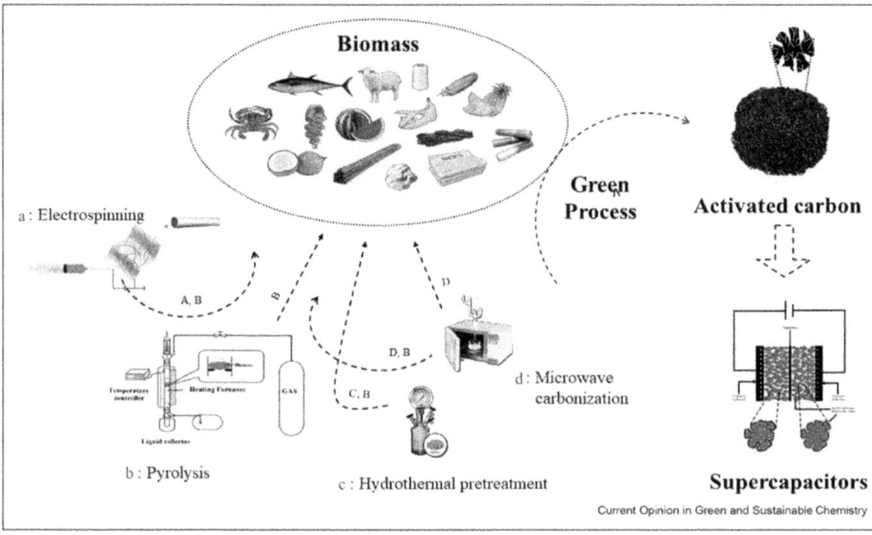

FIGURE 7.4 Different methods for the preparation of activated carbon from biomass sources. (a) Electrospinning followed by pyrolysis conducted by B. (b) Utilization of the pyrolysis process through a traditional furnace. (c) Generation of activated carbon through hydrothermal treatment following pyrolysis by B. (d) Microwave-assisted carbonization for the production of activated carbon from biomass [6].

Carbon-based EDLCs were explored using a physical approach that uses the electrode materials' interface surface area for electrolyte adsorption and desorption. Metal oxide/hydroxide and polymer pseudocapacitors are non-renewable and environmentally harmful. Ginkgo, tree bark, and bamboo-AC materials are used for EDLC electrodes. Figure 7.4 shows how biomass is converted into AC to optimize shape and porosity for various technologies.

7.3 BAST FIBER

Due to their low cost, equipment wear, particular strength and stiffness, and low density, natural fibers are attractive reinforcing materials. Natural fibers' environmental friendliness, compatibility, and biodegradability over petrochemical-based fibers merit their adoption in composites as the supply chain strengthens and lengthens. Plant, mineral, and animal fibers are the principal types. Animal fibers include hair and silk. Main mineral fiber sources are basalt and asbestos.

Natural bast fibers like ramie, jute, kenaf, flax, and hemp have become popular due to their excellent mechanical qualities. Due to their biodegradability and natural availability, bast fibers have promising uses in automotive, structural composites, pulping, and textiles. Bast fibers are renewable, easy to prepare, dampen well, and are low-abrasive on equipment. Thus, bast fibers enhance biocomposites, improving structural stability and mechanical characteristics. Based on their lightweight materials and low environmental impact, bast fiber-reinforced composites are a possible alternative to synthetic ones [7].

Plant fibers are one of Earth's most abundant biomass forms. Industries have employed bast fiber in biocomposites because of its stiffness and strength [7]. Optimized biomass pre-carbonization and activation yield a sheet-like form. Hydrochar is activated with KOH at 700°C, 750°C, and 800°C in an inert atmosphere from hemp fiber hydrothermally pre-carbonized. KOH activates pre-carbonized hemp bast fiber into nanosheets [8].

7.3.1 Activated Carbon-Based Electrode from Biomass of Bast Fibers

Primary biomass components are cellulose, hemicellulose, and lignin. These components can follow different chemical routes after calcination, depending on temperature. Carbon dominates after heteroatom removal. Due to its cost-effectiveness, accessibility, sustainability, and adaptability, biomass waste from natural fiber sources is increasingly used to make carbon products. Despite their high output, typical AC production methods from non-renewable resources like coal, peat, and petroleum are expensive and involve harsh processes.

AC and biomass waste carbon composites have good mechanical and electrical properties due to biomass's natural organization. As supercapacitor electrodes, biomass materials are used due to their appealing carbon framework formed during conversion, complex pore structure, and higher specific surface area. Hierarchical porous carbon with macropores, mesopores, and micropores promotes supercapacitor (SC). Micropores and mesopores improve specific surface area and active site availability, while macropores minimize mass transport diffusive resistance. Various methods are used to convert biomass into AC, depending on the product's application. Carbon activation is traditionally physical or chemical.

Physical activation involves two stages: carbonization below 800°C and activation at 900–1200°C with air, CO_2, or steam. Instead, a precursor is activated using KOH, NaOH, or H_2PO_3 and calcined at the proper temperature. Both techniques yield high specific-surface area AC. Electrolytes cannot diffuse through current microporous pores. Hierarchically porous carbon (HPCs) with macropores, micropores, and mesopores have been studied to address this issue. Macropores reduce mass transport diffusive resistance, while micropores and mesopores enhance active site surface area. HPC synthesis involves a one-pot process to combine carbon precursor and activating agent KOH, followed by a high-temperature treatment at 800–900°C.

The backbone of plant cell walls is lignocellulosic biomass, a complex, hierarchical blend of organic macromolecules and inorganic components. Cellulose, hemicellulose, and lignin create a resistant matrix (Lignin-Carbohydrate Complex, LCC) by proximity and covalent bonding. This complexity makes standardizing porous carbon material production difficult. Simple activation pathways can produce microporous materials with well-defined forms and architectures. As activating agents build HPCs regardless of biomass type, they are advantageous. The biomass precursors, activation methods, and surface functional groups determine the electrode material's surface area, pore size distribution, and performance.

A synthesis mechanism applicable to all biomass precursors is challenging to delineate. Peanut and corncob had comparable atomic surface oxygen concentrations (10.6, 10.3 wt%), although they have distinct oxygen functional groups. Peanut carbon contains approximately twice the number of carbonyl groups compared to corncob carbon, resulting in enhanced wettability with non-aqueous electrolytes.

The method of activation and composite employed to enhance pseudocapacitance and electric double layer (EDL) influenced the EES performance of biomass-derived carbon [9].

7.4 KENAF FIBER

Solar and wind energy are intermittent, complicating the global emphasis on renewable energy as a substitute for fossil fuels. Numerous endeavors were undertaken to fabricate piezoelectric devices utilizing diverse materials. Commercial supercapacitors, known as EDLCs, utilize AC electrodes. Capacitance arises from static charge at the electrode-electrolyte interface. ACs are derived from biowaste such as leaves, flowers, stems, husks, shells, and other organic materials. These composites are designed to enhance energy density by intercalating electrolyte ions at the electrode surface, hence facilitating more charge storage. CuO, Co_3O_4, NiO, and other metal oxides participate in battery-type intercalation reactions due to their measurement of average capacitance rather than instantaneous capacitance. Due to its high cost and environmental detriment, novel ways for improving the energy density of AC are being researched.

The manufacture of biomass-based AC is more straightforward and economical compared to alternative carbon sources. Thermal treatment of natural fibers, such as Kenaf fibers (KFs), can enhance their specific surface area by altering their porosity, surface area, and pore configurations. Chemical activation, at lower temperatures, produces more porous ACs than physical activation. To produce AC with a large surface area and micropore distribution, KOH was recommended as an activation agent.

Capacitance and power density depend on faster adsorption rates. Natural fibers are mostly cellulose, which determines their heat breakdown and physical properties. Hemicellulose decomposes at lower temperatures than cellulose; hence, an increase in hemicellulose heightens the likelihood of ignition of natural fibers, leading to more pronounced thermal degradation. Lignin plays a greater role in char generation compared to cellulose and hemicellulose; hence, fibers rich in lignin enhance carbon yield during decomposition. Kenaf is one of the superior agrofibers for the manufacture of AC. This is due to its abundance, high lignin concentration, and many uses in gardening, papermaking, cordage, bags, and apparel. Kenaf generates 40% useable stalk, almost twice as much as flax, jute, and hemp. Its carbon content yields more than other fiber sources, enhancing its commercial potential.

The fast-growing plant can be harvested in 5–6 months. Textile industry uses carpets and clothing made from kenaf plants, producing 15% of its fibers. KF-AC eliminates wastewater copper. Kenaf AC supercapacitor electrodes may absorb and desorb electrolyte ions during charging and discharging [10].

7.4.1 Activation of Kenaf-Derived Biochar

Activating the biochar powder necessitated chemical and heat processes. Initially, 6 M KOH was combined with a 1:4 ratio of biochar to KOH at room temperature. The amalgamation was agitated for 4 hours and subsequently dried overnight at 105°C in an industrial oven. The vertical reactor within the furnace elevated the temperature of the mixture. The reactor was heated from ambient temperature to 600°C at a rate of 10°C/min while sustaining a nitrogen gas flow of 1 L/min.

AC A600-1 was synthesized by sustaining a temperature of 600°C for 1 hour. The identical heating procedure was conducted for three additional AC samples at intervals of 2, 3, and 4 hours [11].

Kenaf is a promising CO_2 reduction plant, prompting research into its biomass resources. It grows to 2.5–3.5 m in 4–5 months. KFs are used in paper and plastic production. Due to their affordability and lightweight properties, kenaf-based reinforcing fibers have been studied more in recent years, notably for automotive and construction components. Kenaf comprises cellulose, hemicellulose, and lignin, with 30–40% of the fiber being extractable. Applications include kenaf residue biochar. It only treats water, fertilizers, and solid fuels. While biowaste-derived materials have been examined for applications in batteries and capacitors, there has been limited research on conductive additives. Carbon blacks generally have electrical conductivities ranging from 10 to 10^4 S m^{-1}, exceeding those of biocarbons. Composite materials such as carbon nanotubes, carbon nanofibers, and carbon black have been utilized to enhance electrodes [12].

With the increasing demand for natural-based industrial products, there is a necessity for sustainable novel fibers for future applications. *Washingtonia filifera*, commonly referred to as the "Petticoat palm," is a tree indigenous to western Mexico and the southwestern USA, notable for yielding *Washingtonia* leaf stalk fiber (AW). The stipe may exceed 20 m, and the height may approach 23 m. KF, a lignocellulosic material, has gained prominence as a cost-effective reinforcing filler for bio-composites owing to its rapid growth cycle, high yield, resistance to pests and pathogens, and versatility across various climates and soil types. KF is one of the most natural fibers for the enhancement of unsaturated polyester. The low density, non-aggressive handling, enhanced mechanical properties, and biodegradability of KF may contribute to this phenomenon. KF is utilized in pulp and paper manufacturing to mitigate primary pollution at a specified distance from deforestation. The mechanical characteristics of natural fibers render them a viable substitute for polymer composites, owing to their renewable origin, lightweight, affordability, low density, and accessibility. This economical method promotes clean technology and recyclable resources to increase bio-product sustainability and reduce the environmental effect.

Unlike other hybrid natural fibers incorporated into epoxy resins, engineered hybrid biocomposites (KF/WA) are cost-effective materials for construction, household appliances, and automobiles [13].

7.5 HEMP FIBER

Industrial hemp has numerous applications. When cultivated at optimal density for superior structural fiber, it exhibits pest resistance and weed suppression. After planting, it necessitates minimal maintenance until harvest. Hemp fiber competes with glass fibers in terms of strength and rigidity. This renders it a sustainable alternative for numerous applications. The automobile sector was the first to utilize this fiber in composite materials for door panels, trunk linings, spare tire covers, seat backs, and dashboards. Automotive use of natural fibers rose in Europe from 400 tonnes in 1996 to 17,140 tonnes in 2000, with forecasts predicting 60,000 tonnes by 2005, notably hemp fiber utilization. Four hemp fibers struggles to bond with polypropylene, a common molding thermoplastic.

The most successful way is adding maleated polypropylene to increase its hydrophilicity. This alteration enhances natural material compatibility and covalent and hydroxyl bonding. Additionally, fiber improvements can improve interfacial bonding. Chemically treating hemp fiber surfaces improves interfacial bonding, but the composite's qualities are not fully reflected.

Another issue is chemical disposal after treatment. Fungi, albeit understudied, can improve chemistry and morphology cheaply and without damaging the environment. Papermakers have researched fungi for pulping and whitening wood and straw [14].

Aqueous zinc–ion supercapacitors derived from hemp biomass can store energy independently of fossil fuels. These devices utilize renewable, biodegradable, and eco-friendly active components to integrate battery energy storage with supercapacitor power density. ZnO or ZnS, along with aqueous electrolytes, serve as the active components, while hemp biomass comprises zinc ions. Hemp exhibits rapid growth, adaptability, and sustainability. It has a lower environmental impact than many other crops and can be used to manufacture paper, textiles, food, and biofuels. Zinc ions in hemp biomass can store energy in aqueous supercapacitors. Hemp biomass in aqueous zinc–ion supercapacitors may cut carbon emissions, supply sustainable raw materials, and permit local production.

Aqueous zinc–ion supercapacitors made from hemp biomass can store energy without fossil fuels. They can pioneer green energy infrastructure. Multivalent ion energy storage could provide superior specific capacity and energy density. Redox reactions possess more energy storage capacity than monovalent ion systems owing to their abundance of electrons. Aqueous zinc–ion batteries (ZIBs) utilizing a mildly neutral or slightly acidic electrolyte are promise for large-scale energy systems, particularly in grid-scale energy storage. Aqueous electrolyte-based multivalent zinc-ion hybrid supercapacitors are energy storage devices that merge the energy density of batteries with the characteristics of supercapacitors. The devices use aqueous electrolytes and multivalent zinc–ion intercalation materials like zinc oxide or zinc sulfide. Multivalent zinc–ion hybrid supercapacitors provide higher energy density, lower cost, and better safety.

These devices exhibit superior longevity and enhanced environmental sustainability compared to alternative energy storage methods. In conclusion, multivalent zinc-ion hybrid supercapacitors with aqueous electrolytes are promising and will facilitate the shift to a more sustainable and renewable energy paradigm. These designs are effective for portable gadgets, electric cars, and renewable energy systems [15].

Recent research suggests that hemp can be used to make battery materials like carbon nanosheets and supercapacitors, which are being studied as clean, green, renewable energy sources. Since hemp bast, the fiber left after processing, has been discovered, supercapacitor technology has advanced to use it as an effective electrode. Supercapacitors store and discharge electricity like batteries. Supercapacitors have a higher power density than batteries, allowing them to discharge energy quickly, recharge quickly, support more charge cycles, and operate safely in extreme temperatures. These properties make supercapacitors vital to the development of advanced and developing technologies, including electric vehicles, trucks, trains, and wind turbines, which stabilize wind energy.

Graphene, a carbon nanosheet material, is very expensive and requires graphite extraction, which has caused pollution warnings worldwide, particularly in rural China and India. Graphite extraction chemicals are hazardous and pose considerable risks to workers. Thermally flaking hemp bast fibers for 24 hours produces carbon nanosheets. These fibers are byproducts of most industries, making this a great production scenario. Water and air purification beyond supercapacitors may use hemp-derived carbon nanosheets. Mitlin wanted the company to have "graphene performance at activated carbon prices." Hemp carbon nanosheets can generate sustainable energy from trash, benefiting farmers financially [16].

7.5.1 Electrochemical Applications in Buildings

Electrochemistry studies electrical energy and chemical processes. Electrical energy is converted to chemical energy. Fuel cells generate electricity from hydrogen and oxygen. Batteries generate electricity from chemicals. The behavior and performance of electrochemical devices necessitate an understanding of charge transport, electrochemical thermodynamics, and electrochemical kinetics. Electrochemical devices in buildings can alter thermal performance and moisture regulation by modifying surface qualities. Electrochemical devices in buildings function in practical conditions, which may diverge from the ideal [17, 18].

Electricity storage reconciles the disparity between renewable energy generation and load demand. This imbalance arises from fluctuations in hourly energy generation and consumption inside structures. The disparity for solar systems is 63%, whereas for wind energy it is 39%. EES exhibits versatility. Electrochemical applications, such as batteries, provide grid-scale energy storage, automotive use, and building integration to enhance the utilization of intermittent renewable energy. The emergence of electrochemical technologies necessitates careful selection of energy storage solutions for structures. The storage capacity of a system is contingent upon energy storage density, whereas power density pertains to the velocity of energy transfer. Volumetric energy density, expressed in watt-hours per liter, measures energy density. This metric indicates a volume's energy capacity. Gravimetric energy density, expressed in watt-hours per kilogram, quantifies the energy that a mass can retain [19, 20].

Vapor Pressure Deficit (VPD) is measured in W/m^3 or W/m^2, signifying power delivery rate. Energy storage in buildings must be flexible and affordable. Space needed to accomplish a certain storage capacity and the maximal storage capacity without major structural changes are critical. Given that structures require electricity promptly, albeit not as critically as autos, power density must be taken into account. Batteries presently exhibit energy and power density levels that lie between those of fuel cells and capacitors [21, 22]. The continued use of batteries does not exclude the advancement of other technologies. There is a need to evaluate gravimetric and volumetric energy densities while implementing ESSs in buildings. The elevated volumetric energy density enhances the efficiency of storage systems located in certain areas of a building, such as the basement. For buildings requiring lighter components for energy storage, gravimetric energy density will be essential [23, 24].

Electrochemical devices must be assessed based on multiple criteria. The standard parameters include the following:

(1) Energy storage,
(2) Power density,
(3) Response time,
(4) Operational conditions (temperature),
(5) Round-trip efficiency, and
(6) Lifespan.

Additional considerations are required if these devices are intended for application development. Humans use structures, thus safety is paramount. Prioritize fire safety, structural integrity, and safety practices. Building exteriors must be durable due to varied weather conditions. Selection of materials for electrochemical devices in building facades must prioritize resistance to weather, water, and fire. Efficiency, expense, and upkeep must be taken into account when developing applications [25].

EES techniques such as batteries, reversible fuel cells, and supercapacitors are pertinent to construction applications. Batteries dominate the sector, particularly when integrated with solar or wind energy. Lithium–ion batteries have been intensively researched due to their favorable gravimetric and volumetric energy densities. Reversible fuel cells have been evaluated against batteries for the levelized cost of energy storage. Constructing cogeneration systems utilizing fuel cells provides supplementary advantages. Construction materials and components utilize supercapacitors for energy conservation. They function akin to batteries, although they may be cost-prohibitive for certain construction projects. ESSs are likely to gain popularity in building applications as technology advances and costs decrease [26].

REFERENCES

1. Rahman, M. M., Oni, A. O., Gemechu, E., & Kumar, A. (2021). The development of techno-economic models for the assessment of utility-scale electro-chemical battery storage systems. *Applied Energy*, *283*, 116343.
2. Zhang, J., Azari, R., Poerschke, U., & Hall, D. M. (2023). A review of potential electrochemical applications in buildings for energy capture and storage. *Micromachines*, *14*(12), 2203.
3. Del Pero, C., Aste, N., Paksoy, H., Haghighat, F., Grillo, S., & Leonforte, F. (2018). Energy storage key performance indicators for building application. *Sustainable Cities and Society*, *40*, 54–65.
4. Elalfy, D. A., Gouda, E., Kotb, M. F., Bureš, V., & Sedhom, B. E. (2024). Comprehensive review of energy storage systems technologies, objectives, challenges, and future trends. *Energy Strategy Reviews*, *54*, 101482.
5. Manjakkal, L., Jain, A., Nandy, S., Goswami, S., Carvalho, J. T., Pereira, L., ... & Hogg, R. A. (2023). Sustainable electrochemical energy storage devices using natural bast fibres. *Chemical Engineering Journal*, *465*, 142845.

6. Okhawilai, M., & Pattananuwat, P. (2021). Sustainable electroactive materials for energy storage. *Current Opinion in Green and Sustainable Chemistry*, *28*, 100431.
7. Zhu, S., Xie, J., Sun, Q., Zhang, Z., Wan, J., Zhou, Z., ... & Fan, M. (2024). Recent advances on bast fiber composites: Engineering innovations, applications and perspectives. *Composites Part B: Engineering*, *284*, 111738.
8. Dubey, P., Shrivastav, V., Maheshwari, P. H., & Sundriyal, S. (2020). Recent advances in biomass derived activated carbon electrodes for hybrid electrochemical capacitor applications: Challenges and opportunities. *Carbon*, *170*, 1–29.
9. Manjakkal, L., Jain, A., Nandy, S., Goswami, S., Carvalho, J. T., Pereira, L., ... & Hogg, R. A. (2023). Sustainable electrochemical energy storage devices using natural bast fibres. *Chemical Engineering Journal*, *465*, 142845.
10. Subramaniam, T., Ansari, M. N. M., Krishnan, S. G., & Khalid, M. (2024). Kenaf-based activated carbon: A sustainable solution for high-performance aqueous symmetric supercapacitors. *Chemosphere*, *354*, 141593.
11. Razak, M. N. A., Noorden, Z. A., Ani, F. N., Abdul-Malek, Z., Jamian, J. J., & Bashir, N. (2020). Electrochemical properties of kenaf-derived activated carbon electrodes under different activation time durations for supercapacitor application. *Indonesian Journal of Electrical Engineering and Computer Science*, *19*(2), 1105–1112.
12. Eguchi, T., Naganuma, T., Chiba, R., Watanabe, K., Abe, Y., & Kumagai, S. (2025). Application of Kenaf-derived carbon as conductive electrode additive in electric double-layer capacitors. *International Journal of the Society of Materials Engineering for Resources*, *663*, 1–10.
13. Alshammari, M. B., Ahmad, A., Jawaid, M., & Awad, S. A. (2023). Thermal and dynamic performance of kenaf/washingtonia fibre-based hybrid composites. *Journal of Materials Research and Technology*, *25*, 1642–1648.
14. Pickering, K. L., Li, Y., Farrell, R. L., & Lay, M. (2007). Interfacial modification of hemp fiber reinforced composites using fungal and alkali treatment. *Journal of Biobased Materials and Bioenergy*, *1*(1), 109–117.
15. Tekin, B., & Topcu, Y. (2024). Novel hemp biomass-derived activated carbon as cathode material for aqueous zinc-ion hybrid supercapacitors: Synthesis, characterization, and electrochemical performance. *Journal of Energy Storage*, *77*, 109879.
16. Cannabis Tech. Hemp based efficient energy storage solutions. Available online: https://cannabistech.com/articles/efficient-energy-storage-solutions/ (accessed on 19 November 2024).
17. Boxwell, M. (2010). *Solar Electricity Handbook: A Simple, Practical Guide to Solar Energy—Designing and Installing Photovoltaic Solar Electric Systems*. Greenstring: Nairobi, Kenya.
18. Gholami, H., Nils Røstvik, H., & Steemers, K. (2021). The contribution of building-integrated photovoltaics (BIPV) to the concept of nearly zero-energy cities in Europe: Potential and challenges ahead. *Energies*, *14*, 6015.
19. Solar Energy Industries Association. Solar Data Cheat Sheet. Available online: www.seia.org/research-resources/solar-data-cheat-sheet (accessed on 9 July 2023).
20. Solar Photovoltaic Applications Seminar. (1980). *Design, Installation and Operation of Small, Stand-Alone Photovoltaic Power Systems*. PRC Energy Analysis Company: Fairfax County, VA, USA.
21. Roberts, S., & Guariento, N. (2009). *Building Integrated Photovoltaics*. Birkhäuser: Boston, MA, USA.
22. United States Energy Information Administration. Use of Energy Explained. Available online: www.eia.gov/energyexplained/use-of-energy/ (accessed on 6 July 2023).

23. Keoleian, G. A. & Reppe, P. (2023). Residential buildings factsheet. Available online: https://css.umich.edu/sites/default/files/2023-10/Residential%20BuildingsCSS01-08.pdf (accessed on 7 July 2023).
24. U.S. Department of Energy. (2023). 5 Benefits of Residential Solar. Available online: www.energy.gov/energysaver/articles/5-benefits-residential-solarU.S (accessed on 7 July 2023).
25. U.S. Energy Information Administration (EIA). (2022). Homes and Buildings in the West and Northeast Have the Largest Share of Small-Scale Solar. Available online: www.eia.gov/todayinenergy/detail.php?id=54379 (accessed on 7 July 2023).
26. Lvov, S. N. (2021). *Introduction to Electrochemical Science and Engineering*. CRC Press: Boca Raton, FL, USA.

8 Electrical Energy Storage System for Green Buildings

Electrical energy storage (EES) technologies are classified into two distinct categories:

1. Those that deliver high power and
2. Those that provide high-energy storage capacity.

High-energy storage systems provide energy over prolonged durations, whereas high power storage systems deliver energy at exceptionally high rates for periods under 10 seconds. Electrochemical storage systems are applicable in both high-power and high-energy configurations. Furthermore, EES has the capability to store electrical energy through electrical fields, mechanical energy, or chemical energy. Small- and medium-sized enterprises, electric double-layer capacitors/supercapacitors, and flywheels contribute to power quality and grid stability for high-power, short-duration applications. Flywheels utilize mechanical energy, whereas superconducting magnetic energy storage (SMES) systems and electric double-layer capacitors (EDLCs) store electrical energy through electromagnetic and electrostatic means. However, compressed air energy storage (CAES) and pumped water energy storage systems can help balance the load and keep seasonal energy for high-energy uses, which makes the grid more reliable. These systems facilitate the conversion of electrical and mechanical energy. High-power electrochemical storage systems transform chemical energy into electricity on demand.

Germany and the USA operate the two largest CAES facilities globally, featuring power ratings of 290 MW and 110 MW, respectively. However, CAES systems are expected to be adopted in the future, as predicted by the International Electrotechnical Commission, according to which the rated power of CAES applications will exceed 500 MW. The 2009 Chilean 12 MW Li–ion battery project represented the largest initiative of its type. The system has the capability to supply the grid for a duration of 20 minutes at this power level. In 2011, a 32 MW Li–ion plant was installed in West Virginia, USA, capable of delivering its rated power for a duration of 15 minutes. Existing EES systems exhibit a broad spectrum of performance and capacity, tailored to accommodate various application environments and electricity storage requirements. It is expected that all of the individual EES systems that are now accessible in the mentioned ranges would gain significance. Due to the modular

DOI: 10.1201/9781003752837-8

design of all systems, a broad variety of technologies can be used to boost power output and energy capacity [1].

Residential buildings are essential to the development of new flexible power systems. A significant fraction of annual electricity consumption is ascribed to residential use. Every unit in apartment complexes possesses an individual electrical contract, maintained by the occupants. Their use is sometimes so restricted that the chance to engage in demand response or any energy-saving initiatives is greatly limited. Operations within apartment complexes are generally conducted through their aggregate electrical usage, which includes elevators, lighting, and heating systems. Modifying the standard metering becomes a viable option when residents of an apartment block come together to form an energy community. This modification allows for the utilization of photovoltaic (PV) and electrical energy storage system (EESS) to benefit the entire building and its occupants. This modification could lead to various legislative challenges. In countries like Finland, this remains feasible as long as all parties consent to the terms. Participants have the option to leave the community at their discretion [2].

A time-of-use pricing model was implemented, linking electricity rates to designated time intervals to mitigate energy usage during peak hours and promote efficient demand management. This statute delineates two distinct timeframes [3]:

i. peak and
ii. off-peak.

Electricity generation requires extensive energy storage systems. The principal factors influencing these EES technologies are cost, longevity, efficiency, power production, and energy density (ED). Batteries illustrate the essential concepts of electrochemical storage systems. Batteries provide superior efficiency relative to alternative energy storage systems when evaluating factors such as scalability, overall efficiency, longevity, discharge time, weight, and system portability. Although rechargeable lithium–ion batteries (LiBs) are widely recognized as the gold standard in battery technology for storing electricity, they are still mostly used in tiny electronic devices.

The most efficient energy storage systems are engineered to manage high power ratings while sustaining comparatively low energy content, rendering them optimal for guaranteeing power quality and reliability. Considering the attributes of ED in energy storage technologies, it can be inferred that low ED methods encompass flywheel energy storage (FES), capacitors, supercapacitors, and pumped hydro storage (PHS). Conversely, high ED systems include CAES, cryogenic energy storage, thermal energy storage (TES), and batteries.

Ongoing investigations into battery technology reveal that lithium–ion batteries present a viable choice for future large-scale storage systems, whereas sodium–sulfur batteries have already been brought to market in Japan. Batteries are recognized for their desirable attributes as energy storage systems, due to their adaptability in several geographical settings, as well as their power and ED, efficiency, weight, and portability. The primary challenges hindering the future advancement of battery technology for large-scale applications encompass the necessity for comprehensive techno-economic assessments, the scarcity of diverse manufacturers obstructing progress, and the attainment of economies of scale in the short term. The primary considerations for future study and market introduction of stationary batteries are [4]:

1. Capital cost,
2. Determined by the cost per unit of energy divided by cycle life, and
3. The durability of the battery's cycle performance.

8.1 ENERGY STORAGE USING FLOW BATTERIES

In the field of electrochemical systems, redox flow batteries (RFBs) are distinguished as an advanced technology and a notably promising choice for stationary energy storage. This technology offers various attractive attributes:

1. Ensures scalability and flexibility,
2. Allows for independent sizing of power and energy,
3. Attains high round-trip efficiency,
4. Supports significant depth of discharge,
5. Guarantees extended durability,
6. Provides rapid responsiveness, and
7. Minimizes environmental impact. The specified attributes enable diverse operational functionalities and discharge periods, rendering them especially effective for augmenting energy production from renewable sources [5].

Flow batteries can discharge rapidly for a period of 10 hours. Three electrolytes underpin the flow battery technologies presently in demonstration or large-scale project development. Regnesys in the UK supplies sodium bromide (NaBr); VRB Power Systems, Inc. in Canada delivers vanadium bromide (VBr), and ZBB Energy Corporation distributes zinc bromide. In China, the Vanadium Redox Battery (VRB) Power Systems, Inc. is the sole company that has attained a notable degree of advancement.

Vanadium redox couples (V^{2+}/V^{3+} in the negative half-cell and V^{4+}/V^{5+} in the positive half-cell) facilitate energy storage in VRBs. During charge/discharge cycles, the hydrogen-ion-permeable polymer membrane facilitates the exchange of H^+ ions across electrolyte reservoirs. Batteries can achieve 85% net efficiency. Other flow batteries possess distinct power and energy ratings, similar to VRBs. VRBs can store energy for utility companies and industrial applications. Examples include power quality, uninterruptible power supplies, peak shaving, supply security, and renewable energy integration. Stationary applications predominate in VRB development owing to its low ED [6].

RFBs independently store and convert energy. Two-vessel electrolytes accumulate electrical energy. Stacks resembling fuel cells (FCs) convert energy. Vanadium redox flow batteries (VRFBs) are the most prevalent and developed type of RFBs. This is one of the most straightforward electrolyte systems due to the non-mixing of anolytes and catholytes. Another prevalent RFB is the zinc–bromine battery (Zn/Br), a hybrid flow battery with zinc bromide electrolyte contained in two reservoirs. The electrolyte

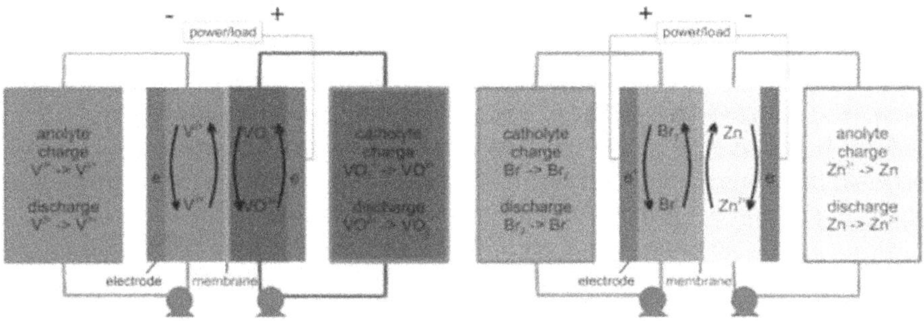

FIGURE 8.1 Left: vanadium redox flow battery, right: zinc–bromine hybrid flow battery [7].

solution deposits metallic zinc onto the negative electrodes of the stack cell during the charging process (Figure 8.1).

The flow battery technology and other new technologies have faced many upheavals and hurdles. A mature technology has been developed by engaged entrepreneurs and developers. Most flow batteries are redox types that use vanadium. Flow batteries are only competed with by the zinc–bromine electrolyte system (Figure 8.1, right). Alternative electrolyte systems remain in the experimental phase. Operational systems exemplify this: Approximately 60% of 72 grid-connected facilities utilize vanadium, whereas fewer than 40% employ zinc–bromine technology. The energy storage market comprises around 2000 MW of lithium batteries and 170,000 MW of pumped hydro facilities, while flow batteries remain a small segment.

RFB technologies, primarily zinc–bromine, are undergoing commercialization, while organic chemistries are under exploration. The anticipated effectiveness of the systems may surpass existing limitations. The foundational ideas of this technology are well established, yet its scalability and future development continue to baffle researchers. The full potential of RFB remains unexploited. The deficiency of electrochemical engineering research regarding reaction environment, cell design, performance, and efficiency hinders advancement.

Energy storage systems are crucial for conventional power generation, distributed energy resource systems, and intermittent renewable energy sources. EES can significantly mitigate industrial challenges like as volatility, reliability, and stability. EES applications encompass scaling, generation, and transmission systems, distribution networks, and the requirements of customers or end-users. The technologies regulate energy, connect and enhance electrical quality, and ensure reliability.

8.1.1 Pros

In RFBs and FCs, power conversion and energy storage are decoupled, enabling the independent scaling of power and energy components. The utilization of larger storage tanks enables virtually limitless capacity. Contemporary designs possess energy capacities ranging from 10^2 to 10^7 Wh, exceeding the majority of ECES by a minimum of one order of magnitude. RFBs and FCs surpass traditional electrochemical devices in storage capabilities exceeding 4–6 hours. The membrane electrode assembly

(MEA) of RFBs, similar to proton exchange membrane fuel cells (PEMFCs), has two catalyzed electrodes partitioned by a polymeric membrane. A porous electrode composed of carbon-based materials such as carbon felt, carbon fiber paper, or carbon nanotubes facilitates the passage of electrolyte to the electroactive areas.

A FC-based storage system typically employs two distinct converters: (1) the FC for power generation and (2) An electrolyzer for energy storage, which converts water into hydrogen and oxygen by the application of electricity.

RFB reactions demonstrate full reversibility, allowing the same cell to function as a converter of electrical energy to chemical energy and vice versa. RFBs function by altering the valence of metal ions while preserving their amount, thereby enabling extended cycle service life. Moreover, the management of gases like hydrogen in FCs presents greater hurdles than the handling and storage of liquid electrolytes in RFBs, which can be effectively confined within two economical tanks. In RFBs, the circulation of electrolytes between the tanks and the cell electrodes can be efficiently regulated using two pumps. The modulation of electrolyte flow facilitates accurate temperature control within cells, permitting operation under optimal circumstances, including peak efficiency.

The state of charge (SOC) can be precisely evaluated by cell voltage measurements, and significant discharges may occur without altering the cell's shape. Separate tanks for the two electrolytes prevent self-discharge, allowing cells to stay fully discharged for extended durations without any negative effects. RFBs demonstrate a rapid reaction capability that facilitates the control of power quality and energy services. Individuals may experience a transient overload syndrome. Moreover, expedited refueling through solution exchange is feasible when necessary, and they necessitate less maintenance.

8.1.2 Cons

The power density (PD) and ED of RFBs are inferior to those of other technologies, rendering them now inappropriate for mobile applications. The extensive active regions and membranes of the cell necessitate larger battery dimensions, resulting in pronounced transverse gradients of the solutions directed toward the electrochemical activity sites inside the electrodes. Electrolytes demonstrate conductivity, rendering them susceptible to shunt currents, which are:

1. Characterized as electric currents that go through distribution and flow channels.
2. This condition results in heightened losses and impacts electrical efficiency.
3. Maintaining precise control of the electrolyte temperature is essential to prevent solution precipitation at temperatures below 15°C and above 35°C.
4. This necessitates the regulation of heat and temperature through specially designed systems.

8.2 FUEL CELL

Significant progress in FC system development was noted during 1985. Electrical power generated from FCs is anticipated to be produced in multi-megawatt installations. A global trend has developed that favors a market-driven and deregulated model for power generation. Consequently, it became evident that small-scale

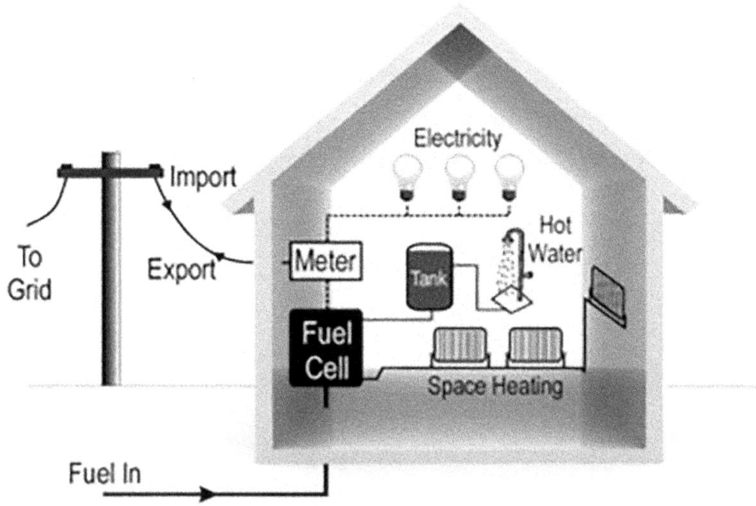

FIGURE 8.2 Fuel cell micro-CHP in a domestic home [9].

electricity generation was economically viable. Consequently, these systems are typically employed for backup power, or when functioning as stationary generators, necessitate expensive redundant capacity to guarantee adequate reliability.

FCs function as an alternative energy source by employing systems that can operate with a reticulated network, bottled gasoline, or liquid fuel. Many problems can be solved with FCs, but their high initial costs mean that they use more energy, which makes them less competitive with reliable traditional power sources. Present market entrance applications necessitate a premium, particularly in distant regions and for those requiring superior power quality. The expenses associated with grid connection in remote regions can be substantial, especially in industries such as agriculture, train signaling, aviation, telecommunications, and forestry [8].

Combined heat and power (CHP) systems entail the simultaneous generation of heat and electricity from a single fuel source, aiming to maximize the efficiency of both outputs. Figure 8.2 illustrates a standard configuration of a residential CHP system. The prime mover technology, namely a FC, obtains fuel from the central network to produce electrical power, simultaneously generating heat as a byproduct. Electricity is utilized directly in home environments and, when grid-interactive, can be imported or exported as necessary. The heat generated during electrical production is harnessed for use in diverse applications, including space heating and residential hot water supply. In domestic built environment applications, enhanced system efficiency leads to a reduction in primary energy consumption, thereby decreasing emissions and operational costs for users.

8.2.1 Model Structure and Fuel Consumption

The reformer's function is to transform fuels such as methane gas, natural gas, or liquefied petroleum gas into hydrogen gas, as pure hydrogen is seldom utilized as a fuel in practical applications. Resources such as water, air, and portions of the

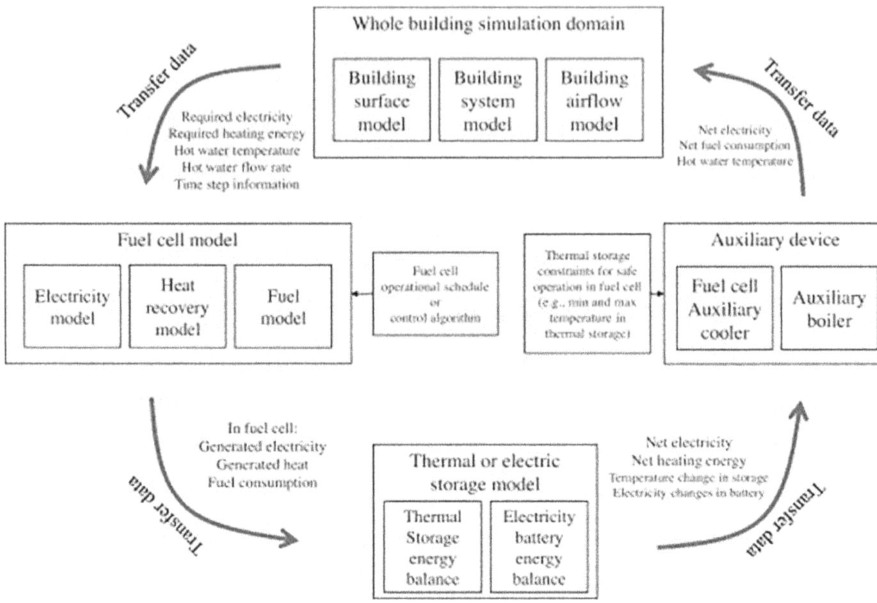

FIGURE 8.3 Model implementation schematic diagram [10].

fuel are utilized in the reformer for conditioning, conversion, and purification of the fuel. The FC stack produces heat and power via electrochemical reactions. The direct current (DC) electricity generated is subsequently transformed into alternating current (AC) power within the power conditioning unit (PCU). To ensure adequate reactions, hydrogen stoichiometry is typically maintained above 1.1, leading to surplus hydrogen fuel that remains unutilized.

Figure 8.3 illustrates the comprehensive methodology for model implementation. After calibrating the FC model, the comprehensive building simulation tool computes the energy consumption of the structure for the present time step. This includes information on power and heating demand, as well as specifics concerning hot water, especially when heating is facilitated through hot water circulation employing a FC system. In grid-connected systems, scheduled FC operation is frequently employed in stationary applications to ensure consistent performance. Temperature fluctuations in the thermal storage require computation, as they influence the efficacy of the FC. There may be minimum and maximum limits for the hot water temperature (stack coolant) in the FC system. The installation of a FC auxiliary cooler or heater results in energy consumption by these devices, which adds to the total electricity and heating energy demand.

Figure 8.4b illustrates the daily heating demand profile generated by the control algorithm employed in the electricity example. The blue, yellow, and gray polygons denote the heating demand within the building, the net heating energy required by the boiler, and the heating energy provided by the FC, respectively [10].

Figure 8.4 illustrates the implementation of the model utilizing hourly time intervals. The estimation is based on the load profile of the exhibited residential

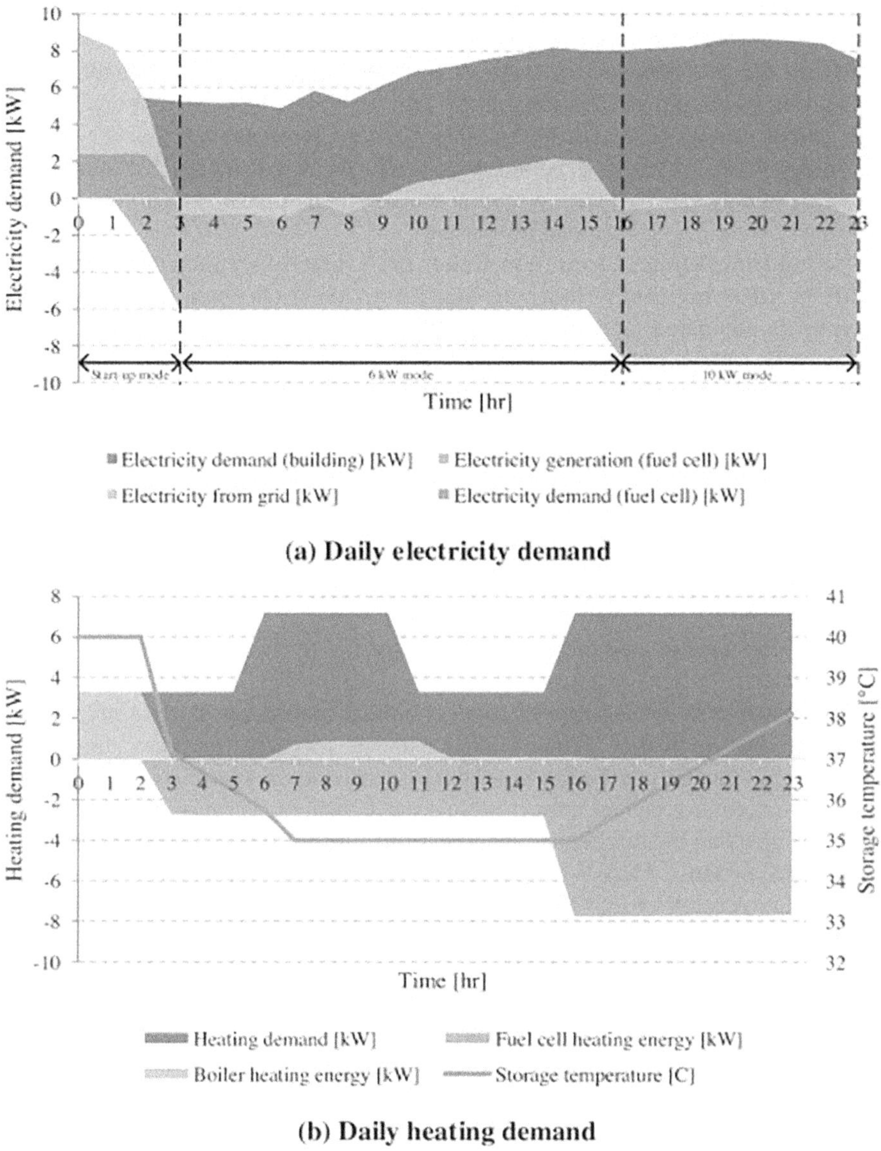

FIGURE 8.4 Example of fuel cell implementation to generate electricity [10].

dwelling. This analysis asserts that one FC system is employed to service three residential units. Figure 8.4a illustrates that the positive Y-axis represents power demand, while the negative section denotes electricity generation inside the FC system. A schedule-based control method is employed for the FC system, as illustrated by the gray polygon in Figure 8.4a. The period from 0:00 to 3:00 is designated as the beginning phase, whereas the span from 3:00 to 15:00 pertains to the 6-kW mode.

Hydrogen is anticipated to play a crucial role in particular decarbonization scenarios, serving as a fuel in future energy systems. This procedure may involve modifying the gas network to function as a hydrogen transporter, enabling low-carbon transportation and heating uses. The development of a hydrogen supply chain enhances the viability of FC technology in the foreseeable future. The storage of hydrogen for electricity generation mitigates the intermittency associated with renewable energy sources. The environmental advantages have been only marginally achieved thus far due to financial limitations.

The implementation of European Union (EU) directives aimed at enhancing air quality in urban regions by reducing nitrogen oxide (NOx) emissions will benefit FC systems compared to competing CHP technologies, such as internal combustion engines. Although there are positive predictions, FC-CHP systems have specific problems that need to be fixed to gain the trust of decision-makers; overlooking these concerns could significantly hinder their potential. The concerns encompass technical, financial, and regulatory aspects, including, but not limited to:

1. Constrained production capability.
2. Ensuring consistent flow distribution is critical to avert cell deterioration.
3. The acquisition of "renewable" hydrogen.
4. An extensive hydrogen infrastructure.
5. Expenditures related to capital technology.

Augmentation of the accessibility of economical combustion-based CHP systems.

Despite the increased interest in expanded research projects, funds, and publications, especially in the USA and the EU, there is still a lack of public outreach.

Moreover, environmental impact evaluations constitute a crucial research domain for comprehending the unintended repercussions linked to the heightened implementation of FC systems. Moreover, although catalysts, steel, and the majority of other materials are generally recycled, graphite plates are frequently disposed of due to the absence of feasible alternatives [11].

Furthermore, it is essential to acknowledge that the operating voltages for both the FC and the electrolyzer are not discretionary. The reference voltage for the electrolyzer should typically be approximately double that of the FC in the reversible solid oxide cells (rSOCs), as shown by multiple researchers. Specified current densities must be employed for the reference voltages. The relationship among the values of these essential variables is referred to as the polarization curve. The ohmic region is characterized by a linear relationship that represents the product of current density and resistance, in contrast to the beginning and final regions. The proton exchange membrane reversible fuel cell (PEM RFC) exhibits increased sensitivity to variations in current density concerning (rSOCs), rendering it advantageous for load-following applications, particularly when integrated with intermittent renewable energy sources [12].

8.2.2 Energy Density and Efficiency

Figure 8.5 illustrates the predicted energy densities of energy storage components in various batteries and FCs. A significant improvement in ED is shown in air-cathode

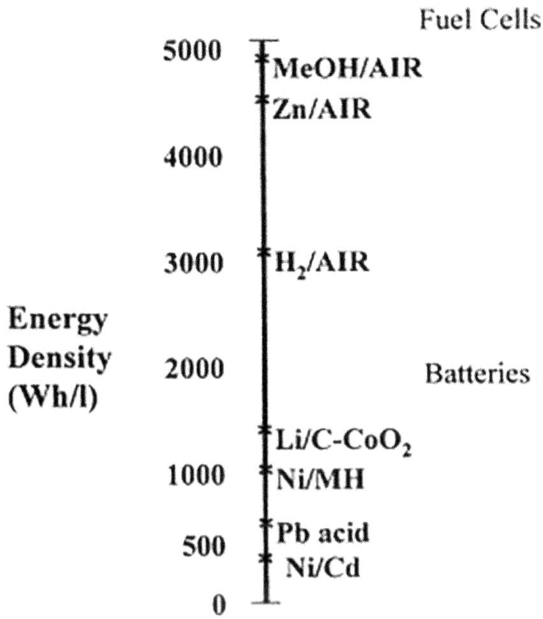

FIGURE 8.5 Theoretical energy densities [13].

FCs utilizing hydrogen, hydrocarbons, and metals as fuel sources. Metal/air systems demonstrate elevated energy densities; however, their constrained rechargeability has impeded broad commercial use, especially in portable applications. A rechargeable battery system is unlikely to develop that can compete with FCs for specific energy.

In practical applications, energy packaging reduces the available energy content of batteries to approximately 25% of their theoretical capacity. Nevertheless, the efficiency of energy conversion remains quite high, exceeding 80%. Inversely, conversion efficiency limits the theoretical FC device ED. Despite their lower energy-conversion efficiency (25–35%), FCs outperform batteries in ED. The advantage is enhanced when significant amounts of energy are required, especially for the forthcoming portable electronic applications.

Figure 8.6 compares the common Li–ion battery to the advantage as energy content increases. If the FC volume is equal to or higher than its portable fuel source, a battery may be better in low-energy scenarios. Devoid of fuel, a 10 W FC with 200 W/l PD occupies 50 cm^3. ED is comparable to a battery with 20 Wh or less of fuel.

A solid oxide fuel cell (SOFC) and double-effect water/lithium bromide vapor absorption system (VAS) tri-generation system was numerically analyzed. Overall system efficiencies of 84% demonstrate the benefits of tri-generation systems in heating, cooling, and power applications. A 300 kW MCFC VAS tri-generation system had a 72% efficiency according to computational research. Combining two commercially available technologies for a tri-generation system was tricky.

Compared to CHP systems, a 520 kW SOFC tri-generation plant with an ORC, heat exchanger, and single-effect vapour absorption system (VAS) saves 20% of

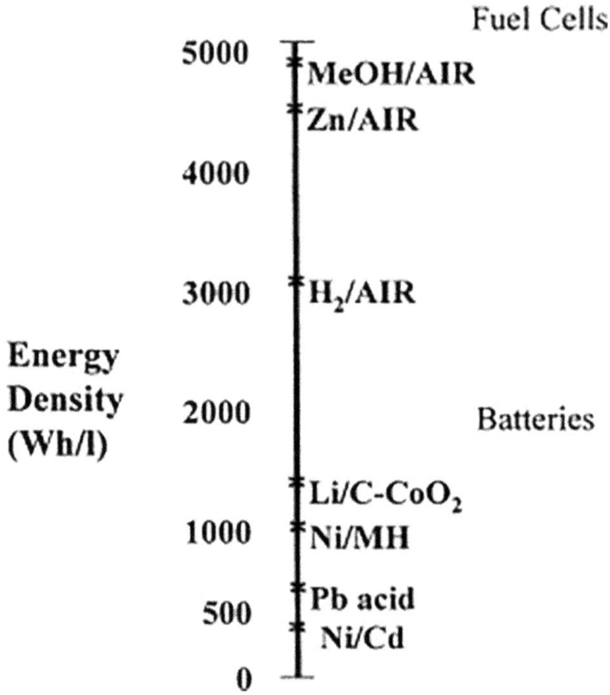

FIGURE 8.6 Practical energy system volumes [13].

energy. Thermal energy for cooling and high-temperature waste heat for ORC power boost performance. Maximum tri-generation efficiency is 74%. The study examined two sizes:

1. Complete peak load-capable SOFC system
2. Partial SOFC system, designed to manage peak loads through the SOFC and grid and achieve net zero grid import in 1 year.

The comprehensive and partial SOFC systems realize carbon emission reductions of 51.4% and 23.9%, respectively, alongside electricity savings of 7.1% and 2.8%, respectively [14].

8.3 SUPERCAPACITORS

Energy storage systems with lift drive DC-links are efficient. Regenerative energy can be stored in these systems to reduce peak loads or power additional equipment. Supercapacitors have high PD, fast charge and discharge rates, many charge and discharge cycles, and long lifespans. Supercapacitors display non-linear characteristics owing to the fluctuation of their capacitance about voltage and frequency. The energy stored in the supercapacitor bank, available for subsequent use, is computed using the following formula [15]:

$$E_{us} = \frac{U_n^2.C}{2}.0.75 \tag{8.1}$$

where

Un = nominal voltage and

C = capacitance.

The energy is multiplied by 0.75 as 75% of the energy stored in the supercapacitor is used. High-frequency constant phase and inserted charge-transfer resistance component define ideal polarizability and low-frequency limiting capacitance boundaries. Increased separator thickness and density decrease electrolyte conductivity, demonstrating electrolyte conductivity. Supercapacitor array charging and discharging cause power oscillations regardless of renewable energy generation and storage. Regulating energy storage and release requires voltage and current stability across all transmission lines. The power grid needs to regulate supercapacitor array input power and energy extraction. Variable wave, solar, and wind energies make electricity generation inconsistent.

Voltage decreases gradually during energy extraction from supercapacitors due to their intrinsic properties. Combining charging and draining might produce voltage swings. Power stability is needed for most energy-intensive devices. Smart power management systems used PVs and supercapacitor-battery hybrid energy storage to predict load energy usage. Hybrid energy storage system (HESS) peak power delay is short due to this predictive system's 100% accuracy and 0.004866 seconds justification time. Responding quickly to power fluctuations reduces battery stress and extends system life. Regulate output power within a range with similar energy storage technologies. Output power exceeding storage capacity can cause mistakes.

A functional supercapacitor stack management system utilizes daily temperature monitoring of an individual supercapacitor. This system oversees the temperature of the supercapacitor stack and regulates both charging and discharging in a supercapacitor-powered urban bus. Dynamically distributing current safeguards supercapacitor stacks from overcharging and discharging. When overcharged, the supercapacitor energizes electric vehicles, and when undercharged, the FC replenishes it. The suggested controller achieves specified objectives with a high-fidelity simulation model. FCs and supercapacitor hybrid energy storage systems employ flatness principles control to regulate output energy components along specified trajectories, applicable in electric vehicles and lifts. This approach ensures dependable control of supercapacitors, preventing overcharging or discharging. Establishing predefined regulations to manage load circuit current and voltage enhances battery longevity and increases energy efficiency owing to the optimum semiactive topological configuration.

8.3.1 Fuel Cell Battery

The HESS superconductor employed two independent control loops to regulate the conditions of the battery and supercapacitor. This strategy significantly decreased hydrogen consumption in the simulation model, and small-scale test bench measurements validated its precision. The tram power control system employs an advanced power-sharing methodology for enhanced safety. This method employs FCs and batteries to energies low-frequency components. A supercapacitor bank employs high-frequency components to fulfil power demands, guaranteeing transient-free, secure operation and prolonging system longevity [15].

8.3.2 Charging and Discharging Mechanism

Powerful supercapacitors can charge from a low current. During charging, supercapacitor layers store energy. Voltage rises with charge density until the supercapacitor achieves its maximum rated voltage. An increase in the rated voltage might cause supercapacitor failure. Management must comply with voltage and thermal boundaries.

ED, PD, capacitance, rated terminal voltage, charging and discharging time, cycle life, and longevity are the supercapacitor's main outputs. Capacitors have higher PD but lower ED than other energy storage solutions. FCs and batteries have higher ED than capacitors because they can hold many charges [16].

Due to their charge storage processes, supercapacitors are electric double-layer capacitors, pseudocapacitors, or hybrids. SC essentials are shown in Figure 8.7a. Schematics of the EDLC, PC, and HSC are shown in Figures 8.7b–d.

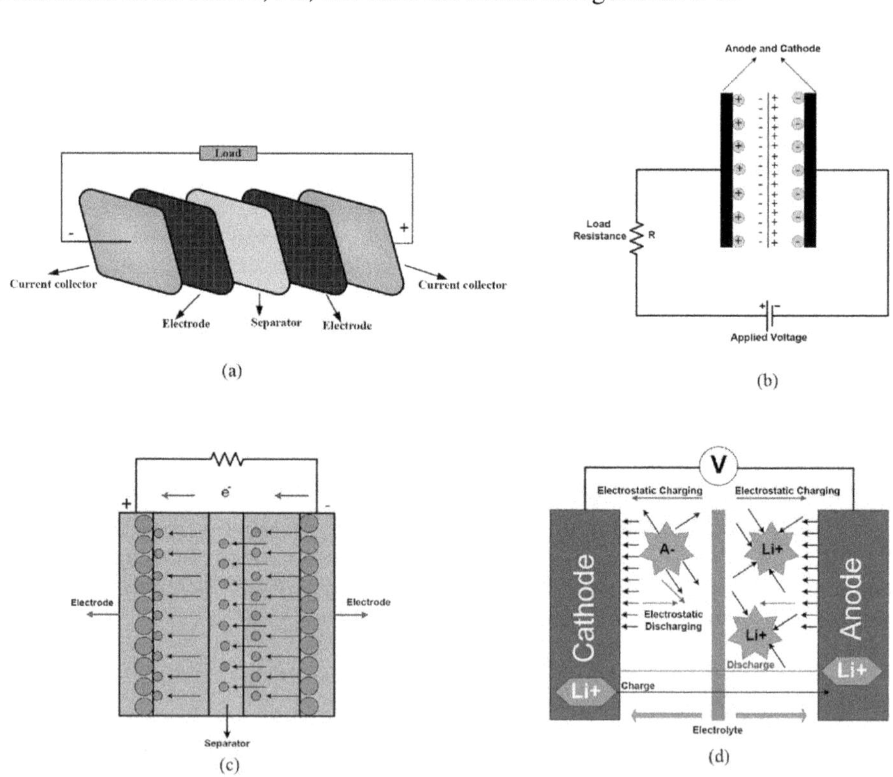

FIGURE 8.7 Schematic structures of the SCs. (a) Basic components of SC (b) EDLC (c) PC (d) HSC [17].

Non-dielectric EDLCs store energy electrostatically. Figure 8.7b shows two electrodes and an electrolyte. Capacitance is directly related to electrode surface area but inversely proportional to electrode distance. A large-surface-area material can increase capacitance. The materials utilized for electrodes and their fabrication procedures are essential for achieving increased capacitance. The interconnected double layers exhibit a significant specific surface area and lowered electrode separation, allowing electric double layer capacitors to attain much greater strength density than traditional capacitors. Furthermore, the properties of the electrolyte substantially affect the performance of supercapacitors. Aqueous and organic electrolytes are commonly employed. Selection criteria must consider electrode and charge collector types.

Electrodes with pseudocapacitance expand when charging and compress when discharging, reducing cycle life and mechanical stability. Supercapacitors and rechargeable batteries are used in many research investigations to develop electrode materials for innovative reaction processes. The requirement for high-energy supercapacitors with long cycle life and high power drives their development.

Certain specialists have observed intercalation pseudocapacitance in several electrode materials for metal-ion batteries by physically manipulating electrode materials. Conducting polymers and metal oxides store energy in pseudocapacitors. Conducting polymers, with their high capacitance, conductivity, and cost-effectiveness over carbon-based electrode materials, are potential electrode materials. Conducting polymer has high strength and energy storage capability. N-doped CP materials prohibit pseudocapacitors from reaching their full potential.

Mechanical stress on CPs during redox reactions considerably influences their stability during many charge–discharge cycles. The limited cycle stability of conducting polymer solar cells has slowed their advancement. The intercalation process, involving the insertion and extraction of protons, augments capacitance by influencing the capacitor's amorphous structure. In its hydrated state, ruthenium oxide (RuO_2) exhibits superior capacitance relative to carbon-based and conducting polymer materials, while also presenting a lower equivalent series resistance than other electrode materials. As a result, RuO_2-based PCs demonstrate enhanced ED and PD relative to other electric double-layer capacitors and carbon-based supercapacitors.

Asymmetric super capacitors (ASSCs) encompass the hybrid super capacitor (HSC). HSC denotes a device that accumulates energy utilizing a battery as the anode and a supercapacitive material as the cathode. ASSCs provide elevatedED and PD across several voltage ranges (up to 2.0 V, 2.7 V, and 4.0 V for aqueous, organic, and ionic liquids) by integrating two electrode materials with distinct potential windows [17].

8.4 SUPERCONDUCTING MAGNETIC ENERGY STORAGE

Contemporary models seek to enhance the efficiency of urban electric network components. Statistics indicate that 54% of the global population resides in urban areas. As the population increases and rural inhabitants migrate to urban areas, this percentage will escalate. DG simulates the electricity system to vary generation methods and tailor them to temporal or geographical requirements. This paradigm promotes low- and medium-power renewable energy production. This pertains to energy storage technologies. SMES systems respond to grid-level energy demands in milliseconds with 95% efficiency. These devices enable UPS, voltage stability, and frequency regulation for integrated networks.

8.4.1 The Superconducting Energy Storage Principle

Superconductivity is one of the few methods for direct electromagnetic energy storage and the optimum solution for electric power storage. The storage system uses a superconducting coil kept at very low temperatures. A converter system transfers off-peak power into DC and stores it in the storage ring for future retrieval. The two components mutually reinforce each other because the energy is maintained in a magnetic field that cycles the current. Without resistive energy loss, the current can continue virtually eternally. The energy in the ring will not dissipate if the system is maintained below a certain temperature.

A niobium–titanium alloy is widely used and economically viable, with a critical temperature of 10 K or –263°C. A coil built from the alloy below this threshold is cooled by liquid helium. Niobium–titanium alloy has good magnetic field tolerance. For applications requiring extremely strong magnetic fields, there are more expensive materials.

Intensive refrigeration is needed to retain a superconducting current in a coil below its transition temperature. Efficient insulating materials can substantially diminish heat transmission rates. Recently, a novel category of ceramic materials has been discovered that demonstrates superconductivity at comparatively elevated temperatures, attainable via cooling with liquid nitrogen. Liquid nitrogen possesses a boiling point of 98 K, corresponding to −175°C. Many materials display significant brittleness, hindering their manipulation; nonetheless, novel strategies are being developed to harness their characteristics. This renders superconductivity a more economically feasible alternative for transmission lines and storage systems.

In the revised electrical grid model, renewable energy sources are of considerable significance. Moreover, renewable energy sources are integrated systems, such as energy storage systems, that enhance optimal performance within the electrical grid. Concerning ESS, it is essential to acknowledge that storage systems can operate in two separate ways. In one capacity, they serve as loads inside the network when charging, while in another capacity, they act as generators during discharging. These systems can be added anywhere in the network. The inquiry concerned MV network transportation. Figure 8.8 shows the SMES System core diagram [18].

The magnetic field of a superconducting coil is maintained by the SMES, which functions as an exceptionally efficient energy storage system.

Integrating a powerful intelligent controller and power electronic converter with the electrical power system improves SMES efficiency. Electric power systems provide reliable energy to electric utility companies, meeting customer expectations [19].

8.4.2 Applications of SMES

SMES devices were originally designed to store energy at 1000 MW or more, as in pumped-storage hydropower projects. This would require a tens- or hundreds-of-km subterranean storage ring. Due to high construction costs and superconducting wire

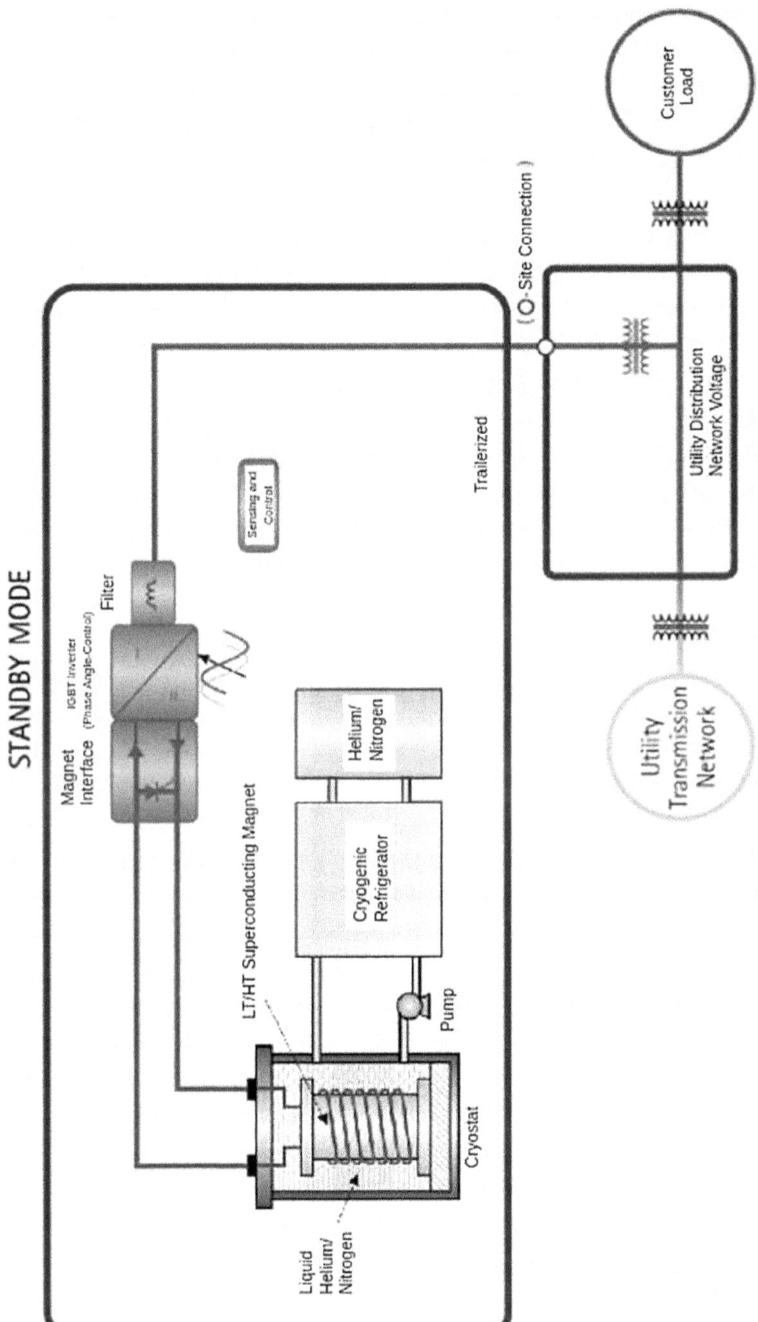

FIGURE 8.8 Basic scheme of an SMES system [18].

prices, the project was impracticable. The US Department of Energy is funding large-scale SMES superconducting energy storage research. Modern improvements enable 100 kW–100 MW small commercial storage facilities. The storage capacity of these commercial devices is 10–30 kWh, which is little for utility storage but important for emergency grid support.

High-quality power is needed in substations and industrial processes. Stability support for scattered generating systems will be improved by larger superconducting coils in SMES development. Micro-SMES systems produce 1–3 MW; distributed generation systems should produce 10 times that amount.

To improve renewable energy reliability, global systems need energy storage. Large SMEs are ideal for this application due to their adaptability. A large SMES system is easier to sit than pumped storage hydropower, the main large-scale energy storage method. Pumped storage hydropower requires ideal locations for two reservoirs, although a large SMES ring can be built in varied situations, such as deserts. These regions have potential for large-scale solar installations, and integrating solar energy with SMES might deliver reliable renewable electricity at low cost.

More cost-effective superconducting materials, especially higher temperature superconductors with greater properties, may affect the growth of major SMEs. Finding room-temperature superconductivity is the goal. That prospect remains distant despite numerous intriguing scholarly journal findings [20].

REFERENCES

1. Chatzivasileiadi, A., Ampatzi, E., & Knight, I. (2013). Characteristics of electrical energy storage technologies and their applications in buildings. *Renewable and Sustainable Energy Reviews*, *25*, 814–830.
2. Koskela, J., Rautiainen, A., & Järventausta, P. (2019). Using electrical energy storage in residential buildings–Sizing of battery and photovoltaic panels based on electricity cost optimization. *Applied Energy*, *239*, 1175–1189.
3. Moya, F. D., Torres-Moreno, J. L., & Álvarez, J. D. (2020). Optimal model for energy management strategy in smart building with energy storage systems and electric vehicles. *Energies*, *13*(14), 3605.
4. Kyriakopoulos, G. L., & Arabatzis, G. (2016). Electrical energy storage systems in electricity generation: Energy policies, innovative technologies, and regulatory regimes. *Renewable and Sustainable Energy Reviews*, *56*, 1044–1067.
5. Arenas, L. F., de León, C. P., & Walsh, F. C. (2019). Redox flow batteries for energy storage: their promise, achievements and challenges. *Current Opinion in Electrochemistry*, *16*, 117–126.
6. Chen, H., Xu, Y., Liu, C., He, F., & Hu, S. (2022). Storing energy in China—An overview. In *Storing Energy,* Second Edition (pp. 771–791). Elsevier.
7. Doetsch, C., & Pohlig, A. (2020). The use of flow batteries in storing electricity for national grids. In T. M. Letcher (Ed.), *Future Energy* (pp. 263–277). Elsevier.
8. Sammes, N. M., & Boersma, R. (2000). Small-scale fuel cells for residential applications. *Journal of Power Sources*, *86*(1–2), 98–110.
9. Elmer, T., Worall, M., Wu, S., & Riffat, S. B. (2015). Fuel cell technology for domestic built environment applications: State of-the-art review. *Renewable and Sustainable Energy Reviews*, *42*, 913–931.

10. Ham, S. W., Jo, S. Y., Dong, H. W., & Jeong, J. W. (2015). A simplified PEM fuel cell model for building cogeneration applications. *Energy and Buildings*, *107*, 213–225.
11. Acha, S., Le Brun, N., Damaskou, M., Fubara, T. C., Mulgundmath, V., Markides, C. N., & Shah, N. (2020). Fuel cells as combined heat and power systems in commercial buildings: A case study in the food-retail sector. *Energy*, *206*, 118046.
12. Chadly, A., Azar, E., Maalouf, M., & Mayyas, A. (2022). Techno-economic analysis of energy storage systems using reversible fuel cells and rechargeable batteries in green buildings. *Energy*, *247*, 123466.
13. Dyer, C. K. (2002). Fuel cells for portable applications. *Journal of Power Sources*, *106*(1–2), 31–34.
14. Elmer, T., Worall, M., Wu, S., & Riffat, S. (2016). Assessment of a novel solid oxide fuel cell tri-generation system for building applications. *Energy Conversion and Management*, *124*, 29–41.
15. Makar, M., Pravica, L., & Kutija, M. (2022). Supercapacitor-based energy storage in elevators to improve energy efficiency of buildings. *Applied Sciences*, *12*(14), 7184.
16. Dissanayake, K., & Kularatna-Abeywardana, D. (2024). A review of supercapacitors: Materials, technology, challenges, and renewable energy applications. *Journal of Energy Storage*, *96*, 112563.
17. Yadlapalli, R. T., Alla, R. R., Kandipati, R., & Kotapati, A. (2022). Super capacitors for energy storage: Progress, applications and challenges. *Journal of Energy Storage*, *49*, 104194.
18. Colmenar-Santos, A., Molina-Ibáñez, E. L., Rosales-Asensio, E., & López-Rey, Á. (2018). Technical approach for the inclusion of superconducting magnetic energy storage in a smart city. *Energy*, *158*, 1080–1091.
19. Khaleel, M., Yusupov, Z., Nassar, Y., El-khozondar, H. J., Ahmed, A., & Alsharif, A. (2023). Technical challenges and optimization of superconducting magnetic energy storage in electrical power systems. *e-Prime-Advances in Electrical Engineering, Electronics and Energy*, *5*, 100223.
20. Breeze, P. (2018). Chapter 5-superconducting magnetic energy storage. *Power System Energy Storage Technologies*, 47–52.

9 Thermal Energy Storage System for Green Buildings

Cost-effective and environmentally friendly thermal energy (TE) storage (TES) devices reduce fuel use. Reuse of excess heat reduces TE loss. Heat is produced by most human activities. Largely, our cooking, vehicle, and other operations lose heat. Innovative waste heat uses are cost-effective.

Due to the cost and effort necessary to collect, store, and use waste heat from many small heat sources over long distances, TES systems are not practicable for every heat source. Buildings use a lot of energy worldwide. Mostly for space heating and hot water production, buildings consume 40% of the world's energy. International Energy Agency states that: space and water heating accounted for 50% of building energy demand in 2021, resulting in 2450 Mt of direct CO_2 emissions.

Renewable energy sources are increasingly used in buildings to reduce energy use and environmental impact. Building integrated photovoltaic/thermal (BIPV/T) systems for fortuitous building energy needs. Industrial buildings, including construction, use more thermal and electrical energy than residential and commercial buildings.

Ensuring safe, healthy, and comfortable indoor conditions in buildings requires significant energy usage. This is due to indirect energy consumption emissions and direct fossil fuel combustion emissions. This requires multiple techniques to reduce building energy usage and greenhouse gas emissions. Since 2010, global direct and indirect building greenhouse gas emissions have increased by 1% yearly.

Building energy consumption decarbonizes rapidly, but all actions must be taken to obtain net-zero emissions by 2050. By 2030, the International Energy Agency recommends 20% carbon-zero buildings. PV/T systems and solar-assisted heat pumps address electrical and TE needs for zero-carbon buildings [1].

Governments and society have worked to develop new building energy efficiency and conservation technology to cut energy consumption without compromising thermal comfort in different weather situations. TES systems reduce peak loads, uncouple energy demand from availability, integrate renewable energy sources, and improve TE management, improving building energy efficiency. Due to their high-energy storage density and stable temperature phase transition mechanism, phase-change materials (PCMs) latent heat thermal energy storage (LHTES) are employed in buildings. PCMs, a TES approach that minimizes building energy demand to

 DOI: 10.1201/9781003752837-9

improve cooling, have been studied in architecture and engineering for 40 years. PCMs in LHTES systems absorb building energy to solve the energy supply-demand gap, an intriguing technology. PCMs heat, cool, and mix at -20°C to 200°C.

9.1 CONCEPT OF PCM HEAT STORAGE

PCMs control temperatures through latent heat. Through condensation and vaporization, LHTES can alter liquid–gas, solid–solid, or solid–liquid phases. Building heating and cooling require solid–liquid PCMs as solid–solid and liquid–gaseous latent heat approaches are technologically limited. The chemical bonds disintegrate when the PCM's ambient temperature reaches its melting point, allowing it to absorb energy and become liquid. PCM charges. Linkages regenerate and exothermically emit heat as the PCM freezes. The volume fluctuates by less than 10% with each charge and discharge [1].

Thermal comfort uses 40% of building energy as it rises. Upgraded building insulation saves energy. With geothermal or solar energy, TES can reduce energy use. For peak demand, TES systems store and recover TE. Heat and cool buildings with TES.

Heat is stored in TES by chemical processes, perceptible heat, and latent heat. Most LHTES research has occurred in the previous two decades. Material phase change powers LHTES, and PCM phases vary with heat. Solid-to-solid, solid-to-liquid-to-gas, or reverse-phase transitions are possible in TES. Many PCMs are solid–liquid and solid–solid. Storage capacity for PCM-equipped LHTES is [2]:

$$Q = \int_{T_i}^{T_m} mC_p dT + ma_m \Delta H_m + \int_{T_m}^{T_f} mC_p dT \tag{9.1}$$

where

a_m = fraction melted,

ΔH_m = the heat of phase transition per unit mass (J/g)

Sensible heat storage (SHS) systems have a wider temperature range and a lower TES density than LHTES devices. The LHTES system relies on PCM, which must meet certain requirements.

Enhancing building insulation and using renewable energy sources are becoming major techniques for energy efficiency and reduction. Due to its affordability and ability to recover energy during peak use, energy storage is essential for achieving these goals during low demand. Different cooling and heating systems have been introduced to save energy.

TES exploits solar TE, reduces energy use, and helps the environment. Latent, sensible, and thermochemical heat. SHS stores TE. Heating a material stores energy with SHS. By heating or cooling a material without altering its phase (such as turning it from a solid to a liquid), sensible heat storage stores thermal energy. To store energy, this approach needs high temperatures. The greatest technique to increase building envelope energy performance and thermal inertia is to store heat energy with PCMs.

The energy is subsequently retained as latent heat due to the fusion of these materials. Subsequently, it is reintroduced into the environment following the process of solidification. The applications of PCMs represent a significant asset in the design of net-zero energy buildings. These materials can absorb, store, and release latent heat. Based on demand, this energy can be used for air-conditioning or free-cooling. Many PCM performance studies have been done with latent heat storage (LHS) applications.

PCMs can be added to the building envelope in several ways:

1. The static and dynamic technologies.
2. The active technique uses a mechanical method to charge or discharge the PCM.
3. Through passive technology, the PCM is only exposed to ambient temperature changes.
4. Maximum PCM efficiency depends on material type, melting temperature range, quantity used, thermo-physical qualities, outdoor climatic conditions, encapsulation method, building placement, and orientation.

Recently, PCMs and their supporting components have advanced, making this topic of study important: Graphite foams are capable of carrying phase change compounds due to their low cost, good performance, and high recovery.

9.2 CLASSIFICATION OF TES SYSTEMS

Energy storage stores TE by heating or cooling a substance. Storage is abundant for renewable energy like solar. Solar energy regulates energy use. The primary solar TES types are shown in Figure 9.1. Capacities, energy charge and discharge speeds, efficiency, storage duration, charge and discharge durations, and costs are system parameters.

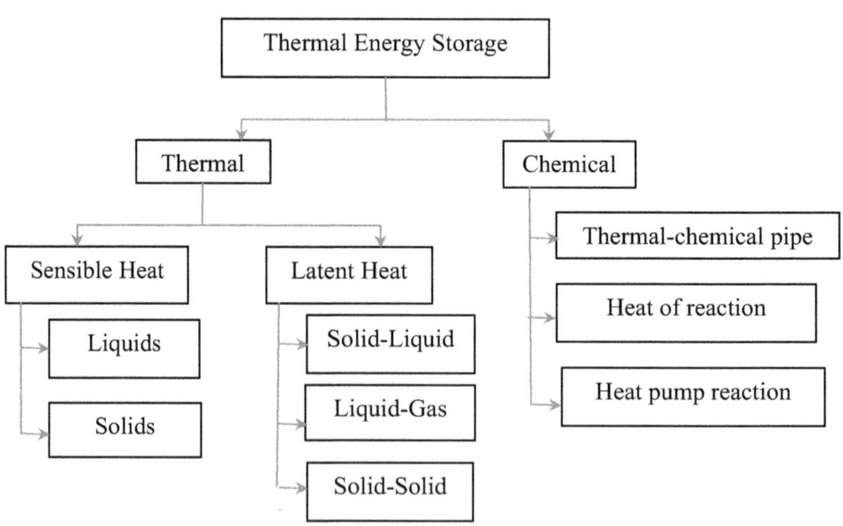

FIGURE 9.1 Types of solar TES [3].

9.3 SENSIBLE HEAT

Over the years, sensible heat TES has gained attention for numerous applications, helping solve the waste or renewable energy time-discrepancy problem. Thermal and physicochemical qualities of materials (density, specific heat, latent heat, etc.), storage capacity, heat loss, and energy supply operational conditions influence TES system selection and optimization. The harmless, accessible, and temperature-compatible liquid water is a basic storage medium.

SHS stores energy by changing material temperatures. Heat storage is proportional to density, specific heat capacity, volume, and temperature change. SHS materials have specific heat, density, and thermal conductivity. Common SHS materials for homes must meet these characteristics:

- Elevated specific heat and energy density;
- Excellent thermal conductivity (>0.3 W/m K);
- Efficient thermal diffusivity;
- Affordable and simple to produce;
- Dependable chemical properties, minimal corrosiveness, and reduced environmental impact, featuring zero ozone depletion potential and low global warming potential.

A well-established technology, SHS has undergone testing in various large-scale residential applications [4].

9.3.1 Active Technologies

TES in active systems can be used to install renewable energy sources in heating, ventilation, and air conditioning (HVAC) for space heating and cooling, upgrade installations, or transfer peak load in new or refurbished buildings. TES integration with the core, floor, walls, external solar façades, suspended ceilings, ventilation, PV, and water tanks is shown in Figure 9.2.

Charge TES with low nighttime temperatures and release it as needed to cool active structures free. PCM is suitable for storage as its phase change energy density is high, and the cold source and application (building comfort) are close. This method has trouble solidifying PCM if summer temperatures do not drop below the phase change range. TES free cools TABS, walls, ventilation systems, and suspended ceilings (Figure 9.3).

9.3.2 Pros and Cons

Usable heat storage, thermally stable SHS, dominates high-temperature TES. Thermal oils and liquid metals cost more than practical heat storage materials. Reasonable heat storage materials' key concern is discharge temperature stability. Thermal discharge lowers the HTF outlet temperature. Material heat is 50–100 times less than latent heat, reducing TES. SHS materials store TE due to their density and temperature range [6].

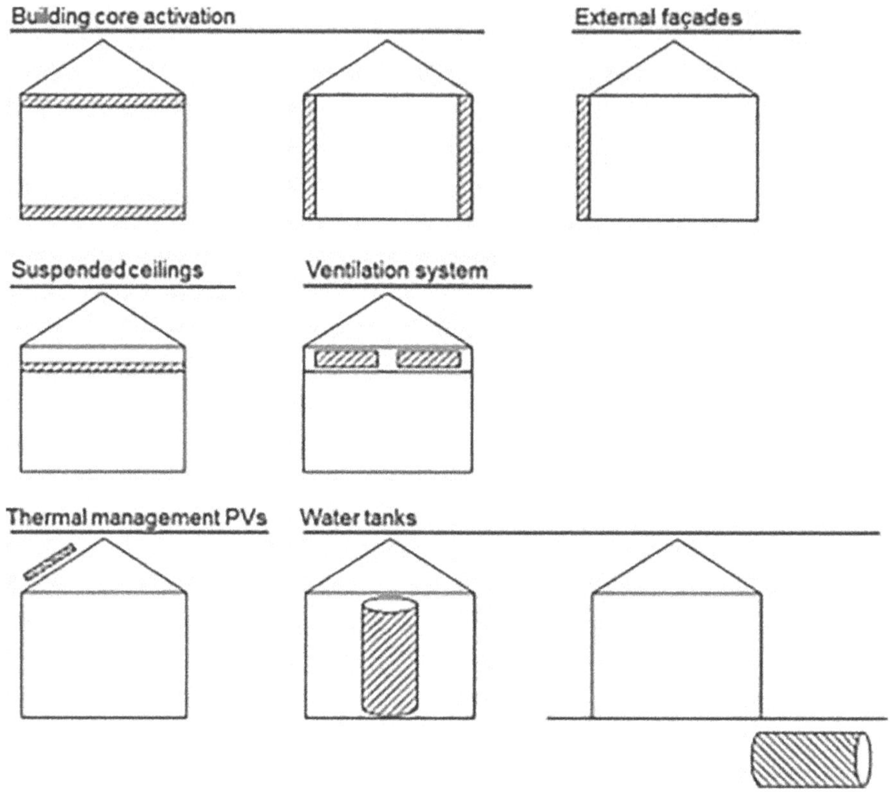

FIGURE 9.2 Integration of the TES in buildings [5].

9.4 LATENT HEAT

LHS and release make PCMs popular for thermal control and energy savings. PCMs have multiple latent heat and melting points, but thermal conductivity limits their use. Building air conditioning is energy-intensive and contributes to climate change. Seasonal temperature and demand make energy conservation challenging. Nearly 40% of global energy demand comes from buildings. Buildings take lots of energy to heat and cool. To satisfy peak electricity demand and reduce costs, utilities provide time-of-use discounts. The time of use market lowers off-peak electricity prices and raises peak prices. TES reduces peak energy demand by storing and supplying excess energy. PCM is a good TES medium because it stores heat and melts and freezes. PCM integrated building envelopes undergo phase transition owing to daily temperature variations, reducing the ambient temperature's impact on inside temperatures, space conditioning energy consumption, and occupant thermal comfort [7].

When LHS PCMs go through isothermal phase changes, they store and release latent heat. Without a phase change, SHS would not be able to make heat. PCMs have high energy levels and small volumetric changes during phase change, so LHS is better than SHS. Solid–solid or liquid–gas PCM recrystallization or pure phase

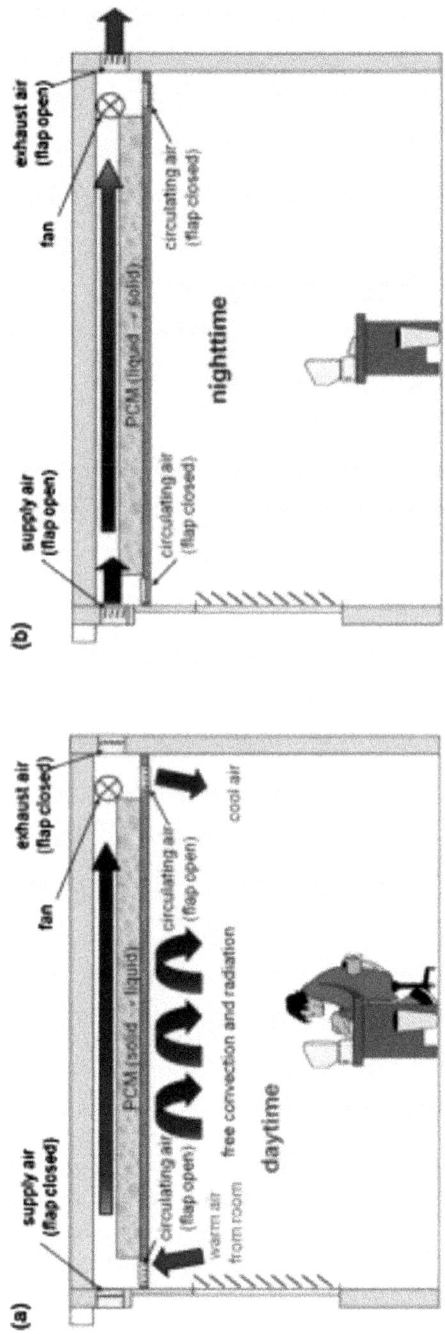

FIGURE 9.3 Cooling (a) and regeneration mode (b) of the ventilated cooling ceiling with PCM [5].

transition can release latent TE. Phase change fusion and solidification cycles occur in liquid–solid PCM TES. PCM masses have liquid–solid phase change storage capacities (Q_s) equal to sensible and latent heat energy:

$$Q_s = m\left[\left[C_{p,s}\left(\Delta T\right)\right]_{solid} + \Delta h_{fusion} + \left[C_{p,l}\left(\Delta T\right)\right]_{liquid}\right] \tag{9.2}$$

Heat energy determination is based on the specific heat capacity of a PCM in solid or liquid state and the temperature gradient (ΔT) in each condition [8].

The integration of PCMs into passive LHTES systems and structures is difficult. This comprises the major design parameters:

1. The PCM's phase-change temperature,
2. Thermal mass quantity, location in the LHTES system, and
3. Passive LHTES system placement in the structure.

Furthermore, these characteristics must be determined for distinct interior loads and climates. Designing residential, service, and high- or low-thermal-inertia buildings requires distinct PCM evaluations. PCM optimization minimizes summer cooling or winter heating loads, determining the method.

Inertia and low thermal mass in lightweight construction reduce thermal comfort and increase overheating risk. Outside and inside stresses make them more vulnerable to sudden changes in temperature. PCM-equipped passive LHTES systems in lightweight residential and business buildings may boost thermal storage capacity and prevent summer overheating from internal loads or solar radiation. A thermoregulated PCM can reduce interior temperature fluctuations and peak cooling demands. Heavyweight construction using PCMs may minimize weight and maintain thermal capacity.

9.4.1 Incorporation of PCMs into Building Elements

To reduce leakage, PCMs must be examined for integration into passive LHTES systems, especially in construction materials or building elements, after selection based on application temperature range and thermophysical parameters. PCMs can be shape-stabilized. Traditional construction materials incorporate PCMs directly or through immersion. During manufacture, liquid or powdered PCM is mixed with gypsum, concrete, or plaster without extra equipment. However, leaks and incompatibility with building materials may occur. Capillary absorption occurs when porous construction materials like gypsum board, brick, and concrete block are submerged in melted PCM.

Encapsulated PCMs can be in pre-integrated construction kits. Protecting PCM requires strength, flexibility, corrosion resistance, and thermal stability. The PCM must be protected from external interactions, have a suitable surface area for heat transfer, and be stable and easy to handle. A continuous polymeric sheet encloses solid or liquid particles or droplets (the core) to form microcapsules.

Microencapsulation reduces process-related PCM leakage, improving PCM integration in cement, lime, concrete, mortar, artificial marble, sealants, paints, textiles, and other coatings.

Microencapsulation increases the heat exchange surface and regulates volume changes during phase change. Research has found supercooling concerns and that microencapsulation affects building material mechanical strength. Package PCMs in tubes, spheres, panels, or other containers for integration into architectural elements using macroencapsulation. The thermal and physical parameters of the container needed for a certain quantity of PCM directly affect heat transport at the PCM, affecting melt time and storage element performance. To increase heat transfer during phase-change processes while limiting material loss, corrosion, and volume changes in PCMs, the container must be optimized.

Macroencapsulation reduces structural damage and mitigates PCM flammability. Its limitations include poor thermal conductivity and edge solidification. Researchers have recently focused on shape-stabilized phase change materials (SSPCMs). Their high apparent specific heat, thermal conductivity, structural integrity during phase shifts, and performance throughout protracted thermal cycling attract attention. A durable composite is made by dispersing PCM, like paraffin, in high-density polyethylene. Paraffin possesses energy like PCMs at 80% mass percentage. SSPCMs have been studied in building walls, ceilings, and floors, showing their intriguing phase transition properties [9].

Figure 9.4 shows the demonstration heating system's performance from 00:00 to 24:00 on February 10, 2019. The LHS system's operating states divide the system's operation into five phases (Phase I–V), with red and blue sections indicating charging and discharging periods. The LHS system can be charged and terminals heated in these conditions. As shown in Figure 9.4b, a heat supply power of approximately 240 kW leads to a heating demand of approximately 59 W/m². This adjustment keeps indoor temperatures between 16.7°C and 18.6°C despite the harsh cold, and 1756.1 kWh were charged to the LHS system, reducing thermal power from 280 to 140 kW. Electric boilers' automatically regulated intermittent heating minimizes thermal power and electric power usage (Figure 9.4b).

In Phase II (07:08–16:38), the electric boilers are turned down to save electricity, while the water pump remains operational to maintain the heating system, as shown in Figure 9.4a. LHS system discharges, as shown by the lower temperature of LHS entry water compared to LHS outflow water in Figure 9.4a, to autonomously heat terminals. Figure 9.4a shows the LHS outflow water temperature and other recorded water temperatures descending in three phases: rapidly, gradually, and sharply over 9.5 hours.

Discharging thermal power of the LHS system ranges from 110 to 200 kW, resulting in a heating load of 28–49 W/m². This capacity can maintain an indoor temperature between 17.7°C and 18.4°C due to the high midday outdoor temperature. The LHS system extracts 1439.4 kWh of TE, with 84.7 kWh from a single storage device, averaging 8.9 kW per storage device.

The current charge–discharge cycle discharges much less TE from the LHS reservoir than the design parameters as Phase II dropped the LHS output water temperature to 40°C.

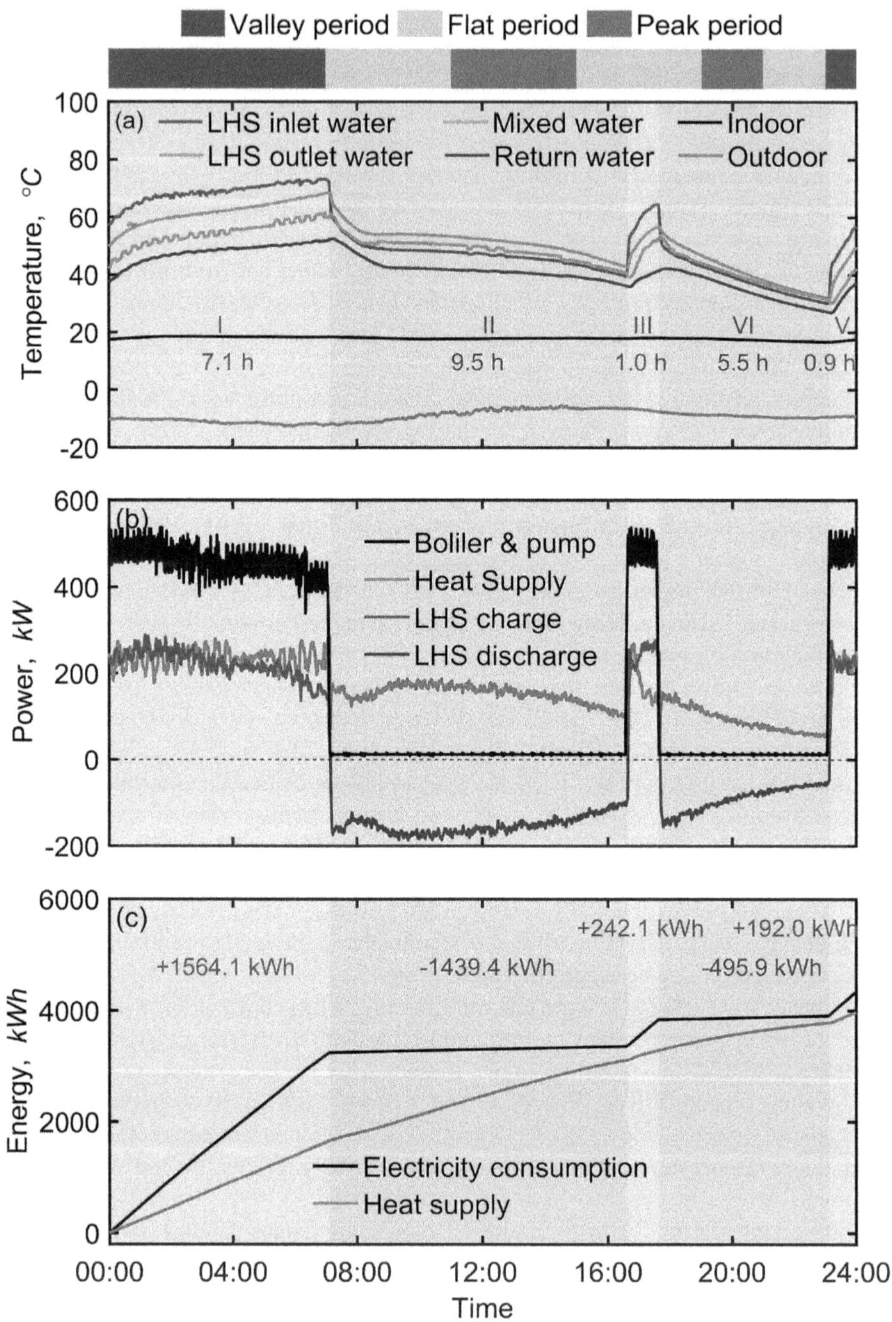

FIGURE 9.4 The demonstration Electric Boiler with Heat (EBH) storage system's performance, integrated with LHS over a single operational day, includes: (a) temperature variations at key measuring points (LHS inlet water, LHS outlet water, mixed water, return water, indoor, and outdoor) over time; (b) transient electric (boiler and pump) and thermal (heat supply, LHS charge, and LHS discharge) powers; and (c) accumulated electricity consumption and heat supply [10].

Before the valley rate phase, the heating system must provide more heat to change building demand. The electric boilers run at full capacity during Phase III (16:38–17:38), the flat power rate time. The LHS system charged 242.1 kWh and maintained a 17.9–18.1°C indoor temperature. Heat directly enhances Phase IV LHS secondary discharge (17:38–23:08). Phase IV water temperatures and heating system heat supply power drop rapidly, indicating LHS system discharge is nearly drained (Figure 9.4). After Phase IV, the LHS outflow water temperature dropped to 30°C, dropping the inside temperature to 16.7–17.5°C.

Total heating system heat supply and electricity consumption are shown in Figure 9.4c. The operative day LHS system consumes 4336.0, 3967.9, 1998, and 1935.3 kWh of electricity, heat supply, heat storage, and heat release. Daily LHS thermal efficiency is 96.8%, and total system efficiency is 91.5%. The operative day's TE dispatched is below the LHS system's heat storage capacity. Improved charging strategies, such as raising the electric boiler ceiling output temperatures, can avoid the second charge–discharge cycle [10].

9.5 WATER TANK STORAGE

Using variable renewable energy sources requires temporal storage solutions only. As seasonal heat storage is mostly done with water, this study focuses on it (Figure 9.5). Water-based and multicomponent filling technologies are the main groups. Tank thermal energy storage (TTES) is created by basins that extend partially or totally above the ground surface, and pit thermal energy storage (PTES) is water-filled sealed pits without structural features for stability. Systems that use multicomponent filling are called gravel–water thermal energy storage (GWTES). Gravel is rarely used for GWTES. Based on filling composition, multicomponent storage can be earth–water or gravel–water [11].

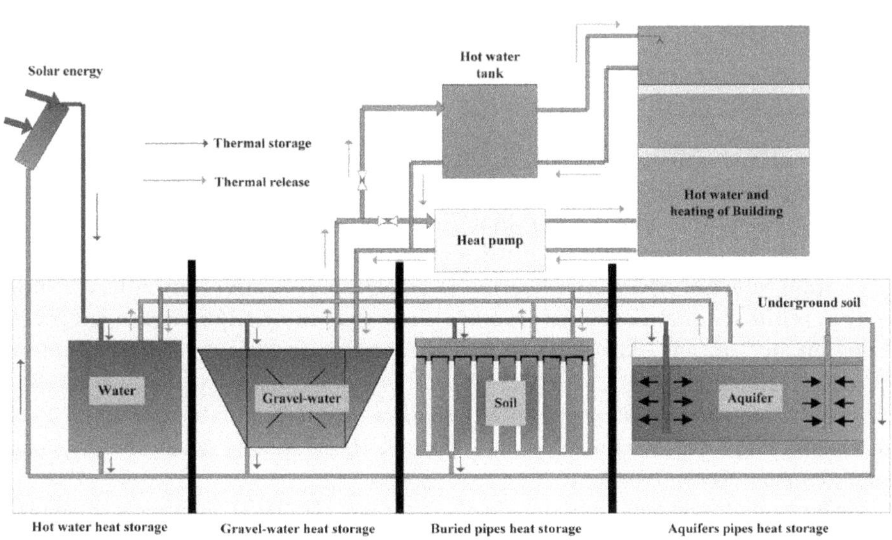

FIGURE 9.5 Typical layout of USTES [13].

9.5.1 Seasonal Cold Storage

Aquifer, borehole, cavern, earth-to-air, earth piles, seawater, rock, and roof pond TES technologies are used for seasonal storage. Seasonal cold and thermal storage cooling research is limited. Winter cold storage uses snow and ice for free cooling, ensuring sustainability.

Efficient use of renewable cold energy for cooling has become more important due to rising energy costs and environmental concerns. Many seasonal cold storage projects have payback times of 5 years or fewer. From conventional systems, Aquifer Thermal Energy Storage (ATES) systems save 90–95% for direct heating and cooling, 80–85% for heat pump-assisted, and 90–95% for industrial process cooling and district cooling [11].

9.5.2 Tank and Pit Long-Term Storages

Concrete or steel seasonal heat storage tanks hold a storage medium. Ground and subsurface installations are possible. Water is usually stored, but other methods exist. Pit storage (P-TES) uses plastic-lined, subterranean pits. Water and gravel make up the storage medium. Direct hot water injection or heat transfer fluid pipelines power the storage system. To reduce thermal losses, tanks and pits have top and side insulation of glass wool, polyurethane, foam glass, or fiberglass.

Minimum 1-m insulation is suggested for long-term storage. Tank and pit installations need firm ground due to geological factors. A drawback is energy deterioration from buoyancy-induced temperature layer mixing. The key benefit is increased capacity rates because of water's high specific heat, especially in T-TES. These TES technologies include solar ponds. These systems have pools with low surface salinity and high bottom salinity, which increases specific volume. Due to its higher density, water toward the bottom is somewhat shielded from solar radiation, limiting mixing [12].

9.6 UNDERGROUND STORAGE

Due to seasonal climate changes, the earth and groundwater passively retain TE. Earth temperature equals seasonal mean air temperature up to 15 m deep. Winter ground temperatures are higher than summer surface air temperatures. Both ground and groundwater are useful for winter heat extraction and summer cold extraction. Summer and winter heating and cooling use extraction. A heat storage system refills in summer. Heat demand may exceed cooling demand, necessitating greater storage. TE is stored underground in underground thermal energy storage (UTES).

Surface temperature changes affect shallow groundwater temperatures to 10 m. The highest depth of annual cyclic variation, 10–20 m, has rather stable groundwater temperatures.

Underground seasonal thermal energy storage (USTES) augments renewable energy and energy conservation. USTES fixes time, space, and strength issues in renewable heating systems. By moving renewable energy heating from summer or transitional seasons to winter, it overcomes short-term thermal storage systems' instability and inefficiency. Due to energy efficiency and environmental concerns, underground solar heat storage has increased worldwide.

Building heating with seasonal thermal energy storage (STES) is excessive in cold climates. In Figure 9.5, STES systems are heat thermal energy storage (HTES), GWTES, borehole thermal energy storage (BTES), and ATES. Storage of solar energy as hot water in a tank or pool is HTES. Given its large specific heat capacity and quick thermal storage and release rates, water is commonly employed in STES systems. To fill the subsurface groove, GWTES uses gravel and water. Walls and bottom are adiabatic and waterproof [13].

9.6.1 Borehole TE Storage

UTES stores heat in the subsurface, transferring heat to and from rocks and ground fluid. Heat pumps and borehole heat exchangers (BHEs) are used in BTES systems to transport heat from a circulating fluid to the subsurface. U-tube or coaxial topologies are used in shallow systems, usually under 300 m. BTES does not require a permeable medium for direct advective heat transfer, unlike ATES, which relies on geological conditions. Thus, it can be used in ATES-unsuitable regions. BTES has low borehole thermal resistance due to thermally efficient boreholes, high conductivity materials, and adequate backfill or grout. It has a high ground volumetric heat capacity and moderate thermal conductivity to reduce conductive heat loss. Maintaining minimum groundwater flow reduces convective heat losses.

The BTES works in most water-saturated rock types with a wide range of ground thermal conductivity. When it comes to array storage, low ground thermal conductivity lowers direct heat loss. On the other hand, high conductivity makes heat rejection and extraction stronger. Optimization of ground thermal conductivity is determined by the BTES dynamic mode. Low thermal conductivity favors long-term storage, while high conductivity benefits short-term storage and frequent charge–discharge cycles. Nearby surface demand and limited groundwater flow will determine BTES sites [14].

9.6.2 Advantages

The advantages of UTES systems over conventional systems are many:

- Nature has large energy reserves that can be stored and used for heating and cooling.
- Reduced operational expenses and increased long-term profitability (almost 75% lower electricity costs for cooling machinery and plants). The recovery of increased investment costs takes less than five years. Environmentalism, that is, resources are recoverable and pollution-free.
- Using unused boreholes for various uses [15].

9.7 PHASE CHANGE MATERIALS

Molecular latent heat-storing material, PCM, melts to a liquid at high temperatures. PCM retains heat in this endothermic process. As the temperature drops, PCM solidifies. PCM releases heat in an exothermic reaction. A substance transfers TE when it changes from liquid to solid or solid to liquid. PCM applications can be used to construct TES systems for peak cooling or heating needs. The PCM uses latent heat to increase building thermal performance and comfort. PCM maintains temperatures within a range during phase change using latent energy (heat).

9.7.1 PCM Major and Common Applications in Building

Buildings have passive and active PCM. The passive system's structure stores and releases heat. Pumps or fans move heat transfer media in active systems.

Three ways PCMs can heat and cool buildings:

1. PCMs used in building walls;
2. PCMs used in floors and ceilings;
3. Storage heat and ice tanks use PCMs [16].

9.7.2 Methods for Using PCMs in Building

PCMs in building envelopes, like walls and roofs, increase heat storage. This improvement improves energy efficiency and lowers electricity use. PCM has a big role in heat storage during phase transition. In high temperatures, PCM melts and absorbs TE, lowering peak temperature and delaying heat transfer into indoor spaces. In cold weather, solidification releases energy, improving heating and thermal comfort. Day and night, the PCM absorbs heat to cool and release it to heat over 24 hours. Heating and cooling efficiency improve. Building PCMs work during the day, as seen in Figure 9.6.

Some PCMs are unsuitable for building thermal storage. Two factors must be addressed when choosing a building PCM:

1. melting point and
2. latent fusion heat.

Physical and chemical parameters must be met for PCMs. Buildings can use passive or active PCM systems. Their differences are shown in Figure 9.7. Passive thermal storage systems (PTSSs) incorporate PCMs into building materials, removing the need for heat transfer equipment. However, active thermal storage systems (ATSSs) need mechanical or electrical power to operate.

9.7.3 PCM on Wall

PCM wallboards are made from concrete and gypsum with PCMs. Microencapsulation, macroencapsulation, vacuum impregnation, and direct mixing are integrated. To evaluate its application, many experiments and practical tests were done. The PCM–concrete combination's heat storage capacity increased while maintaining thermal conductivity. PCM was scarce and had low thermal conductivity, therefore thermal conductivity varied little.

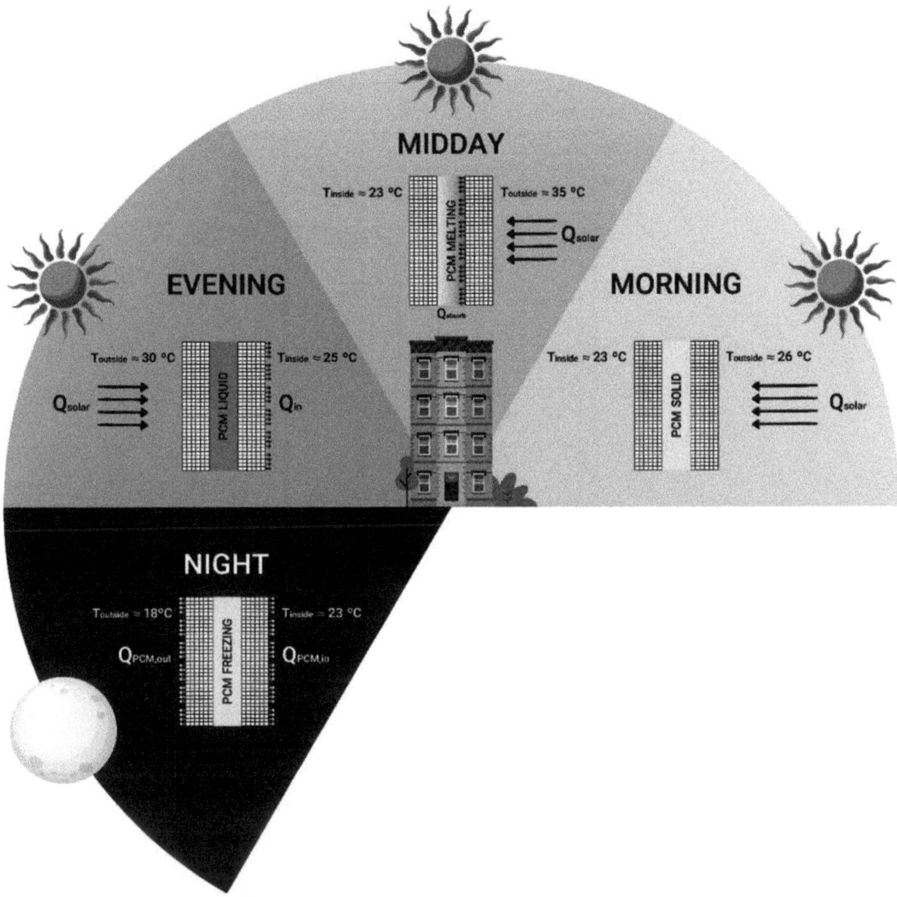

FIGURE 9.6 The operating principle of PCM incorporated into a building, over a typical daytime [17].

A year-long simulation explored PCM wallboard performance versus a room without them. Summer thermal testing was performed on office building PCM wallboards. The aluminum honeycomb wallboards contain 60% 5 mm micro-encapsulated paraffin. Wallboard PCMs reduced peak surface temperatures and temperature fluctuations, improving thermal comfort.

It also delays peak heat fluxes with heat insulation and PCM. The selected salt hydrates appear to considerably affect building passive cooling. A clay brick wall with horizontal hollows held PCM macrocapsules. RT18, commercial organic paraffin, was the experimental PCM. Steel macrocapsules in plastic tubes contained liquid paraffin phase transition material. Paraffin was chosen for its safety, non-corrosivity, and metal compatibility. For minimum and highest peak temperatures, walls with PCMs were one hour behind those without [18].

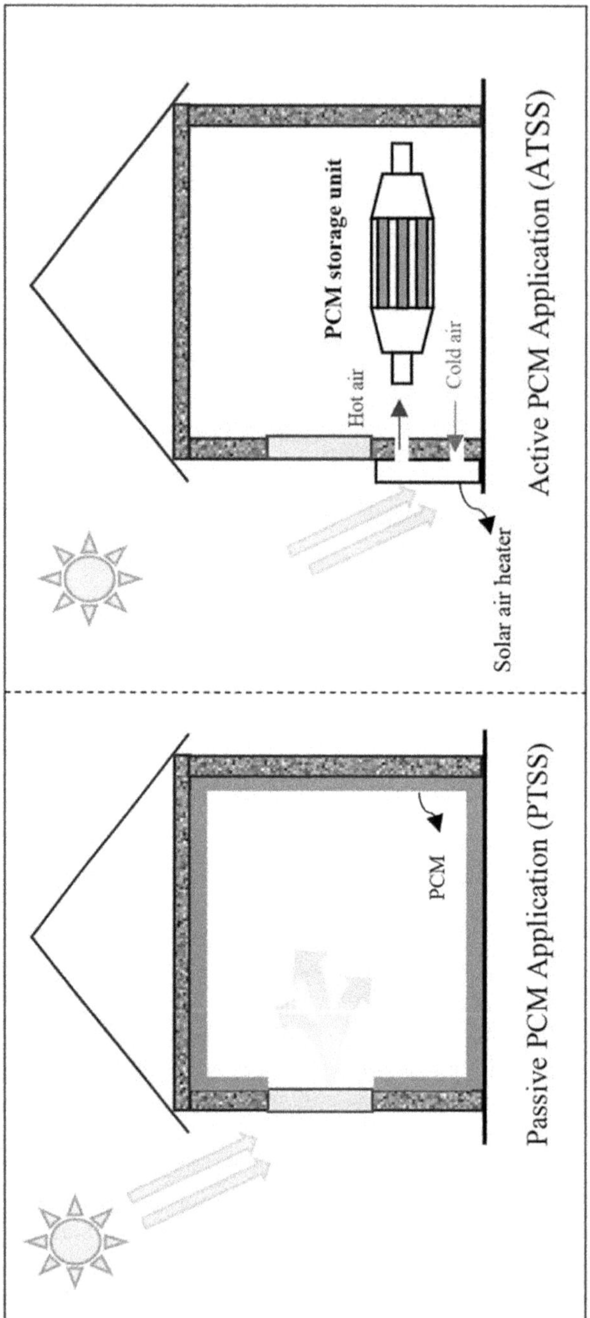

FIGURE 9.7 Passive PCM system versus active PCM system [17].

9.8 PASSIVE TECHNOLOGIES

Low heating and cooling in buildings ensures year-round thermal comfort. Building envelope insulation and storage limit outdoor temperature swings. Thus, indoor air temperatures can be comfortable without heating or cooling. Environmental and internal factors affect indoor air temperature. Air temperature, wind speed, and sun radiation affect external solicitations. Internal solicitations result from heat and loads. A building's envelope must function better to save energy. TES in exterior and partition elements improves building energy efficiency. It may benefit from PCMs. Phase-change latent heat loads allow PCMs to store more energy in the building envelope than sensible materials. Effectively uses solar energy.

Figure 9.8 shows a thermodynamic building. The system's envelope displays the balance between building (or adjacent) energy generation, consumption, and storage during operation. The balance depends on internal and external loads, envelope thermal resistance, and heat capacity. Building elements using PCMs store energy, lowering heating and cooling energy usage passively and sustainably. Energy balancing requires lower energy generation. PCMs permit net-zero energy buildings. As climate, design, orientation, and PCM amount, position, and type affect thermal benefits, these studies require a thorough dynamic modeling of the building's thermal behavior under projected use conditions [19].

In hybrid PV/T systems, water or air heat extraction devices increase solar radiation energy conversion. Sloping roofs and façade solar chimneys improve ventilation (Figure 9.9). A new PV-Trombe wall is shown in Figure 9.10.

The chimney effect and electricity from the PV module heat the space in winter by circulating air through vents. Figure 9.9 shows how the PV-Trombe wall uses the chimney effect to circulate outside air to cool the room and remove heat gain. Cool air is drawn air from the shadow side into the room and passed through the chimney to create the PV-Trombe wall [20].

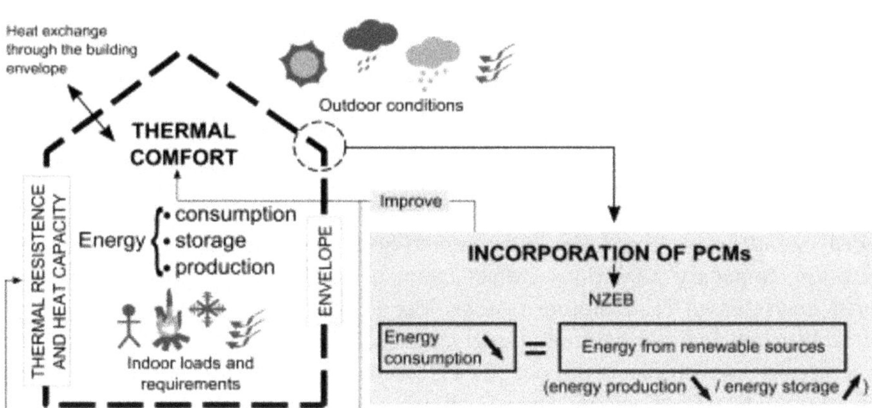

FIGURE 9.8 Net-zero energy building (NZEB) as a thermodynamic system and PCMs' possibility to minimize energy use and improve interior comfort [19].

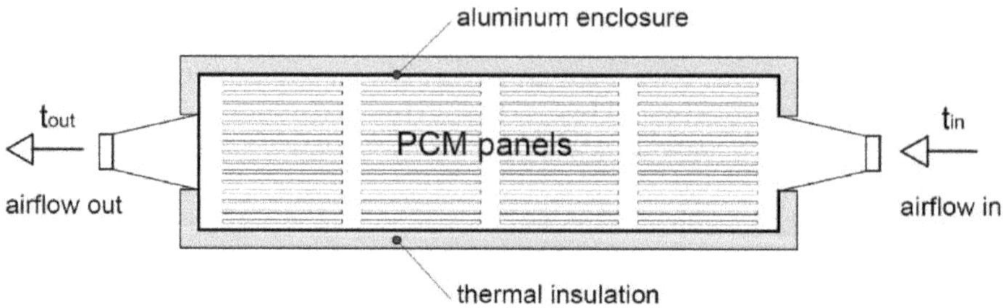

FIGURE 9.9 Ventilation unit with PCM heat storage configuration [22].

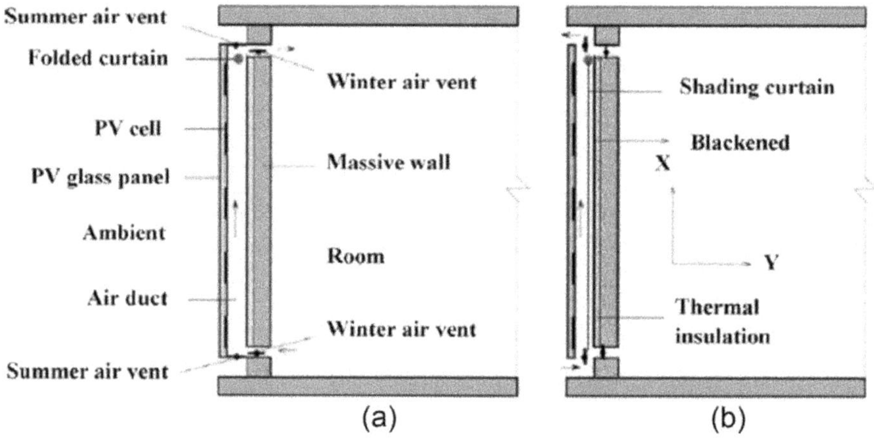

FIGURE 9.10 PV-Trombe wall for (a) winter heating and (b) summer cooling [20].

Figure 9.11 compares LHS and SHS media under the same temperature and building wall thickness. Figure 9.11 indicates that PCM maintains three times more energy than water and six times more than concrete.

9.9 ACTIVE TECHNOLOGIES

A typical method of dense material forming a TES based on SHS and an innovative approach of lightweight material with LHS may be suitable for building integration. Recent decades have seen LHS TES research flourish. During PCM's internal phase transition, boundary conditions collect latent heat. This physical system absorbs, stores, and releases TE. Building systems lose PCM energy potential due to incomplete charging and discharging. Integrating PCM maintains internal temperature during phase transition, a major benefit. PCMs switch phases to store a lot of TE and release more heat of phase transition (LHE) at a small temperature range.

Heat and cooling energy demand is heavily impacted by internal air exchange. A decentralized ventilation system, the integrated ventilation unit benefits interior spaces that can be ventilated solely or with a central plant system. Heat can rise, especially on bright days. PCM stores and releases TE. The ventilation unit's air-PCM

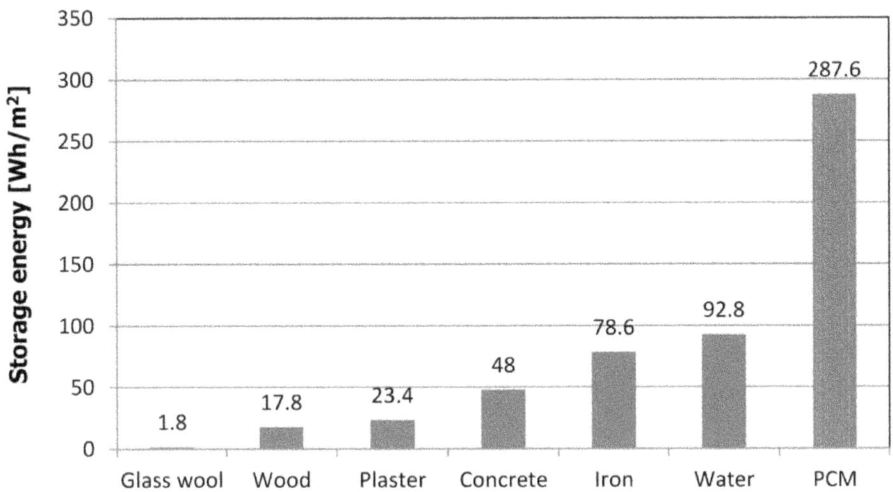

FIGURE 9.11 Comparative construction material storage capacity under comparable temperature and wall thickness conditions [21].

heat exchanger (Figure 9.10) may increase ventilated air thermal stability from exterior to inside.

Value-added designs and heat storage materials increase TES system performance and building sustainability. Figure 9.12 compares energy source availability and building energy efficiency to sustainability criteria, and technological, socio-economic, environmental, and cost-effective techniques link these factors.

9.10 FREE COOLING TECHNOLOGIES

Global warming increases natural ventilation despite higher ambient temperatures and lower urban temperature swings. Without artificial cooling, natural airflow cools buildings at night and improves indoor thermal comfort during the day. Windows and ducts provide nighttime ventilation. Free cooling overcomes night ventilation's shortcomings. Overheating and cooling applications benefit from free cooling systems, especially night ventilation systems. Only the fan and storage unit utilize energy, making these solutions superior to mechanical cooling. Energy consumption and supply can be saved by moving building interior cooling to off-peak periods. Off-peak energy is stored and used during peak demand.

Free cooling has gained popularity due to rising energy demand worldwide. As more sustainable energy sources are found, governments and administrations can reduce human reliance on natural resources. Compared to thermal benefits, free cooling uses little fan power. If free cooling is insufficient, active (mechanical) cooling may save energy. Based on the findings, this consequence is expected. Technologies do not control indoor relative humidity better than free cooling, according to research. Free cooling's energy savings may offset air conditioning's higher starting expenses, according to an economic study. Figure 9.13 shows free cooling basics. The elements affecting building free cooling are shown in Figure 9.14.

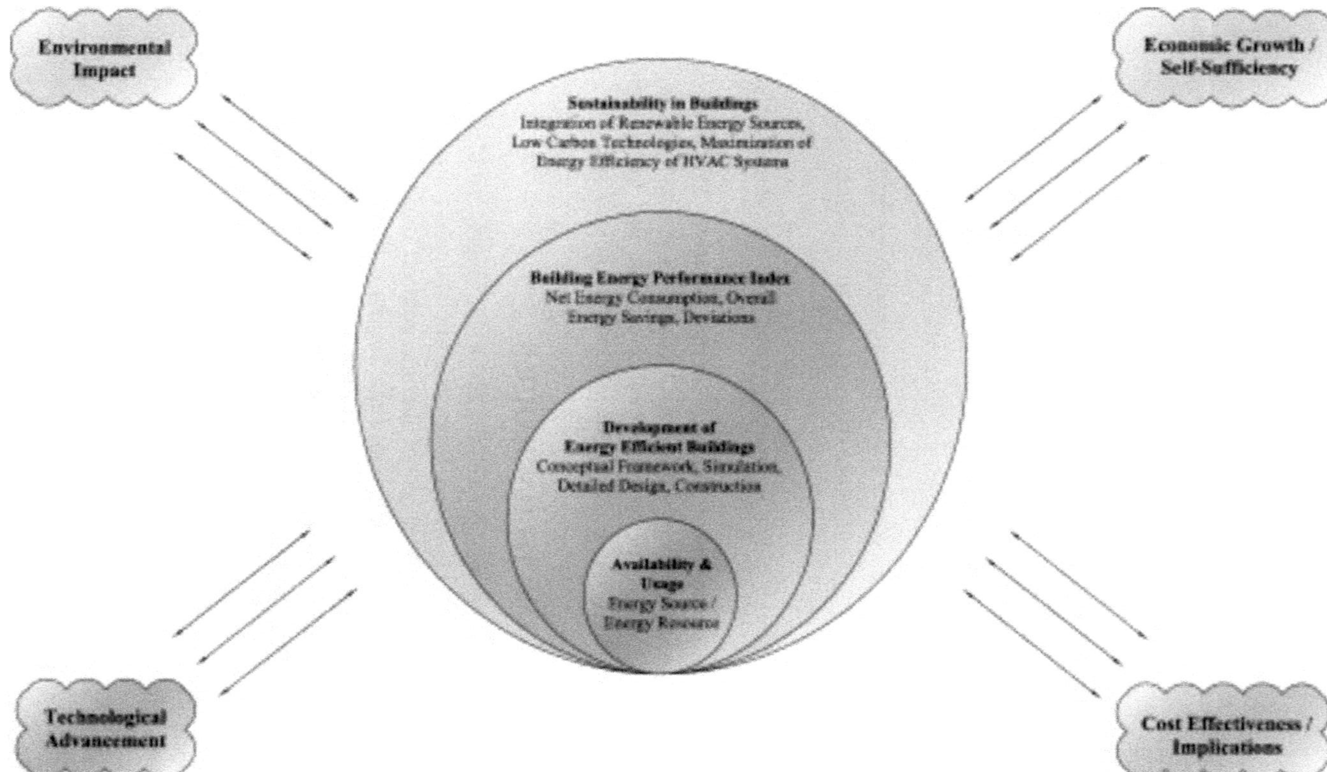

FIGURE 9.12 Sustainability chart for the development of energy efficient buildings [23].

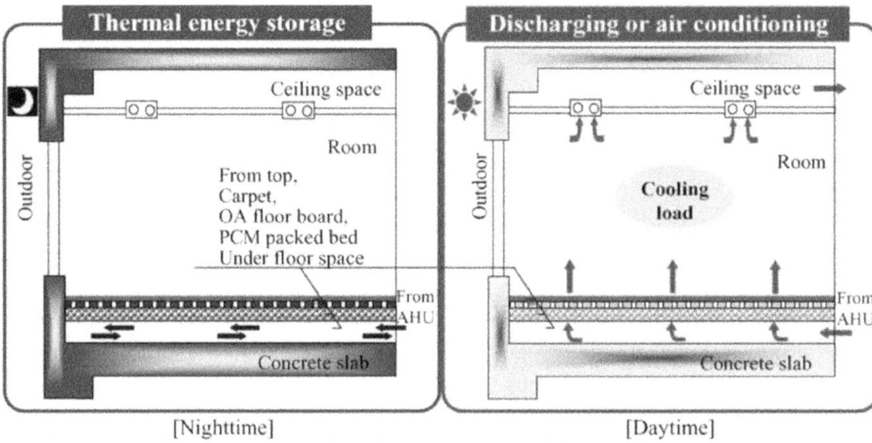

FIGURE 9.13 The concept of free cooling [24].

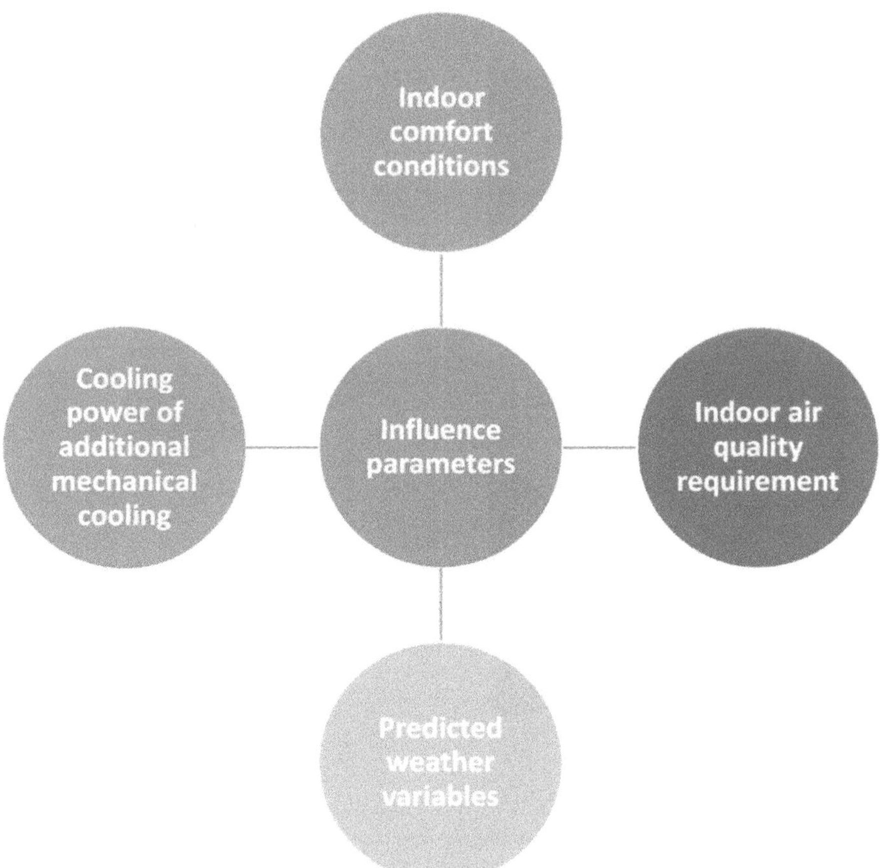

FIGURE 9.14 Affecting parameters on free cooling of building [24].

9.10.1 Economical and Environmental Feasibility of Free Cooling Systems

Figure 9.15 shows the free cooling system's economic viability. About 17% of the free cooling system is storage material (PCM). A viability analysis shows that the PCM storage system (free cooling) costs 9% more than the standard refrigeration system with similar power and has a payback period of 3–4 years. The free cooling system used 9.4 times less electricity than the split-type air conditioner.

Although PCM-based free cooling systems cost more than conventional air conditioning units, they have a shorter payback period and use 10 times less electricity. Free cooling systems drastically cut building CO_2 emissions by saving electricity. Alternative air conditioning systems like PCM-based free cooling can minimize CO_2 emissions by 430 tons/year.

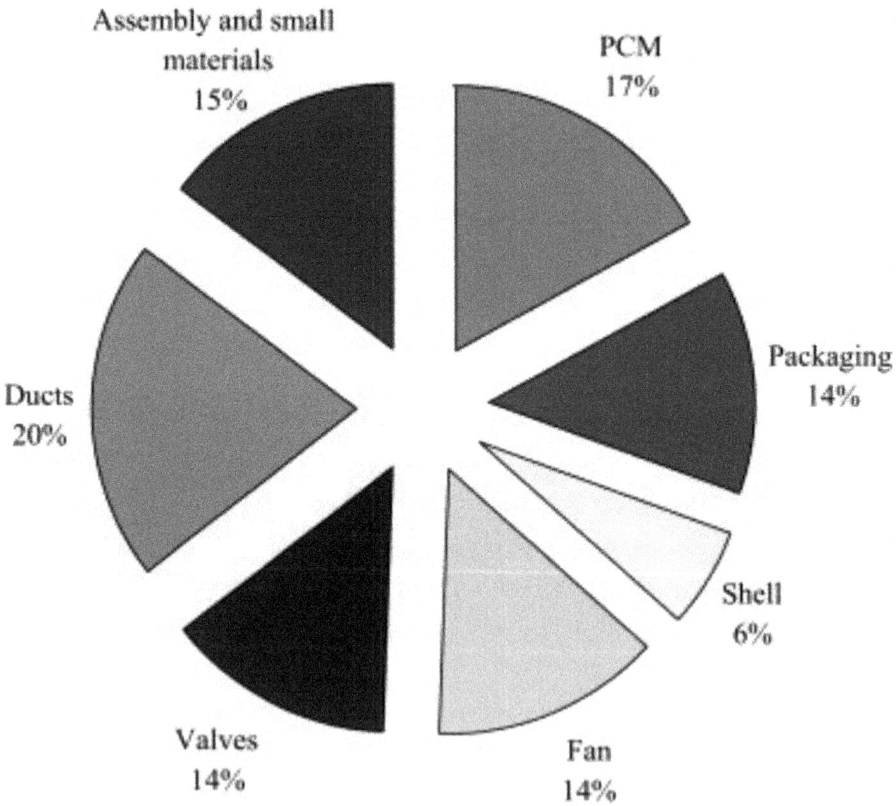

FIGURE 9.15 Percentage of free cooling system cost [25].

REFERENCES

1. Jouhara, H., Żabnieńska-Góra, A., Khordehgah, N., Ahmad, D., & Lipinski, T. (2020). Latent thermal energy storage technologies and applications: A review. *International Journal of Thermofluids*, *5*, 100039.
2. Zhang, S., Ocłoń, P., Klemeš, J. J., Michorczyk, P., Pielichowska, K., & Pielichowski, K. (2022). Renewable energy systems for building heating, cooling and electricity production with thermal energy storage. *Renewable and Sustainable Energy Reviews*, *165*, 112560.
3. Romdhane, S. B., Amamou, A., Khalifa, R. B., Saïd, N. M., Younsi, Z., & Jemni, A. (2020). A review on thermal energy storage using phase change materials in passive building applications. *Journal of Building Engineering*, *32*, 101563.
4. Li, G. (2016). Sensible heat thermal storage energy and exergy performance evaluations. *Renewable and Sustainable Energy Reviews*, *53*, 897–923.
5. De Gracia, A., & Cabeza, L. F. (2015). Phase change materials and thermal energy storage for buildings. *Energy and Buildings*, *103*, 414–419.
6. Alva, G., Lin, Y., & Fang, G. (2018). An overview of thermal energy storage systems. *Energy*, *144*, 341–378.
7. Hassan, F., Jamil, F., Hussain, A., Ali, H. M., Janjua, M. M., Khushnood, S., ... & Li, C. (2022). Recent advancements in latent heat phase change materials and their applications for thermal energy storage and buildings: A state of the art review. *Sustainable Energy Technologies and Assessments*, *49*, 101646.
8. Carnie, J. T., Hardalupas, Y., & Sergis, A. (2024). Decarbonising building heating and cooling: Designing a novel, inter-seasonal latent heat storage system. *Renewable and Sustainable Energy Reviews*, *189*, 113897.
9. Soares, N., Costa, J. J., Gaspar, A. R., & Santos, P. (2013). Review of passive PCM latent heat thermal energy storage systems towards buildings' energy efficiency. *Energy and Buildings*, *59*, 82–103.
10. Zhao, B. C., Li, T. X., Gao, J. C., & Wang, R. Z. (2020). Latent heat thermal storage using salt hydrates for distributed building heating: A multi-level scale-up research. *Renewable and Sustainable Energy Reviews*, *121*, 109712.
11. Yan, C., Shi, W., Li, X., & Zhao, Y. (2016). Optimal design and application of a compound cold storage system combining seasonal ice storage and chilled water storage. *Applied Energy*, *171*, 1–11.
12. Guelpa, E., & Verda, V. (2019). Thermal energy storage in district heating and cooling systems: A review. *Applied Energy*, *252*, 113474.
13. Zhou, X., Xu, Y., Zhang, X., Xu, D., Linghu, Y., Guo, H., ... & Chen, H. (2021). Large scale underground seasonal thermal energy storage in China. *Journal of Energy Storage*, *33*, 102026.
14. Brown, C. S., Kolo, I., Lyden, A., Franken, L., Kerr, N., Marshall-Cross, D., ... & Diamond, J. (2024). Assessing the technical potential for underground thermal energy storage in the UK. *Renewable and Sustainable Energy Reviews*, *199*, 114545.
15. Lee, K. S. (2013). *Underground Thermal Energy Storage*.
16. Cui, Y., Xie, J., Liu, J., Wang, J., & Chen, S. (2017). A review on phase change material application in building. *Advances in Mechanical Engineering*, *9*(6), 1687814017700828.
17. Xiong, Q., Alshehri, H. M., Monfaredi, R., Tayebi, T., Majdoub, F., Hajjar, A., ... & Izadi, M. (2022). Application of phase change material in improving trombe wall efficiency: An up-to-date and comprehensive overview. *Energy and Buildings*, *258*, 111824.

18. Zhou, D., Zhao, C. Y., & Tian, Y. (2012). Review on thermal energy storage with phase change materials (PCMs) in building applications. *Applied Energy*, *92*, 593–605.
19. Soares, N., Costa, J. J., Gaspar, A. R., & Santos, P. (2013). Review of passive PCM latent heat thermal energy storage systems towards buildings' energy efficiency. *Energy and Buildings*, *59*, 82–103.
20. Chel, A., & Kaushik, G. (2018). Renewable energy technologies for sustainable development of energy efficient building. *Alexandria Engineering Journal*, *57*(2), 655–669.
21. Khadiran, T., Hussein, M. Z., Zainal, Z., & Rusli, R. (2016). Advanced energy storage materials for building applications and their thermal performance characterization: A review. *Renewable and Sustainable Energy Reviews*, *57*, 916–928.
22. Junaid, M. F., ur Rehman, Z., Čekon, M., Čurpek, J., Farooq, R., Cui, H., & Khan, I. (2021). Inorganic phase change materials in thermal energy storage: A review on perspectives and technological advances in building applications. *Energy and Buildings*, *252*, 111443.
23. Parameshwaran, R., Kalaiselvam, S., Harikrishnan, S., & Elayaperumal, A. (2012). Sustainable thermal energy storage technologies for buildings: A review. *Renewable and Sustainable Energy Reviews*, *16*(5), 2394–2433.
24. Alizadeh, M., & Sadrameli, S. M. (2016). Development of free cooling based ventilation technology for buildings: Thermal energy storage (TES) unit, performance enhancement techniques and design considerations–A review. *Renewable and Sustainable Energy Reviews*, *58*, 619–645.
25. Waqas, A., & Din, Z. U. (2013). Phase change material (PCM) storage for free cooling of buildings—A review. *Renewable and Sustainable Energy Reviews*, *18*, 607–625.

10 Low-Energy IoT and Appliances

In low-energy buildings, good insulation, heating and cooling systems, and renewable energy (RE) sources reduce primary energy demand. Award-winning architects created a net-zero building with energy-efficient windows, heating, and lighting. Photovoltaic panels and an energy-efficient heat pump on the roof help the building reach its energy targets.

Energy efficiency and RE generation enable zero-energy buildings to utilize solely the energy that they can produce on-site. Sustainable buildings improve occupant comfort and well-being while conserving or replenishing natural resources and ecosystems by considering all environmental impacts, including material use and water contamination.

Smart city data was collected and analyzed using sensors, lighting, and meters. Community utilities, infrastructure, and services benefit from data. Internet of Things (IoT) is widely used, making smart city efforts viable. Meters, appliances, and IoT help power users manage energy. IoT architecture allows real-time sensor data analysis. IoT offers real-time alerts, energy demand estimates, consumption trends, and energy-saving tips. IoT can detect electric motor vibration/temperature concerns before they become problematic. Real-time data is collected, consolidated, and analyzed in smart cities to improve services.

IoT data helps cities with traffic, pollution, and safety. Companies employ IoT for remote monitoring and innovative administration of dispersed tasks. The IoT can help cities reduce pollution, manage traffic, and keep people safe. IoT infrastructure tech helps RE systems. IoT connects smart roofs, solar panels, and rainwater harvesters. IoT data collection helps smart grid providers find service faults faster [1].

In smart settings, IoT can smooth device/appliance interaction without human intervention. In 2050, 170 billion IoT devices are expected to span buildings, city blocks, and cities. Energy efficiency was examined at multiple levels in IoT. Smart cities use IoT. Smart cities conserve and manage energy. Using multiple electrical devices and sensors to collect city data is "smart." Technology connects physical devices to networks for information and services through IoT. City and building energy savings may be possible using IoT. IoT in buildings collects, transmits, stores, and analyzes data.

DOI: 10.1201/9781003752837-10

Recent research shows that IoT and energy efficiency methods have brought academics to building energy profiling. Smart buildings and advanced detection technologies have also made IoT a key aspect of smart cities. IoT applications in smart buildings to enhance energy efficiency and lessen environmental repercussions are prevalent. More research shows that good system communication can cut building energy use. IoT is vital to smart object design and operation in the built environment because of advances in networking, processing, and sensing.

Smart gadgets can compute and communicate, from sensors to appliances to phones. Smart parts let diverse networks establish an IoT base. Smart buildings that integrate IoT tech reduce energy use, environmental impact, RE use, and user adaptability, fostering smart, sustainable communities. Smart buildings can connect and control systems and spaces for optimal integration due to developments in networking infrastructure, wireless technologies, and smart algorithms. Smart buildings are interactive, adaptable, multifunctional, automated, and efficient [2].

10.1 PRINCIPLES OF GREEN IOT

Smart devices and IoT networks help broadcasters. Manage data, devices, and connections sustainably with the green IoT (GIoT). It promotes energy efficiency and eco-friendly design, production, and use with wireless sensor network (WSN), cloud computing (CC), machine-to-machine (M2M), direct current (DC), and advanced metering infrastructure. Increased signal bandwidth enhances data processing and communication in GIoT.

Breakthrough technology like IoT is improving communications and global living. The IoT can enhance global growth. All radio frequency identification (RFID) devices need power. In a world where billions of network devices use energy daily and massive data centers collect sensor data, energy-intensive computing and analytics are needed. Thus, it is vital to analyze energy use in large IoT systems. "GIoT" is a new field that addresses these energy issues. This effort aims to reduce IoT-related CO_2 emissions and boost energy efficiency. The IoT's economic and environmental effects are becoming clearer. As energy use soars and conventional sources decline, GIoT has become more significant to scholars and industry suppliers [3].

In the last decade, several studies on energy-efficient IoT have appeared across various disciplines, motivated by the possible integration of IoT into daily life and the enormous energy savings projected (Figure 10.1).

Energy-efficient, IoT-focused methods use customized simulation environments to evaluate performance. These methods differ in hardware, architecture, metrics, topology, network conditions, and simulation software. There has been no comparative performance evaluation of current approaches. Each method is tested against common hardware, sensor, processing, and networking settings. Without a standard platform, comparing energy use and harvesting efficiency is difficult [4].

The smart world is a time when items can work together to help people. The smart ecosystem has many IoT components. The basic principles of green information and communications technologies (ICTs) and the enabling ICTs for GIoT are studied. The latest advancements and future vision of sensor cloud, a revolutionary paradigm

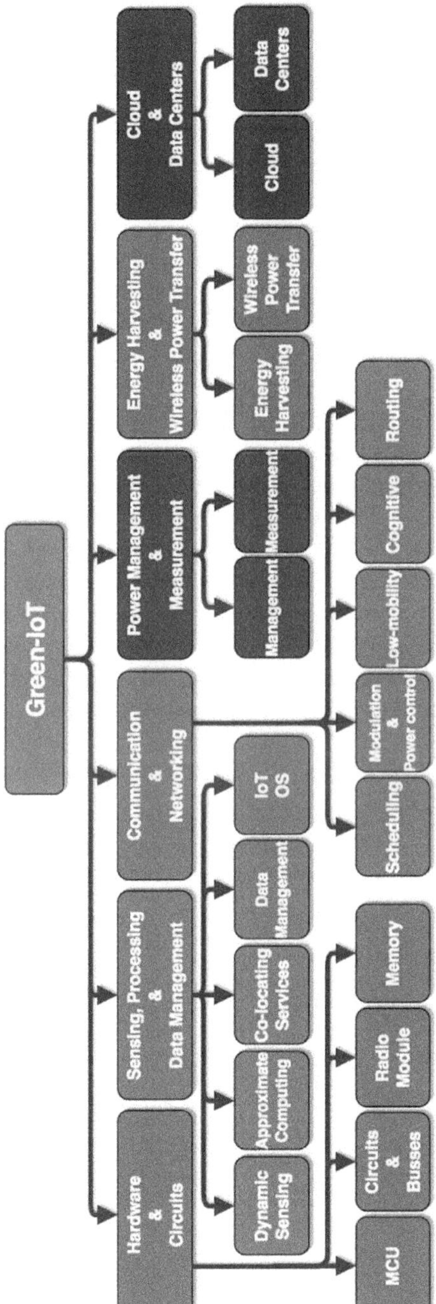

FIGURE 10.1 Green-IoT technical taxonomy [4].

in GIoT, are evaluated and introduced. Classifying IoT components into core categories for easier connecting is summarized below:

1. hardware,
2. middleware, and user-end visualization,
3. actuators,
4. embedded devices, and additional communication components. Physical devices include sensors, actuators, embedded devices, and
5. communication devices.

Middleware is a collection of technologies enabling on-demand storage of data acquired by sensors, processed by embedded systems, and used in data analytics. User end visualization is a set of tools for visualizing and interpreting data across platforms, allowing users to track events triggered by data from multiple sensors [5, 6].

Figure 10.2 shows a modular design based on Figure 10.3, which includes enhancements from turning the highlighted concerns and limits into recommendations for improving building information modeling (BIM) and IoT system interactions. Left side of the figure shows the two main sources of information used to integrate IoT and BIM technologies. The modular blocks allow data extraction and placement into the model. Thus, data is current and consistent. The blocks' independence allows private or open-source commercial solutions to be integrated without affecting other system components.

The two bottom-left sub-blocks include IoT sensors and gadgets. This first one encompasses sensors and all devices that automatically collect field data and execute orders as actuators. Second, IoT middlewares can report sensor data to the Central Control Unit and send commands to sensors/actuators from other blocks. Barriers

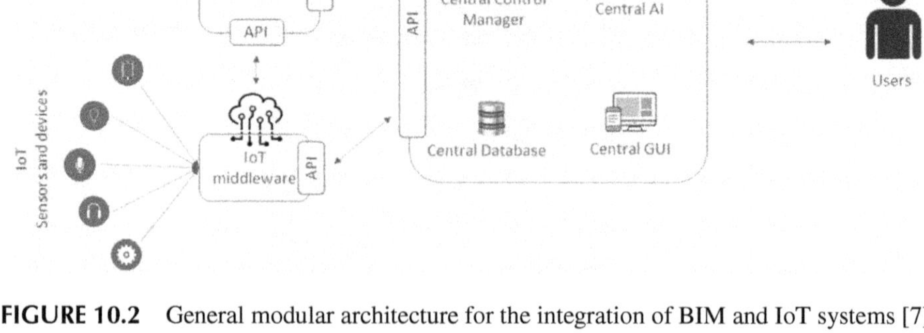

FIGURE 10.2 General modular architecture for the integration of BIM and IoT systems [7].

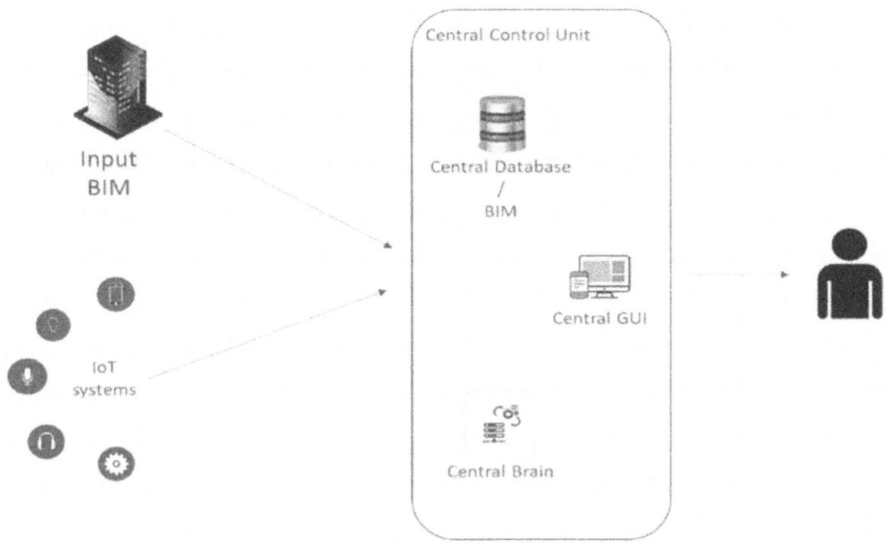

FIGURE 10.3 Simple overview of common information flow [7].

prevent context, protocols, and standards from entering systems and services. Add new IoT sensors without affecting other blocks, and propose new middleware contributions like energy optimization without modifying sensors. Modularity provides security. Each block is separate, hence security problems should be addressed at least once per block to understand risks and update solutions. New technologies like Edge and Fog Computing and blockchain, which speed up and secure IoT communications, can be adopted with block independence. The Central Control Unit receives sensor and BIM data after information flow.

A Central Control Unit contains four primary sub-blocks. The Central Database holds all field-generated static and dynamic data. The Central Control Manager gathers and shares data with all unit blocks. Instead, the central artificial intelligence (AI) sub-block employs AI to extract more inferred data. Finally, the central graphical user interface (GUI) displays collected data and, when present, user-intervening methods. BIM systems are versatile enough to replace the Control Central Unit or one of its subsystems. "Static Data Import Management," the third component of the proposed architecture, covers all Central Control Unit Application Programming Interface (API)-based data import from static solutions. Data is injected from static files through APIs.

Any company in the sector can produce a plugin for this block to enable a new proprietary format due to its independence. Building permissions, authorizations, bills, consumption evaluations, and life cycle assessment data can be imported.

BIM and IoT technologies can be used together because it is flexible. Eco-friendly ideas would come up because of it. IoT experts may be able to use BIM solutions to make sensors or middleware that are better for the environment or use less power.

10.2 IOT FOR GREEN BUILDING MANAGEMENT

Cutting down on energy use without lowering quality of life is a top priority in civil engineering and building. Building costs and energy efficiency must be equal. This challenge needs a multi-objective decision-making tool that incorporates all of them. Many nations strive to cut CO_2 and increase energy. To achieve this goal, energy production must use more RE [8].

Smart city big data management and CC do not take into account how much power and battery drain media streaming uses. Build a medical sensor-based electrocardiogram (ECG) analysis wearable and power-saving transmission power management system. IoT, communication, and network kinds are studied, but not smart city power and battery-efficient media transmission. Provide a smart city IoT-machine learning semantic framework for urban data modeling.

For energy-efficient data transmission with low packet loss, develop defect detection, error repair, and delay tolerance methodologies. However, smart cities ignore power- and battery-efficient media delivery. This technology changes radio access networks to optimize IoT platform end device streaming bandwidth. While transmitting media, smart city IoT sensor devices ignore power and battery drain [9].

10.2.1 DATA MINING

Data mining (DM) is commonly utilized in the green building (GB) sector to swiftly locate and evaluate building automation system (BAS) energy usage facts. DM uses advanced algorithms to find hidden facts in semi-structured and unstructured data. This method outperforms statistical models in showing energy use and occupant choices. Artificial neural networks (ANNs) and support vector machines (SVMs) are common DM methods for complex, non-linear system dynamics.

DM uses building automation system (BAS) data, pre-processes time-series data, resolves missing values, uses k-means clustering, and interprets cluster analysis results to understand energy usage patterns (Figure 10.4). DM has been employed in studies to find energy-saving building methods. By looking at how much power was used, ubiquitous computing was used to find habits that used less energy. They used DM algorithms in a smart home testbed to prove that energy use and occupant behavior are linked in a way that is interactive.

It comprehensively explored the usage of ANN in Heating, Ventilation, and Air Conditioning (HVAC) systems, spanning dynamics modeling, indoor environment optimization, chiller performance, and heating/cooling load prediction. In addition, DM and machine learning improved a building's human behavior database. MLP and supervised learning were utilized to construct an occupant interaction framework.

The unsupervised DM approach K-means clustering was applied to assess school energy use trends.

The research employed DM to show how GB occupant behavior and energy utilization are related. When situations deviate from training parameters, performance diminishes, and neither DM technique can build a model. Application for these programs online is time-consuming [10].

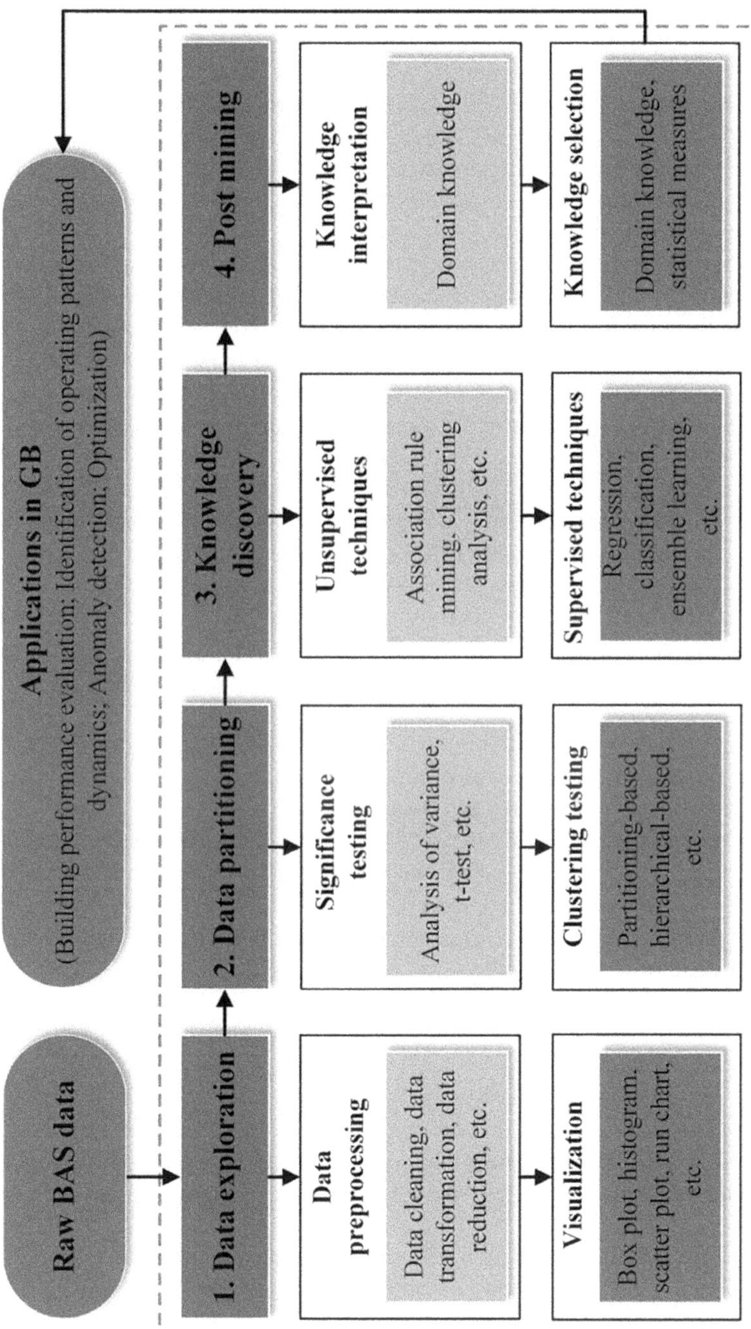

FIGURE 10.4 The GB DM framework [10].

10.2.2 Smart Technologies

Russian, Chinese, Indian, and American CO_2 emissions account for 66% of global emissions. The residential sector consumes 27% of global energy and emits 17% of CO_2. The worldwide construction industry is embracing GBs as climate change worsens. GBs integrate sustainability, energy efficiency, environmental protection, and livability into every aspect of their design and operation. RE, high-efficiency energy-saving solutions, green construction materials, and intelligent building design optimization. Green construction is becoming more popular, creating several chances for intelligent technology.

Smart building technologies are mostly AI extensions. AI has numerous intriguing subfields. To determine supervised learning, training data must have labels. Without example data, reinforcement learning learns by trial and error. Construction companies employ logistic regression, K-nearest neighbors, SVMs, and expert systems for material inspection, regression, classification, and anomaly detection. Reinforcement learning finds the optimum option. A GB uses it for building management, resource scheduling, etc.

Simulating the human brain with multi-layer neural networks is its key component. The strong generalization and real-time object identification of neural networks like you only look once (YOLO) and Single Shot MultiBox Detector (SSD) make them useful for construction site and other building safety monitoring. The networks also aid in building safety management. Architectural applications include market forecasting and emotional input policy research.

The actual operation mechanism of the on-site deployment is mostly divided into the following modules:

1. The front-end information-gathering module captures temperature, sound, and image data utilizing sensors and devices.
2. To ensure data transfer security and efficiency, the gateway module can connect several networks, improve equipment compatibility, translate data between protocols, and encrypt and compress data.
3. *Storage Module:* An edge server or cloud will store the transferred data for later processing and analysis.
4. The decision module processes and evaluates data using a number of ingenious algorithms before formulating and executing the relevant functional decisions.
5. The feedback module facilitates human–computer interaction. Users alter the system's assessment procedures or decision parameters, examine and optimize the outcomes of decision execution, and make other modifications to enhance building utilization management and data analysis.

The trend toward the construction of smart buildings may be seen in large part in the development of AI technology and other technologies, as well as the derivative combination [11].

10.3 IOT-BASED HUMAN ACTIVITY DETECTION AND BUILDING EFFICIENCY

As people in modern societies spend more time inside, buildings are made to be more energy efficient to improve people's health and help meet decarbonization goals. Stakeholders must be informed about the use of new technology in building design and operation. Figure 10.5 shows the reviewed components that can be included to improve building performance for humans. People may be able to solve well-known problems with new buildings if they all know about the technology that can help them. Building designers, operators, managers, and occupants must understand how equipment works to maximize performance, together with technology developers and vendors.

Building performance must consider all of the elements, including users, building systems, and AI, as buildings are turning into Cyber–Physical Social Systems (CPSSs). Consequently, it was found that passive sensors can be used for many things, like keeping track of whole automation systems or individual behaviors like eating habits.

Additionally, active sensors can investigate a variety of subjects, including specific behaviors and the functioning of entire systems. Nonetheless, a major advantage of using energy is its capacity to change into a variety of forms. This is true for nanogenerators, which can be used to make smart clothes and other things that people can use. These kinds of apps would help us learn a lot more about how people's physical traits affect their behavior and how to plan and run buildings to provide user-centered services.

Furthermore, kinect technology has several features, such as depth, skeleton, microphones, and RGB cameras, that are helpful for the human component of building performance. Therefore, the optimal feature (or combination of features) can be used based on the needs of the stakeholders. For instance, some would consider utilizing skeletal scanners, while others might use microphones to operate building

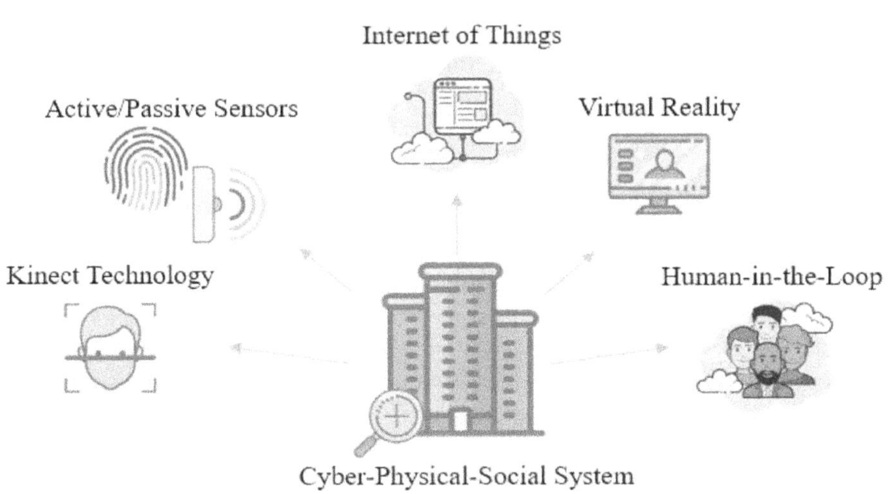

FIGURE 10.5 Transformation of buildings into Cyber–Physical Social Systems technologies [12].

systems. In this instance, such a decision may be influenced by the specific needs of a user or a group of users.

The behavioral sensor methods are beneficial, but combining them is even better. As many sensors as possible can be linked using IoT technology to learn about building performance by connecting common things.

The Human-in-the-Loop (HITL) component allows for low energy use and improved indoor environmental quality based on human needs. This is true no matter what is touted as good for IoT systems. Even fully automated structures can include humans in decision-making. This allows users to share decision-making and have more control over the building's systems. Feedback is vital to adding an HITL component to various buildings. Feedback can come from users to building systems and from building systems to users. Combining human and machine intelligence can yield better results. The best results may be achieved if human capacities are considered while generating feedback, as stakeholders can learn which input is most helpful for each group [12].

Neural architecture search (NAS), which automates deep learning network creation, outperforms manually designed architectures in many computer vision tasks. This differs from deep learning methods that rely heavily on domain expertise. Although inventive and efficient, these searching areas are difficult to design, understand, and train. Reduced inner structure components make models smaller and more efficient. The optimization process is mostly evolutionary. Evolutionary algorithms (EAs) can help solve difficult problems with an easy assessment of various solutions. Evolutionary optimization is a way for Ensemble Deep Learning (EDL) to find the most accurate and efficient deep model setups. For human activity recognition, EDL evolves a deep model, resulting in an easy human activity learning. Therefore, the model's configuration is an individual, and its purpose is performance. Precedent research suggests that EDL improves deep model learning flexibility.

AUTO-convolutional neural network (CNN) employs a genetic algorithm (GA) to find the best topological constraints for CNN-based classification methods, but it is easy to evaluate (though rather computationally costly). Users can build many designs, test their effectiveness, and pick the best one without knowing CNN design or dataset parameter selection [13].

10.3.1 Occupancy Detection

Sensor DM for context-aware modeling, including air quality monitoring, and the estimation of occupancy of indoor facilities, including workplaces, residences, apartments, and labs, utilizing a range of sensors have been the topic of a lot of recent research. For instance, various tests on occupancy detection have been conducted with conventional sensors like light, temperature, humidity, and carbon dioxide (CO_2) gas. A method for recognizing occupancy in a genuine dorm room and a basic test room using CO_2, passive infrared (PIR), and volatile organic compound sensor data. Inferring latent variables from indoor CO_2 levels and proxies for occupancy detection. Tests indicate proxy sensing counts people better than machine learning. To detect occupants, a mass balance equation and interior CO_2 concentration are used in homes and offices.

Statistical and AI technologies like machine learning and pattern recognition have been utilized to estimate or predict occupancy. Uncertainty in neural network models that assess occupancy using CO_2 concentrations rather than variable airflow rate shows that occupancy variation frequency greatly affects estimation accuracy.

As most previous research used living people, occupancy detection systems for elderly people in homes are unsuitable. Humans, plants, and other organic bodies in early decomposition produce CO_2 and volatile organic chemicals such as hexane, toluene, and dimethyl disulfide.

Thus, CO_2 may impact occupancy detection after death. Because oxygen consumption and CO_2 production relied on weight, age, and activity, occupancy was difficult to quantify in prior investigations. Many sensors to improve occupancy detection accuracy waste power and resources for real-time systems' calculations. Due to noise detection, filtering, and vector representations, computer vision-based systems are sensitive to occlusion and light-related brightness.

Wearable location-based technology helps lost elders prevent senile dementia. Power use analysis alerts seniors to an anomaly. This system relies on KEPCO's AMI, which presents scaling concerns. Power use changes a resident's behavior, but the system needs prior knowledge. As power utilization drops below the threshold, the algorithm fails to determine context information like elder occupancy and aliveness.

Due to health risks, such as respiratory difficulties, particulate matter (PM) concentration is a major worry in Korea. Much research has been done to enhance air quality monitoring system PM concentration estimates and determine the causes and effects of growing PM concentrations. It tracks health and absence data from 70 schools to examine how indoor environmental quality variables affect student performance. Absenteeism and indoor environmental quality are unrelated; however, temperature, hygiene, and classroom ventilation rate may affect students' performance and health [14].

Numerous studies suggest that knowing individualized occupant actions could help building management systems minimize energy use, provide context-aware user support, and improve occupants' thermal comfort, acoustic insulation, and other aspects of life. In recent decades, vision-based, PIR, and integrated sensors in wearables and mobile devices have been created to identify human activity. These systems either need extra infrastructure to measure activities or need residents to wear uncomfortable, cumbersome, and intrusive equipment. However, the vast and widespread availability of WiFi infrastructure in buildings makes it possible to employ WiFi signals to achieve occupancy sensing, including occupancy localization, occupancy counting, and occupancy detection. This is predicated on the premise that an occupant's movement can change how WiFi signals propagate, as viewed at receiving terminals. As the IoT expands, billions of WiFi-enabled devices—like smart TVs, sound bars, and thermostats—are on their way to being widely utilized indoors.

Some new channel state information (CSI)-based activity recognition algorithms have been presented; however, traditional classification methods like decision trees or SVMs are still needed. The proposed machine learning framework still requires tedious feature engineering, which entails manually detecting time/frequency domain patterns. Such an extraction may not generate informative information and may neglect some intrinsic temporal connections among the CSI sequences that are pertinent to human activity.

10.3.2 Application in Smart Buildings

Figure 10.6a and b show occupants' receptive responses in two typical interior settings. The office tenant arrived at 8:35 a.m. (Figure 10.6a). The first detection of his walking condition in his cubicle at 11:35 a.m. and the absence of activity from 11:55 a.m. to 12:30 p.m. suggest he had lunch for 35 minutes. Commercial building management systems can cut energy use by regulating lighting and ventilation based on tenant behavior. In contrast to Figure 10.6b, the resident woke up around 9:50 a.m. on weekends. His Saturday afternoons and nights were spent in the living room, possibly watching TV or movies. As illustrated in Figure 10.7, this occupant lounged in

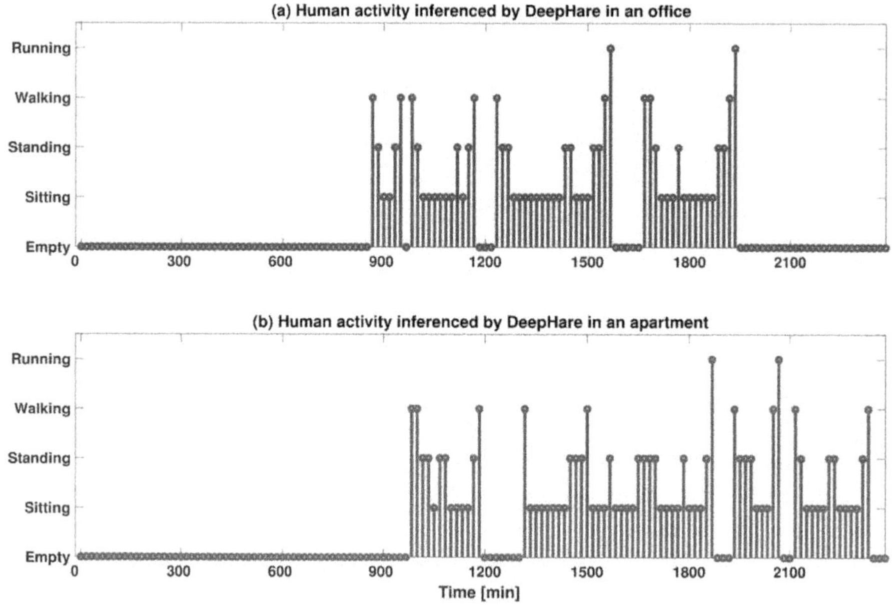

FIGURE 10.6 One-day DeepHare occupancy estimates in typical indoor spaces [15].

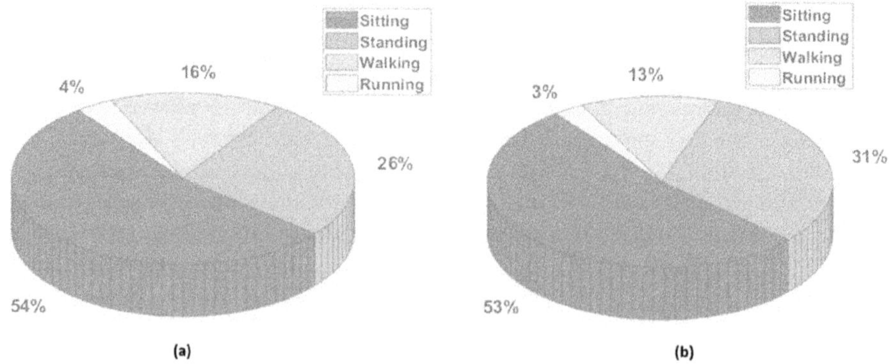

FIGURE 10.7 DeepHare occupancy profile in normal indoor situations. (a) Office occupancy behavior profile on one weekday, (b) living room occupancy behavior profile on one Sunday [15].

the office and flat over 53% of the time. He should exercise more to be healthier and more productive.

10.4 IOT IN SMART CITIES

Wireless tech's rapid growth has changed citizens' daily life. They use modern technologies with cutting-edge technology for daily home tasks. This useful resource is mostly for residents of modern urban areas worldwide.

A complex network of physical objects can connect freely via the internet as the IoT. Insufficient attention has been paid to defining AI. In many situations, AI will do as well as a human. No single definition of a smart city exists.

Smart city concepts differ by city and nation. In India, it possesses a distinct connotation compared to Europe and other regions. This may be perceived as a compact region dedicated to inclusive, sustainable development, establishing a replicable model that acts as a reference for other communities. The research pertains to the revised information system success model. A decision-making process often involves a series of processes, such as recognizing and addressing challenges, examining alternatives, and evaluating potential advantages. This stage pertains to data assimilation.

Users utilize tools to assess decisions. Insufficient information regarding possibilities complicates the user's ability to navigate the situation. Consequently, the caliber of information is paramount. Researchers advise users to verify the service characteristics of their selected option. A user depends on the system's quality, for which they pursue information. User engagement and satisfaction are influenced by the quality of information, systems, and services. "Intention to use" pertains to attitudes, while "usage" relates to behavior. In a procedural context, "use" precedes user pleasure, whereas in a causal one, "use" leads to user contentment.

Similarly, enhanced customer satisfaction with the IoT is expected to foster a more positive disposition, hence increasing IoT utilization. The concept of "net benefit" is challenging to elucidate without contextual understanding. Nonetheless, net benefit may also refer to net gain or net loss. The net benefit can be either individual or national. Therefore, context is essential for accurately assessing net benefit. All of these difficulties relate to the exploitation of information systems that enable the IoT by users. Upon engaging with the government's IoT services, users will utilize IoT-enabled devices, thereby creating substantial data. Consequently, the utilization of IoT devices by spinal cord injuries (SCI) residents presents an opportunity to gather substantial data.

The government might then evaluate this data using appropriate AI techniques. This would allow for quick, reliable decision-making, which would benefit the consumers of these services. Figure 10.8 shows the full cycle.

10.4.1 Smart Mobility and Transportation

Mobility is frequently conflated with transit in intelligent urban environments. Smart mobility and smart transportation are often used interchangeably to denote analogous

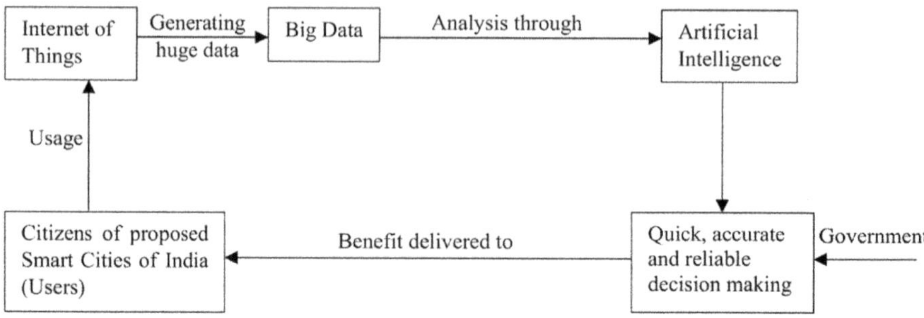

FIGURE 10.8 Flow chart of IoT implementation in smart cities [16].

services, particularly the implementation of IoT networks to enhance human mobility inside smart cities. Mobility and transportation encompass public transit, private automotive commuting, recreational travel, and other modes of mobility, each contributing to pollution to differing extents. Smart mobility and transportation initiatives seek to enhance accessibility and efficiency for smart city inhabitants, irrespective of physical, sensory, or cognitive impairments.

Public spaces like community gardens and parks are vital to urban life and the growing urbanization process. However, in certain rapidly growing urban regions, open spaces have been repurposed over the last 10 years. As open areas become scarce, intelligent mobility can help citizens use, share, and manage them.

The energy and reserve markets have developed demand-responsive, intelligent parking facilities for plug-in electric vehicles (PEVs). The operational dynamics of PEV parking facilities have been analyzed to develop incentive- and price-based demand response schemes. Figure 10.9 illustrates the concept of intelligent velomobility within smart e-bike monitoring systems (SEMS). The SEMS platform acquires real-time historical data from the e-bike via sensors attached to the bicycle. The technology can independently monitor the E-bike's battery without user intervention [17].

10.4.2 Challenges in Smart City Implementation

Smart cities, like other tech advances, have unique issues. Most connections made by RFID and similar technologies are hackable. To create hack-resistant smart communities, improvements are needed. Smart cities' adoption connects all personal gadgets to a single network, raising privacy and security concerns, especially about hacking. In addition, a lot of money is needed to build smart cities. Only careful planning and efficient use of equipment can yield benefits instead of drawbacks. As machinery's uses grow, simple jobs are more at risk. Smart cities may raise unemployment [18].

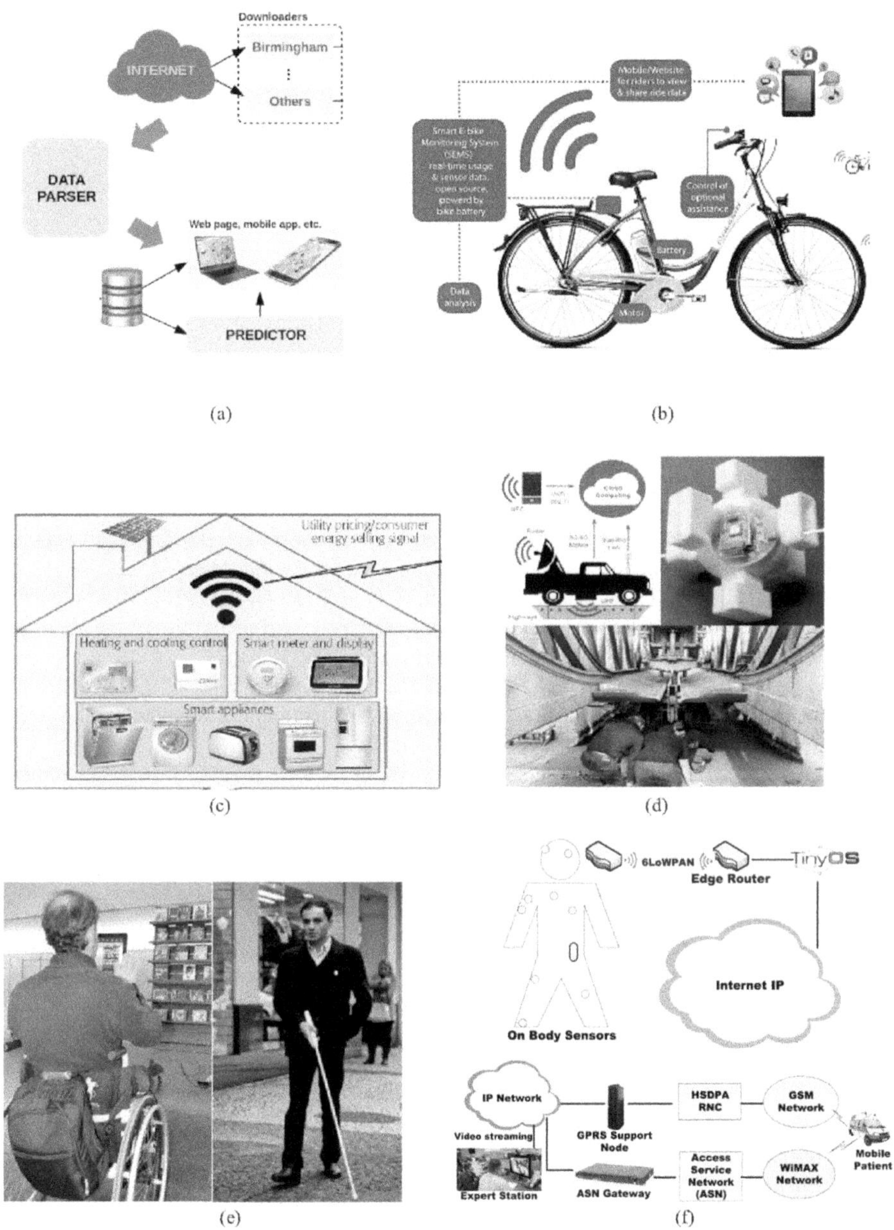

FIGURE 10.9 (a) Data parsing system for intelligent parking solutions. (b) SEMS. (c) Depiction of an IoT-enabled intelligent residence. (d) Representation of advanced civil infrastructure incorporating self-sustaining wireless sensing technology, compatible with IoT integration. (e) A wheelchair user engaging with a self-interactive augmented reality application, accompanied by a demonstration of an intelligent electronic walking cane. (f) Remote surveillance in ubiquitous technology and individualized healthcare [17].

REFERENCES

1. Zhao, W., Chen, J., Hai, T., Mohammed, M. N., Yaseen, Z. M., Yang, X., ... & Xu, Q. (2022). Design of low-energy buildings in densely populated urban areas based on IoT. *Energy Reports*, *8*, 4822–4833.
2. Al-Obaidi, K. M., Hossain, M., Alduais, N. A., Al-Duais, H. S., Omrany, H., & Ghaffarianhoseini, A. (2022). A review of using IoT for energy efficient buildings and cities: a built environment perspective. *Energies*, *15*(16), 5991.
3. Alsharif, M. H., Jahid, A., Kelechi, A. H., & Kannadasan, R. (2023). Green IoT: A review and future research directions. *Symmetry*, *15*(3), 757.
4. Tuysuz, M. F., & Trestian, R. (2020). From serendipity to sustainable green IoT: Technical, industrial and political perspective. *Computer Networks*, *182*, 107469.
5. Yogi, M. K., & Bhavani, K. G. D. (2018). Green IOT: Principles, current trends, future directions. *International Journal*, *6*(3), 156–158.
6. Pandiyan, P., Saravanan, S., Kannadasan, R., Krishnaveni, S., Alsharif, M. H., & Kim, M. K. (2024). A comprehensive review of advancements in green IoT for smart grids: Paving the path to sustainability. *Energy Reports*, *11*, 5504–5531.
7. Malagnino, A., Montanaro, T., Lazoi, M., Sergi, I., Corallo, A., & Patrono, L. (2021). Building Information Modeling and Internet of Things integration for smart and sustainable environments: A review. *Journal of Cleaner Production*, *312*, 127716.
8. Wang, Q., Chen, G., Khishe, M., Ibrahim, B. F., & Rashidi, S. (2023). Multi-objective optimization of IoT-based green building energy system using binary metaheuristic algorithms. *Journal of Building Engineering*, *68*, 106031.
9. Sodhro, A. H., Pirbhulal, S., Luo, Z., & De Albuquerque, V. H. C. (2019). Towards an optimal resource management for IoT based Green and sustainable smart cities. *Journal of Cleaner Production*, *220*, 1167–1179.
10. Qiang, G., Tang, S., Hao, J., Di Sarno, L., Wu, G., & Ren, S. (2023). Building automation systems for energy and comfort management in green buildings: A critical review and future directions. *Renewable and Sustainable Energy Reviews*, *179*, 113301.
11. Yin, S., Wu, J., Zhao, J., Nogueira, M., & Lloret, J. (2024). Green buildings: requirements, features, life cycle, and relevant intelligent technologies. *Internet of Things and Cyber-Physical Systems*, 4, 307–317.
12. Bavaresco, M. V., D'Oca, S., Ghisi, E., & Lamberts, R. (2019). Technological innovations to assess and include the human dimension in the building-performance loop: A review. *Energy and Buildings*, *202*, 109365.
13. Ismail, W. N., Alsalamah, H. A., Hassan, M. M., & Mohamed, E. (2023). AUTO-HAR: An adaptive human activity recognition framework using an automated CNN architecture design. *Heliyon*, *9*(2), 1–20.
14. Jeon, Y., Cho, C., Seo, J., Kwon, K., Park, H., Oh, S., & Chung, I. J. (2018). IoT-based occupancy detection system in indoor residential environments. *Building and Environment*, *132*, 181–204.
15. Zou, H., Zhou, Y., Yang, J., & Spanos, C. J. (2018). Towards occupant activity driven smart buildings via WiFi-enabled IoT devices and deep learning. *Energy and Buildings*, *177*, 12–22.
16. Chatterjee, S., Kar, A. K., & Gupta, M. P. (2018). Success of IoT in smart cities of India: An empirical analysis. *Government Information Quarterly*, *35*(3), 349–361.
17. Alavi, A. H., Jiao, P., Buttlar, W. G., & Lajnef, N. (2018). Internet of Things-enabled smart cities: State-of-the-art and future trends. *Measurement*, *129*, 589–606.
18. Janani, R. P., Renuka, K., & Aruna, A. (2021). IoT in smart cities: A contemporary survey. *Global Transitions Proceedings*, *2*(2), 187–193.

11 Efficient HVAC Systems

Building Heating, Ventilation, and Air Conditioning (HVAC) systems can be made more energy efficient by optimizing the control and operation of the plant and by enhancing the designs of the various mechanical and electrical components that make up the system. In many developed countries, houses use more than 40% of all power. This means that better HVAC systems will save a lot of power. In developed countries, most people stay inside most of the time.

This has led to the design of commercial and residential buildings with occupant thermal comfort in observance, which influences energy demand. Buildings are responsible for 38% of CO_2 emissions and 35% of global energy use, much of which is needed to keep occupants comfortable. Thermal comfort has progressively affected HVAC energy use, contrary to the trend of reducing energy use in buildings and clusters. This means current thermal comfort models need work to be more accurate.

The framework's building design guidelines mostly use a constant temperature setpoint, with comfort zones specified by air temperature limitations (for office buildings, these limits are imposed between 20°C and 26°C; available ranges are 21–23°C). The growing interest in this area among scientists over the past decade is more proof that, despite advancements in energy efficiency and HVAC systems, there is a need for more sophisticated thermal comfort models to account for escalating heating and cooling demands. An important component of building design is providing thermal comfort indoors. Due to the reference standards' steady-state assumption that ignores the human body's thermal adaptability, HVAC system control strategies misestimate occupants' thermal demand and building energy usage.

In Figure 11.1, scatter plots show all the suggested dynamic controls for indoor air relative humidity and temperature set annually. These scatterings are shown in psychrometric diagrams (Figure 11.1). These come from the correlations of the individual comfort indicators that were evaluated and modified for the human body's multi-node model. These scatter plots show annual air temperature and relative humidity pairs.

Energy use and pollution from industry affect society. Despite its efficacy, developing nations are hesitant to adopt renewable energy due to infrastructure costs and technology maturity. Improving end-use devices' energy efficiency can reduce energy consumption.

DOI: 10.1201/9781003752837-11

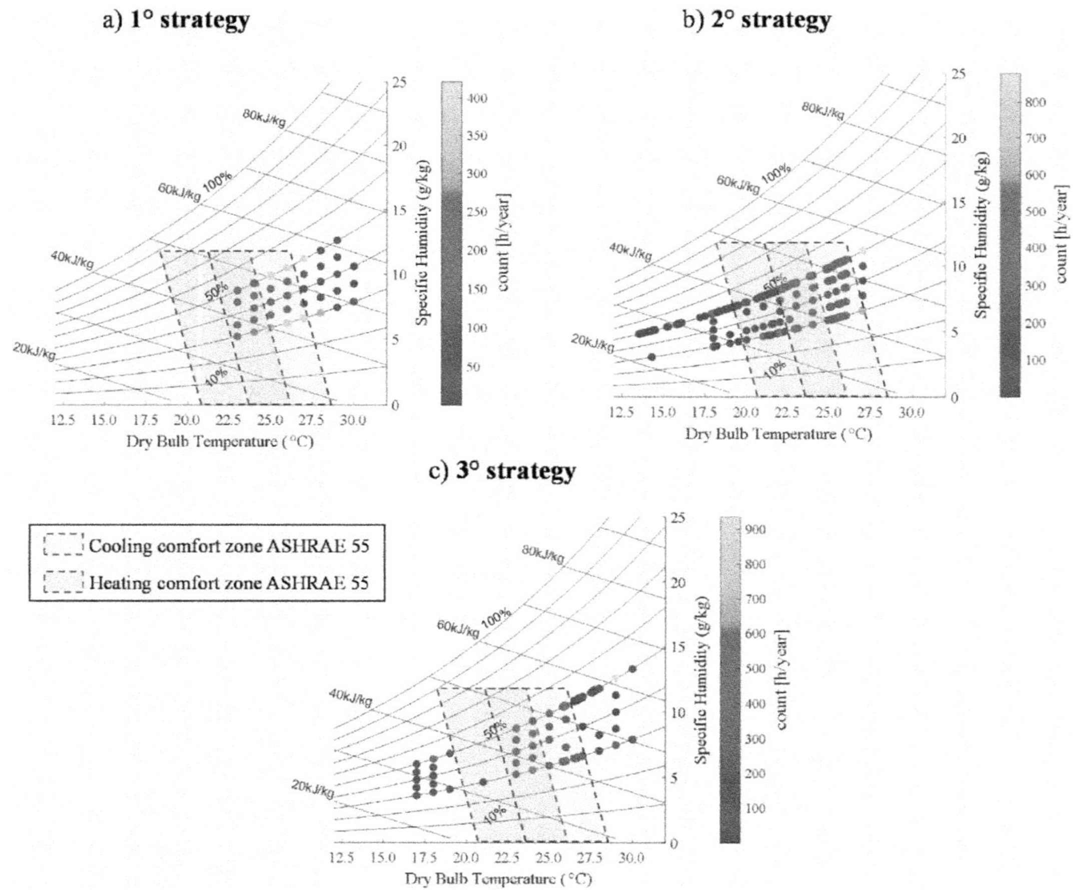

FIGURE 11.1 Annual temperature and humidity scatter plot [1].

Manufacturing, transportation, building automation systems, etc. are all pushing for more efficient energy consumption. Among the various components that make up a building's automation system, the HVAC system is one of the major consumers of energy. HVAC systems are used in both developed and developing countries, increasing their annual energy consumption. Energy efficiency is improved several times in the building process for commercial and residential structures. Building technologies that allow outside air to flow can help achieve a more comfortable temperature. These technologies greatly reduce the load on HVAC systems. HVAC systems are in operation, but can improve their energy efficiency.

Such systems optimize by selecting and redesigning components during manufacture. This technique is better for new public and residential installations because replacing the entire system with such enhanced components would require more expenditure. Control methods include closed-loop decoupled Proportional-Integral (PI) controllers, advanced model-based control, and open-loop heating curves. Traditional control methods like the heating curve and decoupled PI aim for closed-loop stability, and tweaking is based on user knowledge. Model-based control strategies ensure thermal comfort and optimize energy use. To keep control during peak hours, supervisory scheduling uses cost per unit of energy to achieve lean operation.

These new methodologies offer a world of possibilities for better system design that maximizes efficiency relative to cost per unit of energy consumed [2].

11.1 GEOTHERMAL ENERGY INTEGRATED WITH HVAC

The rise in living standards and developing urbanization has made energy use in buildings a global concern. The energy demand for HVAC systems will rise as they are vital for interior thermal comfort. Energy saving will depend on developing energy-efficient, non-fossil-fuel HVAC systems. The HVAC sector favors shallow geothermal for sustainable energy. Ground Source Heat Pumps (GSHPs) have long been thought of as a low-cost method to obtain shallow geothermal energy.

The borehole-free cooling (BFC) technology has been proposed to reduce HVAC energy use for interior cooling. Directly linking the Ground Heat Exchanger (GHX)'s discharge water to the cooling terminal cools the space without a borehole. BFC can considerably improve the GSHP system's yearly cooling effectiveness, and the water that leaves the GHX system matches the interior air parameter. A BFC system that uses ground source absorption heat pumps (GSAHPs) in cold areas. BFC increases summer cooling and interior comfort while avoiding GSAHP system degeneration. BFC uses the subsurface-to-indoor air temperature difference.

Some researchers suggest employing natural energy sources, like cooling tower or GHX water, to intercept heat transfer across building envelopes through water pipes in walls and windows.

Regular double-paned window blinds may absorb 60°C of solar heat. Shallow geothermal energy efficiency should increase with a higher temperature differential. Heat can also be used from the soil–air temperature difference and the envelope temperature gradient. For most of the heating season, the soil-to-air temperature differential reached 20°C. Thus, winter direct usage of shallow geothermal energy seems promising. As building envelopes and fresh air contribute loads, shallow geothermal is best [3].

11.2 BUILDING ORIENTATION

A building's layout on a plane perpendicular to the horizon or the sun's trajectory at an azimuth (AZM) angle ranging from 0° to 135° is provided by BO. The conventional values for N, E, S, and W are 0°, 360°, 90°, 180°, and 270°, respectively. An efficient BO uses winter daylighting and summer insolation shielding. This strategy is more obvious in rectangles. Therefore, several climates have used well-oriented elongated designs. Their shorter façades face the overheated time, preventing summer insolation, while their longer façades face the underheated period, enhancing winter daylighting. The sides of a building that face northwest get twice as much sunshine in the winter as those that face east or west. In the summer, they get at least four times as much. In Figure 11.2, there is a rectangle-shaped building that is in the direction of the summer sun in both the Northern and Southern Hemispheres.

Building orientation (BO), site layout, and design greatly impact energy use. Building alignment with the lot can also alter the "solar envelope," the physical limit of adjoining properties, and how long they get sunshine.

Figure 11.3 shows how three block orientations affect solar envelope size and shape, which is relevant to building and development efficiency. Passive solar absorption, essential for passive solar functioning, is a complex dynamic process involving BO, geometry, glazing ratio, heat-flow channel, inter-zone heat transfer mechanisms, thermal mass, and interaction. The temperature and geography of the area change the amount of passive solar energy that is available, which makes the problem more difficult to solve. One important first step in using passive solar design concepts in building and development is to think about the orientation. A simple and inexpensive way to save energy is to make sure the building is oriented correctly. This must be done during the planning phase so that more advanced passive solar systems can be added later [5].

All day, the module exploits the northern wall's intense solar radiation. The amount of shade a building requires to block summer heat and let in winter sun depends on its orientation and the weather. Shading greatly affects a building's thermal performance. Shading comes from internal sources (blinds, roller shades, drapes) and external ones (overhangs, nearby buildings, trees).

Energy consumption is determined by convection, conduction, and radiation through the roof, floor, walls, and windows. Wind direction and speed dominate convection heat transfer. Higher wind speeds increase module heat loss and gain and windward wall impact, raising energy expenditures. Most of the wind came from the east and west, hitting those two walls harder. Airflow cools the body by dissipating heat and evaporating sweat. To enhance ventilation, build above ground level to catch cooling breezes. Windows should be placed on opposing sides of the room for better entry and departure, with larger inlets and outlets.

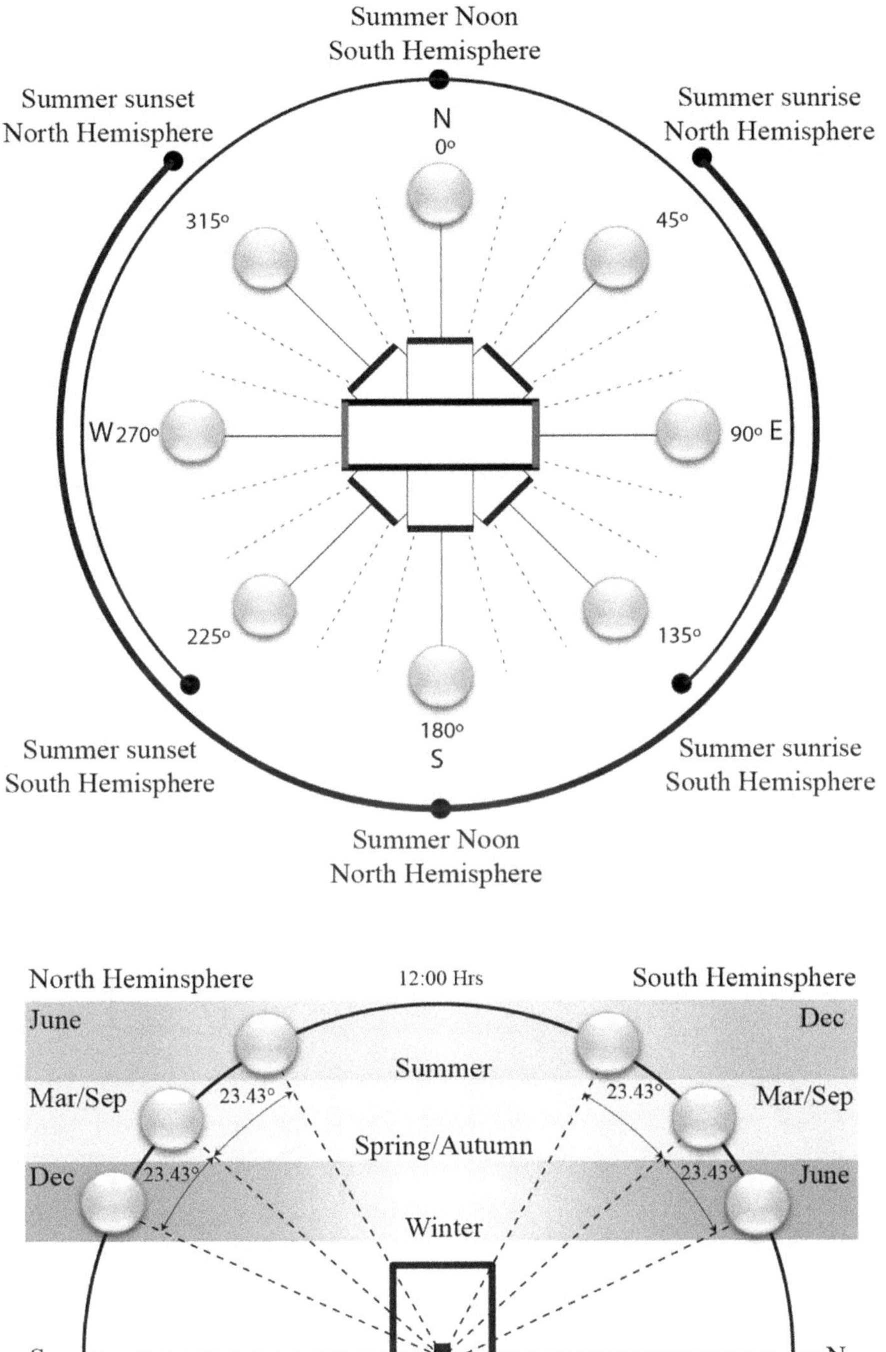

FIGURE 11.2 A two-dimensional plan and section depicting the trajectory of the summer sun as it crosses the Northern and Southern Hemispheres, displaying a rectangular structure [4].

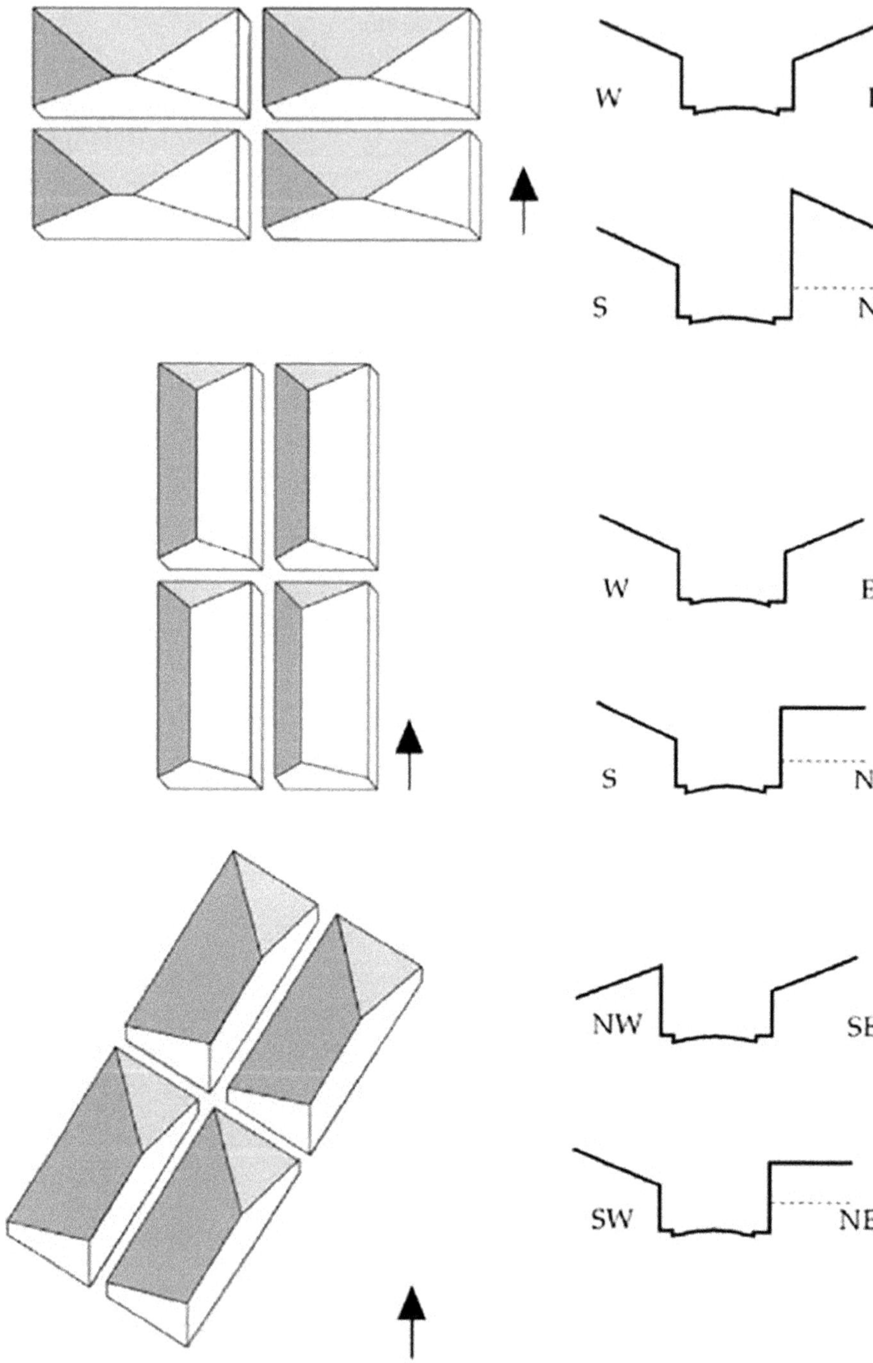

FIGURE 11.3 Effect of block orientation on size and shape of the solar envelope [5].

Energy savings and comfort can be achieved cheaply by orienting. Windows facing east or west lose more heat than they gain in winter, causing the structure to overheat in summer. Sun gain and winter heat loss through windows are reduced in south-facing rooms [6].

11.3 BUILDING ENVELOPE

Comfort and energy savings might result from proper alignment. East- or west-facing buildings are overly hot in summer and frigid in winter. The structure's side windows face the cold wind stream in winter and the hot stream in summer, speeding heat loss and gain. Winter heat loss through windows is low in south-facing buildings [7].

11.3.1 Green Roof

Green wall systems can defy gravity and reduce building needs with permeable, lightweight materials. City streets' building façades have many reflections. The green wall absorbs some acoustic energy when it interacts with something. Installing green wall systems in city streets could considerably reduce the noise increased by building façade reflections. This amplification effect is particularly visible in tight roadways.

For lower green roofs, courtyards and non-exposed façades are best. In enclosed courtyards, green roofs and appropriate roof shapes lower noise by 7.5 dBA. Roof edge low screens only work with absorbing faces. The effect of wall vegetation is greatly affected by assumptions about the reference case's material qualities. Street façade vegetation is best at decreasing road traffic noise when the untreated source canyon is almost stiff. To be most efficient, treat the upper floors while outside. Treating the whole receiving canyon with vegetation has an added benefit above simply the top half. Greening the building envelope has less impact than any measure alone. The best combo seems to be roof edge screens with green roofs or wall plants.

In some situations, like noise from traffic, green roofs and walls do not seem to work. It was determined that green roofs and street canyon wall plants considerably lowered noise regardless of the receiver's location in the courtyard. The acoustical efficiency varies greatly throughout space because of the courtyard's roof screens and wall plants.

Long-distance contributions are rare but can be large downwind. When distant traffic predominates, the courtyard is unaffected by surrounding street canyon façade modifications. The calculations below show the highest noise reduction possible with building envelope measures. Substrate water concentration may greatly alter acoustic absorption.

Water is essential for plant growth. Understanding porous material water dispersion is necessary to predict acoustic absorption. The mechanics of water movement in the emerging substrate and rain runoff need further study to determine long-term noise reduction benefits.

11.3.2 Green Walls

Green walls have been considered for canyon façades, upper or lower. Use reference case assumptions and brickwork properties to estimate insertion loss. Green walls work with 0.82-reflective bricks below 2 dBA. Lower absorption coefficients than bricks may cause low-frequency discrepancies in wall vegetation. For fully vegetated source canyon façades, a 0.95 reflection coefficient yields the best insertion loss of 4.4 dBA.

Noise reduction increases with frequency due to absorption in insertion loss spectra. Sound frequency sensitivity alters with street canyon greening [8].

11.3.3 Indoor Driving Temperature

Thermal behavior can be reflected by several indices that practically reflect the indoor environment. To reduce energy use and temperature discomfort, many studies treat buildings as complex, dynamic systems in which technology, human behavior, and space elements must be examined simultaneously. It was also claimed that studying the actual thermal environment rather than only modeling gives a better understanding of a building's adaptability. Indoor temperature helps assess thermal comfort during a monitoring campaign. For health reasons, the UK Department of Environment and the World Health Organization (WHO) set minimum indoor temperatures of 18°C and 16°C for bedrooms and other places and 21°C and 18°C for living rooms. The harsh reality of cold homes in Europe has led to several empirical studies on indoor temperature in winter. One of the most debated issues in the field is reliably measuring indoor temperature in different construction conditions.

The energy usage of 227 Italian social housing complexes was evaluated using data-driven methods, and standardized performance indices were created to compare comparable structures in different temperature zones. Undertook a 2- to 4-week monitoring effort to directly measure the indoor temperatures of 1600 low-income households and discovered that social and physical factors, including building age, thermal efficiency, and population, might affect temperature. Based on technological and sociological factors, including income, age, and occupant count, predictive analytics algorithms explained 45% of the daily average interior temperature fluctuation. Although time-consuming and laborious, tools that track empirical data on indoor temperature might improve energy plans and help identify what renovations are needed for existing structures [9].

11.3.4 Applied Passive Solar Design Techniques

Building materials are also affected by passive solar design. This includes glazing, insulation, and structural materials. The passive design strategies are as follows:

1. the building envelope;
2. shade from balconies or overhangs;
3. shadowing from surrounding structures;
4. natural ventilation (operable windows);
5. winter solar heating.

Passive solar design was used (1–5). For nighttime-occupied homes, automatic daylighting control is unsuitable.

However, the orientation of the hotel buildings is crucial, since, aside from solar access, it also interacts with

1. views;
2. privacy between the guest room and neighbors;
3. building's functional and social ties.

Simple floor plans allow easy accessibility of diverse activities in the hotel, and also lower square footage of the floor helps increase the efficiency.

Corollary shading elements are key energy-saving components, and shading tactics help balance the temperature of the space during the day in hot summers and temperate climes.

Awnings and other fixed shade solutions can better control solar access when the structure faces south. This concept keeps the current building's rectangular shape, with the longer sides facing north and south. This allows fixed overhangs on south-facing windows to exclude direct sunlight.

The following tactics were utilized and compared:

1. adding shade components;
2. increasing or upgrading outside wall insulation and glazing system;
3. increasing the number of exterior windows [10].

According to Figure 11.4, 62.5% of the 50 research documents studied were about road traffic noise. However, 21 publications (n = 21) emphasized street geometry and façades, not noise. Street, courtyard, and building façade layouts may affect noise nuisance. Multisensory integration approaches have also been studied to improve outdoor sound perception. Positive auditory stimuli, visual features, thermal comfort, and smell variables affected soundscape quality, including acoustic comfort and noise discomfort. Natural noises have improved traffic noise. Traffic noise nuisance has been investigated. Lilac aroma has been added to studies on how individuals perceive road traffic noise. Temperature is thought to affect the soundscape.

Noise reduction, level differences, noise attenuation, and insertion loss are used to study the noise level differences between treated and untreated building envelopes. Throughout the analysis, the phrase "insertion loss," which was most usually used in the gathered studies, was applied consistently [11].

11.3.5 Air Layer Utilized in External Walls

Photovoltaic walls, double-skin façades, green façades, Trombe walls, solar wall systems, wall-based solar chimney systems, and naturally or mechanically ventilated façades have interior air layers. Their structures and activities affect their thermal capabilities. Wall air layers ventilate like mechanical or natural ventilation. The buoyancy effect, caused by the air temperature differential (density difference) between the entrance and outflow, controls airflow in natural ventilation. Mechanically ventilated walls use fans to circulate air.

The Trombe wall captures solar radiation to provide heating and ventilation energy to improve building comfort. Engineer Edward Morse patented the Trombe wall in 1881. The Trombe wall was popular in France thanks to Felix

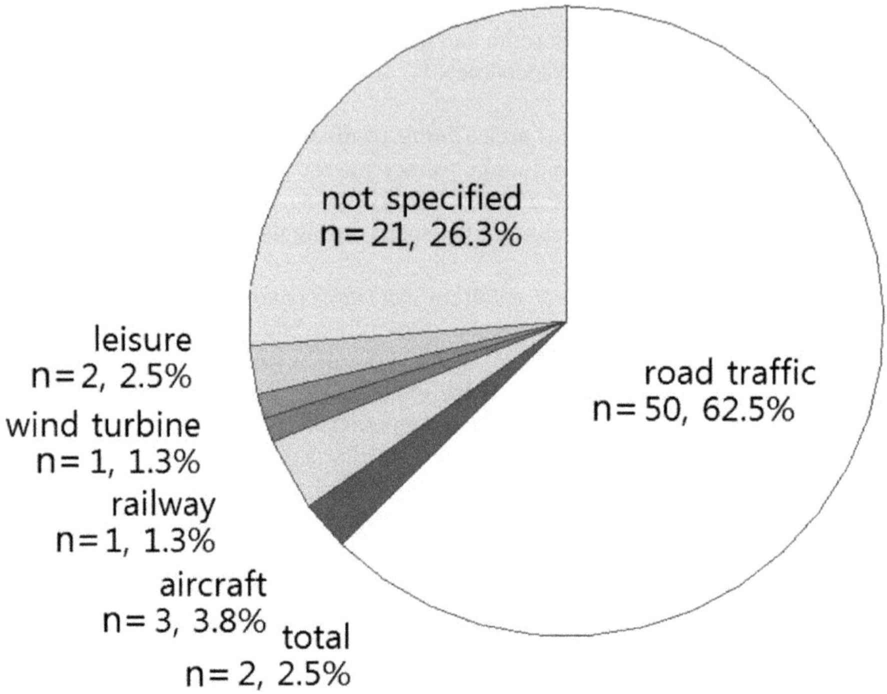

FIGURE 11.4 Noise sources [11].

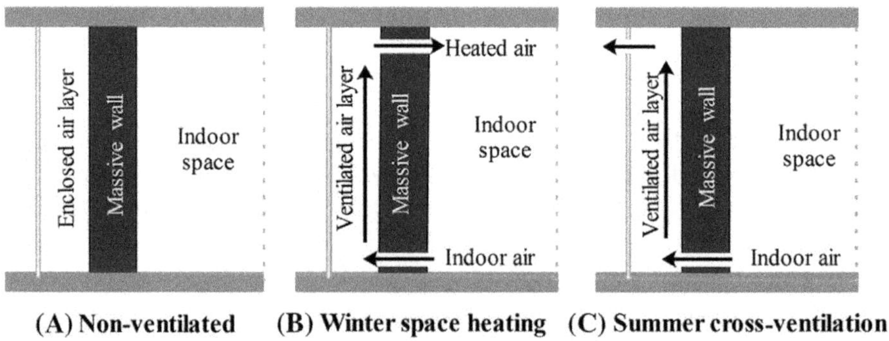

FIGURE 11.5 Different working modes of a Trombe wall [12].

Trombe and Jacque Michel. Trombe walls have a large wall, a glass cover, and an air passage between them [12].

Trombe wall function has been improved through research. Enhancing the air channel, adjusting the entrance and outlet air vents, and adding thermal insulation improve the Trombe wall. The most important step is replacing fixed vents with movable ones and adding dampers to external windows. This alteration helps the conventional Trombe wall adapt to diverse climates, purposes, and seasons (Figure 11.5).

11.4 SELECTION OF CHILLER

In offices and commercial buildings, air conditioning systems typically utilize the most energy. Building cooling needs vary by activity type, occupancy level, and ambient temperature. Multi-chiller systems can operate in extremely changeable conditions as they can be turned on and off as needed. They also utilize less energy than single-chiller systems under partial load. Engineers can inform their design decisions using the comprehensive load profile instead of merely the peak load, which is frequently an atypical scenario, by evaluating the annual variations in thermal load. Dusty HVAC's challenge is cooling buildings and industrial facilities efficiently at a cheap cost. Exterior temperature and partial load ratio (PLR), which is calculated by dividing the chiller's actual cooling capacity by its maximum capacity under the same conditions, are known to affect performance. Chiller selection and operation must be considered in cooling plant design [13].

A chiller plant schematic is shown in Figure 11.6. There are chillers, cooling towers, condensing water pumps, chilled water distribution pumps, storage tanks, and distribution pipes in chiller plants.

Chillers can be in series or parallel. Cooled water distribution pumps, condensing water pumps, cooling tower fans, and chiller compressors use energy.

Data-driven chilled water plant scheduling was developed. Reliable energy use prediction models were built using multi-layer perceptrons. A two-level intelligent method solved the optimization problem due to the model's non-convexity and nonlinearity. Computational results showed the recommended technique saved 14% energy. Optimization of the chiller's on/off state and cooling water temperature can save a chiller plant a lot of electricity [14]. Chillers 03, 04, and 06 have PLR-coefficient of performance (COP) curves and 80% PIs in Figure 11.6, with condenser intake and evaporator output temperature differentials of 3.3°C and 10.8°C. In Figure 11.7, chillers perform best with a PLR of 0.7–0.9. Chiller 06 outperforms Chiller 03 at 3.3°C, more so than Chiller 04, and a similar pattern was seen at 10.8°C. At 10.8°C, Chillers 03 and 04's performance differences are less noticeable than at 3.3°C due to their PIs' overlap, notably in the PLR range of 0.3–0.4. Comparable performances complicate the selection between Chiller 03 and 04. Further research is required to ascertain whether the performance variations of these chillers are attributable to unequal working hours to offer effective maintenance recommendations.

Instead of utilizing the seasonal energy efficiency ratio (SEER), which is used in the USA, the chillers and plant were evaluated using efficiency (kW/ton) and COP, which are used worldwide. Cooler plant efficiency is the kilowatts utilized by chillers, condenser water pumps, and cooling towers to produce one refrigerant tonnage of cooling. Refrigerant tonnage, a metric for the rate of cooling energy, corresponds to the removal of 12,000 Btu of heat per hour. Increased energy consumption for this operation enhances the efficiency of the chiller plant [16].

11.4.1 Cooling Strategies for Air Conditioning in Buildings

Refrigerant tonnage, a measure of cooling energy capacity, equates to the extraction of 12,000 Btu of heat per hour. The augmented energy demand for this activity improves the efficiency of the chiller plant. Solar gains and lights drive most building

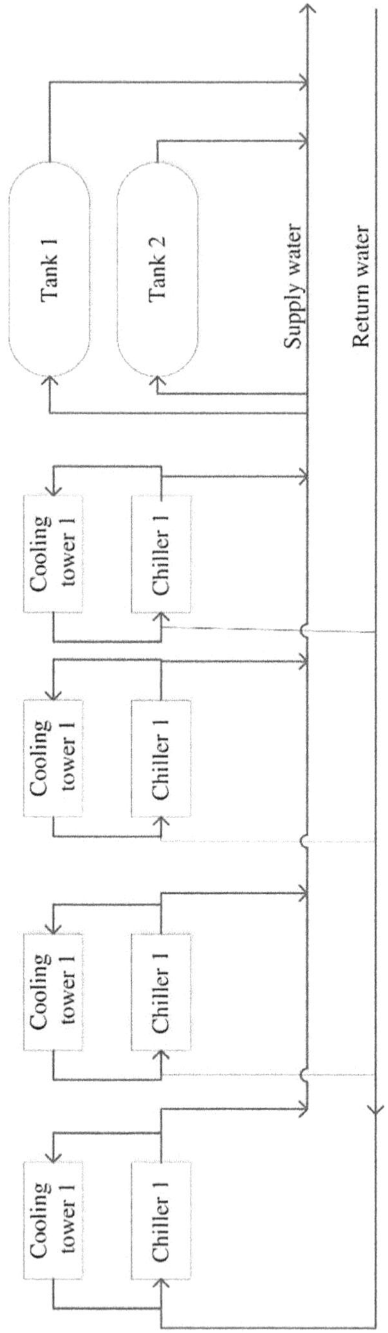

FIGURE 11.6 Schematic diagram of a typical chiller plant [14].

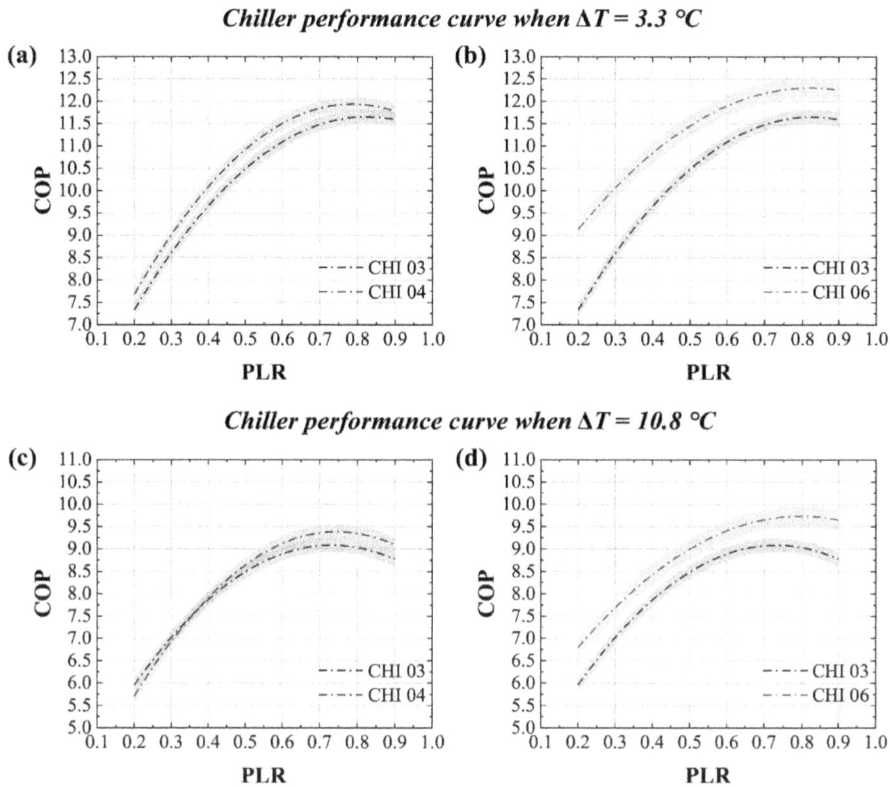

FIGURE 11.7 Chiller PLR-COP curves with PIs (80% prediction): (a) Chiller 03 and Chiller 04 when ΔT = 3.3°C; (b) Chiller 03 and Chiller 06 when ΔT = 3.3°C; (c) Chiller 03 and Chiller 04 when ΔT = 10.8°C; (d) Chiller 03 and Chiller 06 when ΔT = 10.8°C [15].

cooling. Minimizing solar heat gains via insulation and overhangs can boost building energy efficiency and minimize cooling demand. Commercial building natural ventilation solutions for solar gain reduction are shown in Figure 11.8.

Conversely, active cooling methods should be considered if a structure cannot be cooled passively. Electrically powered compression chillers are the most common cooling systems on the market today, with installed systems typically having an average COP between 3.0 and 5.0.

11.4.2 Data-Driven Analyses and Knowledge about the Chillers

Data from each chiller was reviewed to determine how they responded to outside variables like chiller load ratio and external temperature. Maintenance personnel checked and repaired chiller control and protection components (valves, solenoids, relays, sensors, fuses, etc.) before the research. Earth difficulties, blocked valves, damaged solenoids, blown fuses, control relay faults, and faulty wiring can cause anomalous chiller behavior. Analyses followed chiller element repairs. Variations raised the condensing pressure. Peak electricity usage induced by high pressure hampered chiller performance. It harmed air-cooled chiller components, such as fan

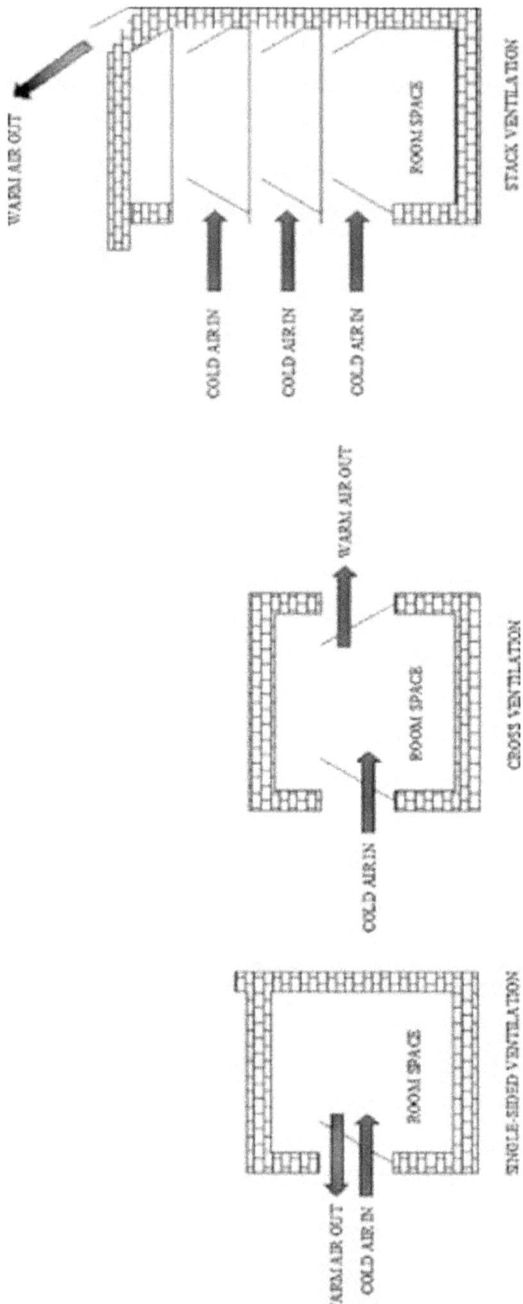

FIGURE 11.8 Common concepts of natural ventilation [17].

bearings, compressors running in hard conditions that affected the compression ratio, and some loading and unloading valve solenoids.

11.5 VARIOUS COMPONENTS IN HVAC SYSTEM

Fresh air, and proper indoor temperatures, and humidity are major areas of HVAC systems. Solar radiation and other heat/humidity sources affect indoor cooling. Figure 11.9 shows how the building outside (passive channel) or HVAC system (active channel) can pull the interior cooling load outdoors. The temperature difference between inside and outside drives heat transfer in the building envelope. The HVAC system must use input power to maintain the interior–exterior temperature differential when the outside temperature is too high.

The second law of thermodynamics makes the heat pump in an HVAC system a heat-work conversion process that can be studied using energy analysis. Interior cooling and outdoor heat exhaustion are heat transfer processes that obey equations; therefore, entransy analysis may investigate them. Heat-work conversion and heat transport are linked by condensing and evaporating temperatures. A greater evaporation temperature reduces heat pump work and rises when entransy dissipation drops. Cooling structures raise temperatures. Here are several HVAC entransy dissipation reduction strategies:

1. *Promote Heat Transfer from Heat Sources to Cooling Sources:* Although heat transfer causes entransy dissipation, transporting heat straight from heat sources to cooling sources reduces it in HVAC systems. Radiant cooling transmits heat directly, unlike air conditioning.
2. *Avoid Cold–Hot Fluid Mixtures:* Entransy dissipation occurs. Thus, minimizing fluid mixtures in water systems and interior conditions like huge buildings with stratified air conditioning will help dissipate energy.
3. Increase thermal resistance by reducing the unpaired heat transfer flow rate ($\xi = -1$). For example, multistage heat pipes or chillers in sequence reduce heat transfer area or entransy dissipation by reducing mismatched flow rate.

HVAC system design is essential for comfort and air quality. Air quality and thermal comfort must be balanced to improve Indoor Air Quality (IAQ) in medical

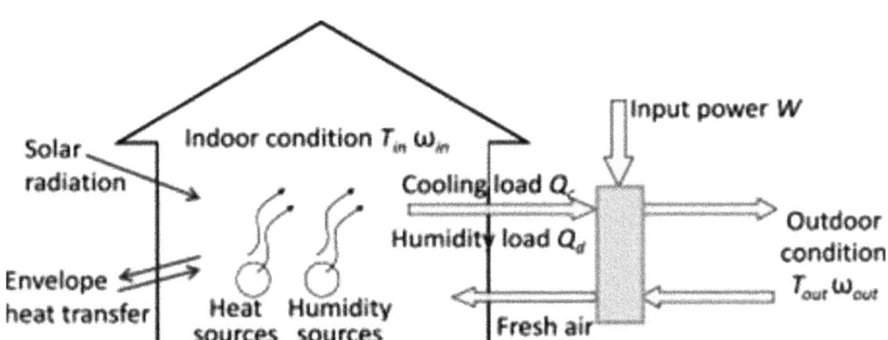

FIGURE 11.9 Concept diagram of an HVAC system [18].

settings. Air conditioners do not always filter and transmit outside air efficiently. Thus, a few minute particles may penetrate the cleaned air and infect the public with COVID-19.

The study also evaluates air conditioners' existing state, future needs and expectations, difficulties, and pandemic compatibility. However, given the coronavirus epidemic, it is important to study the topic from all angles. Reviewing prior research helps the current study determine the importance of key factors that group features with the worldwide coronavirus epidemic's death rate and prevalence and its relationship to air conditioners and air filters. These include how air pollution affects

1. the global coronavirus pandemic, how airborne particles spread the infection, and
2. how medium-temperature and humid regions affect the epidemic [19].

A typical centralized HVAC system with in-building and out-building parts is shown in Figure 11.10. Five categories are as follows:

1. indoor air loops;
2. chilled water loops;
3. refrigerant loops;
4. condenser water loops;
5. outdoor air loops.

Small improvements in operating efficiency can save a lot of energy. The intimate coupling of thousands of rooms and hundreds of cooling coils in a large-scale HVAC system makes it difficult to ensure optimal performance.

Several energy conservation studies have examined the efficiency of specific elements or entire systems. Stoecker's cooling coil model made many empirical assumptions and assumed constant air and water flow. Many experts are interested in how variable speed drive (VSD) pumps can improve energy usage in chilled water loops.

While several researchers investigated fundamental duct and pipe networks, others analyzed all cooling coils with similar cooling loads, which is impractical. With more than two zones, derivative-based approaches would struggle to optimize a two-zone variable air volume (VAV) heating system. After assuming all coil cooling loads were

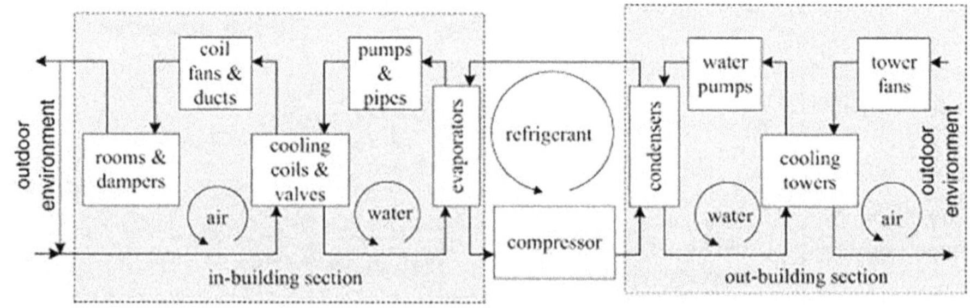

FIGURE 11.10 Schematic of a typical HVAC system [20].

equal, the chilled water rate was optimized as a function of air-water and refrigerant-water coil sizes. Using genuine systems for tests and a summary of generic notions that are hard to apply to other systems for exact control, the entire system was optimized. Integrated Direct Digital Controls (DDCs) improve comfort and energy efficiency in conventional facilities in innovative ways. Large volumes of data strain DDC systems' communication networks and processing capacities.

Despite their popularity, conventional derivative-based optimization algorithms cannot handle mixed-integer nonlinear optimization. Recently developed evolutionary computation approaches enable genetic algorithm implementation in multiple domains. By simplifying the initial optimization problem to reduce independent variables and calculation time, DDC systems are less stressed.

Simulated and actual HVAC systems are used to evaluate computational models of HVAC components in model-based energy analysis. Using component models, the dual-duct air handling unit (AHU) simulates real systems despite its tiny size.

To accurately assess energy supply and product, HVAC energy fluxes must be described. Electricity and mechanical labor are energy supplies. The energy supply should include less energy from secondary flow streams that heat, chill, or humidify or dehumidify primary flow streams.

Energy flow from the source is relevant. Even though the increase in primary flow stream exergy is an exergy product, the relevant energy flow is a service energy flow because it supplies other systems. Figure 11.11 shows the classification principle schematically.

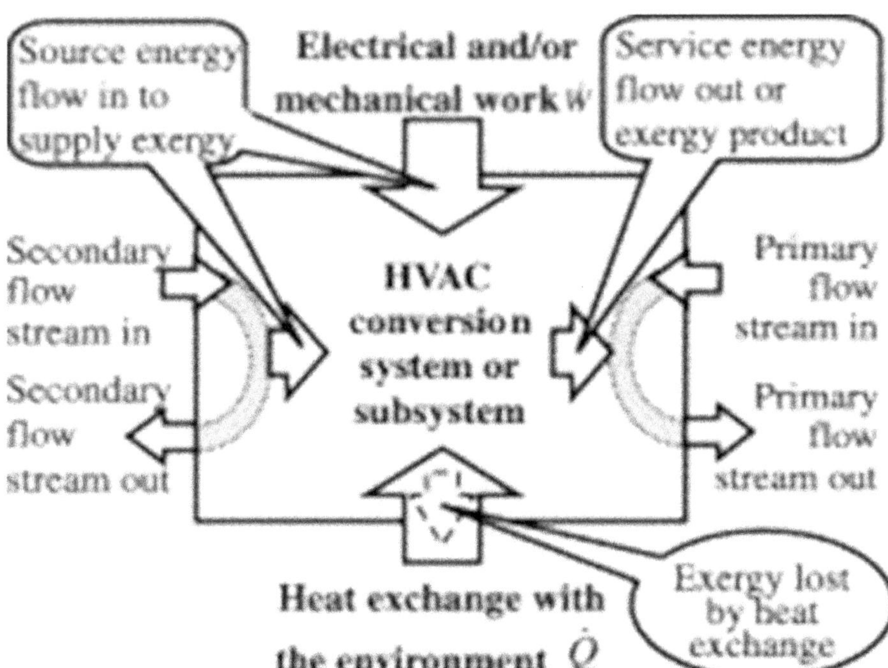

FIGURE 11.11 Schematic of energy flows [21].

REFERENCES

1. Barone, G., Buonomano, A., Forzano, C., Giuzio, G. F., Palombo, A., & Russo, G. (2023). A new thermal comfort model based on physiological parameters for the smart design and control of energy-efficient HVAC systems. *Renewable and Sustainable Energy Reviews*, *173*, 113015.
2. Kirubakaran, V., Sahu, C., Radhakrishnan, T. K., & Sivakumaran, N. (2015). Energy efficient model based algorithm for control of building HVAC systems. *Ecotoxicology and Environmental Safety*, *121*, 236–243.
3. Lyu, W., Li, X., Yan, S., & Jiang, S. (2020). Utilizing shallow geothermal energy to develop an energy efficient HVAC system. *Renewable Energy*, *147*, 672–682.
4. Valladares-Rendón, L. G., Schmid, G., & Lo, S. L. (2017). Review on energy savings by solar control techniques and optimal building orientation for the strategic placement of façade shading systems. *Energy and Buildings*, *140*, 458–479.
5. Morrissey, J., Moore, T., & Horne, R. E. (2011). Affordable passive solar design in a temperate climate: An experiment in residential building orientation. *Renewable Energy*, *36*(2), 568–577.
6. Albatayneh, A., Alterman, D., Page, A., & Moghtaderi, B. (2018). The significance of the orientation on the overall buildings thermal performance-case study in Australia. *Energy Procedia*, *152*, 372–377.
7. Sadineni, S. B., Madala, S., & Boehm, R. F. (2011). Passive building energy savings: A review of building envelope components. *Renewable and Sustainable Energy Reviews*, *15*(8), 3617–3631.
8. Van Renterghem, T., Hornikx, M., Forssen, J., & Botteldooren, D. (2013). The potential of building envelope greening to achieve quietness. *Building and Environment*, *61*, 34–44.
9. Ferrari, S., Zoghi, M., & Cardelli, R. (2025). A monitoring campaign introducing the indoor driving temperature to assess the building envelope performance. *Applied Energy*, *377*, 124650.
10. Sozer, H. (2010). Improving energy efficiency through the design of the building envelope. *Building and Environment*, *45*(12), 2581–2593.
11. Yang, W., & Jeon, J. Y. (2020). Design strategies and elements of building envelope for urban acoustic environment. *Building and Environment*, *182*, 107121.
12. Zhang, T., Tan, Y., Yang, H., & Zhang, X. (2016). The application of air layers in building envelopes: A review. *Applied Energy*, *165*, 707–734.
13. Pargas-Carmona, L. A., Da Silva, J. A., Sant'Anna, Â. M., & Risco-Martín, J. L. (2022). An optimization scheme for chiller selection in cooling plants. *Journal of Building Engineering*, *49*, 104066.
14. Wei, X., Xu, G., & Kusiak, A. (2014). Modeling and optimization of a chiller plant. *Energy*, *73*, 898–907.
15. Wu, S., Yang, P., Chen, G., & Wang, Z. (2025). Evaluating seasonal chiller performance using operational data. *Applied Energy*, *377*, 124377.
16. Abd Rahman, N. M., Lim, C. H., & Fazlizan, A. (2021). Optimizing the energy saving potential of public hospital through a systematic approach for green building certification in Malaysia. *Journal of Building Engineering*, *43*, 103088.
17. Hughes, B. R., Chaudhry, H. N., & Ghani, S. A. (2011). A review of sustainable cooling technologies in buildings. *Renewable and Sustainable Energy Reviews*, *15*(6), 3112–3120.
18. Zhang, L., Liu, X., & Jiang, Y. (2013). Application of entransy in the analysis of HVAC systems in buildings. *Energy*, *53*, 332–342.

19. Elsaid, A. M., Mohamed, H. A., Abdelaziz, G. B., & Ahmed, M. S. (2021). A critical review of heating, ventilation, and air conditioning (HVAC) systems within the context of a global SARS-CoV-2 epidemic. *Process Safety and Environmental Protection*, *155*, 230–261.
20. Lu, L., Cai, W., Xie, L., Li, S., & Soh, Y. C. (2005). HVAC system optimization—in-building section. *Energy and Buildings*, *37*(1), 11–22.
21. Chengqin, R., Nianping, L., & Guangfa, T. (2002). Principles of exergy analysis in HVAC and evaluation of evaporative cooling schemes. *Building and Environment*, *37*(11), 1045–1055.

12 Design of Green Buildings

Industrial resources are overused due to global economic growth. This increases greenhouse gas emissions, environmental damage, and resource loss. Global concerns include climate change, energy issues, and environmental degradation. Construction is a major sector and buildings increase greenhouse gas emissions and environmental impact. Buildings account for 40% of global energy use, whereas 30% is attributed to greenhouse gas emissions. "Building emissions" denotes the resource utilization and greenhouse gas emissions over a building's lifecycle.

Many diseases and serious health issues can result from building construction materials' discharge of toxins into indoor and outdoor surroundings. Consequently, lowering greenhouse gas emissions and building energy use is a major concern in the construction industry. In constructing, renovating, and other facilities, ecological environmental designs are used. Environmental preservation, waste and pollution reduction, worker productivity, residents' health, and energy efficiency are achieved. Green building costs are controversial despite reducing energy use and pollution.

Green supply chain participants prioritize realistic and feasible private profits. Green building value is not reflected by slow development. Thus, promoting green buildings is essential to the construction industry's future. Green buildings are essential to construction sector sustainability [1].

Using sustainable principles, Green Hemp Bast (HB) designs minimize energy and pollutants throughout the building life cycle. Green HBs consume less energy than normal buildings; hence, energy efficiency is the main goal. Green architecture may save energy costs by 30–80% and improve worker productivity and well-being. Using eco-friendly materials, installing green roofs to boost interior thermal comfort and reduce Heating, Ventilation, and Air Conditioning (HVAC) system demand, and strategically placing windows to optimize natural light are some practical alternatives. Using green design construction methods and materials during construction may increase project costs by 10–20% or running costs, notwithstanding the benefits. In new and rebuilt architecture, green design is often used. Use Green Building Rating Systems (GBRS) to assess building performance. The building's design, construction, use, and maintenance are discussed. Green construction criteria are usually listed in a building rating system to assign scores to factors. GBRS commonly fall into three categories:

DOI: 10.1201/9781003752837-12

1. indoor environmental quality,
2. energy, and
3. material [2].

12.1 NEED FOR BUILDING INFORMATION MODELING

Building information modeling (BIM) allows facility managers to access, review, and process building data in a digitalized three-dimensional (3D) environment, improving Operation and Maintenance (O&M). The rapid advancements in BIM require researchers and industry specialists to have a current knowledge of BIM application and research in facility O&M.

BIM adoption O&M principles, including data requirements, inefficient areas, and process improvements, need further survey research. These arguments require an enhanced life cycle cost analysis approach and more research on innovative systems' return on investment to demonstrate BIM–O&M's value [3].

In a digital symbolic representation, the connection between symbols and implementation techniques is reversed. For example, in text processing, a character is not drawn using the strokes that comprise the alphabet, but is inserted directly as a Unicode symbol. The standard process for adding a wall or other building feature to BIM is as follows:

1. Choose the proper symbol to represent the type of wall from a menu or palette. Among other intrinsic features, the symbol's several layers of materials create a stable cross section.
2. Specify the location, length, and other permissible features of a wall type instance. The following factors could cause some confusion at this point:
    a. The task is usually performed in a display that resembles an analog representation, such as a floor plan.
    b. Users usually "draw" the wall's geometry and location by clicking on the points where the axis endpoints should be.
    c. The axis is shown for a brief moment before a fully detailed drawing of the wall instance appears, very similar to what one would draw on paper.

BIM's use of interfaces (views) that mimic familiar analog representations results in a floor plan that looks as though a wall has been added. Importantly, the final detailed drawing of the wall instance is a product of the type that was chosen. Instead of using the type description to deduce the type, BIM uses it to create this graphical appearance.

BIM contains large libraries of predefined symbols to help with this procedure, much like text processing software has enormous catalogs of Unicode characters. The BIM symbols come in a variety of abstraction levels; they can represent a generic external wall, a general wall, a specific kind of external wall, or even a wall manufactured by a specific company. The basic interactions between symbol sorts generate hierarchies that facilitate abstraction and decision deferral. Early on in the design process, a wall might be generic; later on, it might be specific. Therefore, rather than being imprecise, generic types are a fuzzy sum of their subtypes. Moreover, if a selected wall type presents issues, the symbol hierarchies allow for a visual reassessment of prior, more abstract type selections before progressing to other subtypes.

BIM activities can be broken down into three distinct areas, each with two subfields: participants and results. These areas are Technology, Process, and Policy (TPP) (Figure 12.1). Applying scientific information in real life is what technology is all about. Software, gear, tools, and networking systems made by many tech companies help the Architecture, Engineering, Construction, and Operations (AECO) sectors work better, make more money, and be more efficient. If you want to help people make choices, a policy is a list of written rules or ideas. Within the policy field, a group of people work together to train practitioners, do study, share benefits and risks, and keep business conflicts to a minimum. A method is "a structured ordering of work activities across time and space, with a clear beginning, end, and inputs and outputs that are easy to see." People who

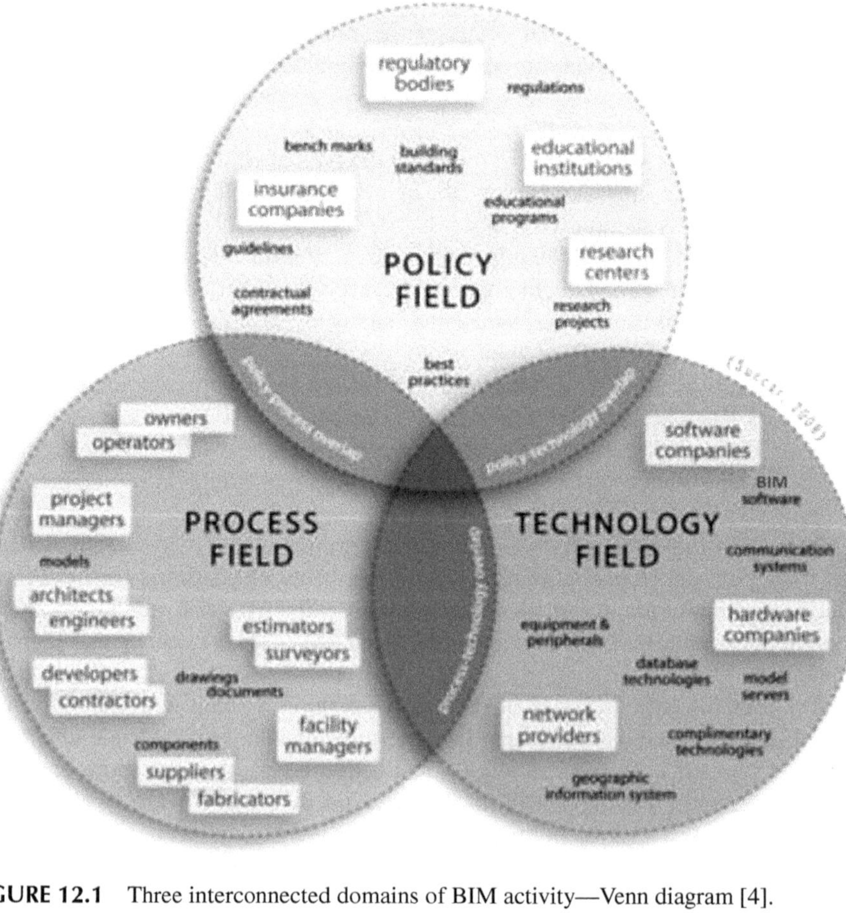

FIGURE 12.1 Three interconnected domains of BIM activity—Venn diagram [4].

buy, plan, build, modify, use, control, and maintain structures work in the process field.

While BIM does adhere to traditional geometric displays, it is not limited to that domain. Symbols have several properties beyond their geometric qualities. These include properties related to their materials and, somewhat redundantly, different performance attributes like acoustic and thermal insulation. Some perspectives, like schedules, make nongeometric properties explicit and represented alphanumerically, whereas geometric views make them implicit in the visual appearance of symbols. Instead of working on separate representations and periodically syncing them, all disciplines work together on the same shared model in BIM. Even when they focus on different traits or symbols, they can still access the same data that shows how a design or building is currently doing [4].

Advantages and Disadvantages of Using BIM: In a nutshell, here are some ways in which the AEC industry might benefit from BIM:

1. *Improve and Optimize Planning:* With a BIM method, decision-makers can evaluate digital site plans and choose the optimal one from among several possibilities for project planning.
2. *Foster Greater Imagination in the Design Process:* 3D models provide more room for detail and imagination than traditional 2D drawings.
3. *Minimize Rework and Enhance Quality:* BIM technologies provide conflict detection, which helps find faults that might be missed during design. This prevents unnecessary rework and improves the quality of work done throughout the building process.
4. Reducing errors and rework saves time, which means the construction duration can be shortened. Therefore, the total amount of time needed to construct can be cut in half.
5. *Reduced Overall Expenses:* BIM enables better methods for design, construction, operation, and maintenance. So, it is possible to lower the conventional expenses of money, materials, and people while maintaining economic viability.
6. Architecture, civil engineering, facility management, mechanical, electrical, and plumbing (MEP) engineering, and other fields may all be involved in a single project, so it is important to foster collaboration between these teams.
7. *Master Full Lifecycle Management and Multidimensional Visualization:* With BIM, Architecture, Engineering, and Construction (AEC) projects can be managed digitally from start to finish. Stakeholders can see the big picture at every stage to multidimensional visualization, which also provides a foundation for sustainable management.

Furthermore, BIM implementation could be hindered by a multitude of issues. Some of these challenges stem from economic factors, social mores, technological limitations, and educational factors. The primary technical hurdles are the technical requirements for supporting hardware and the complexity of learning BIM-based applications. Functioning as advanced BIM tools such as Autodesk Revit, Bentley,

Tekla, and ArchiCAD, the program centers on the building design and construction process. For many seasoned professionals in the AEC field, the learning curve for these tools may be too steep. Interoperability between BIM software could be a technological challenge, as BIM relies on models that are accurate and consistent over their entire lifespan. Consequently, there may be problems with information transmission and sharing due to the software's varied system structures and data formats.

Investments in BIM can be costly at the outset due to the high price of software, hardware, and technician salaries. We are facing a significant obstacle here. Big AEC companies or groups that have used BIM before might not have any issues with the expense.Nevertheless, BIM is not cheap, especially for Small and Medium-sized Enterprises (SMEs).

Issues of morality also provide significant challenges. The success, security, and dependability of the project at every level depend on the adoption of industry standards and market regulations for BIM. Achieving uniform industrial standards can guarantee accurate and consistent data storage, exchange, and sharing across all BIM tools.

According to the needs of the AEC business, standardized BIM techniques and rules can help professionals and stakeholders work together. Building SMART's Industry Foundation Classes (IFCs) is the only publicly available BIM data format worldwide. Unfortunately, there is no national or global BIM standard for the AEC business.

BIM education is the catalyst for this industry-wide shift in the AEC sector. It is arguable that BIM trainings for elite engineers and architects are not as effective as BIM education in schools for producing the required BIM abilities. By enrolling in BIM classes and gaining experience with BIM-based tools, students can gain early access to emerging technologies. Given the dynamic nature of the employment market, this increases their employability upon graduation [5].

12.2 SUSTAINABLE ARCHITECTURE

To maintain this intricate system functioning properly, humans must practice sustainable growth. As the annals of human social progress are endless, there can be no universally accepted formula for long-term prosperity. It provides insight into the system's performance and overall trend. As a society grows, it should improve everyone's quality of life. Growth should not go beyond what resources and the world can handle, though. Beyond the outside, "the healthy built environment" includes the inside as well. Most people in industrialized nations face traffic snarls and, more specifically, vehicle-related air pollution when they consider urban pollution. Although it is their responsibility to ensure that towns have adequate ventilation, they might be able to contribute to a decrease in urban air pollution. Fighting urban air pollution, however, mostly boils down to transportation policy and technology. However, architects and the professionals they employ bear a disproportionate share of responsibility for the indoor environment's quality. For the past ten years, researchers all across the globe have been concentrating on this last issue. This led to the concept of "the healthy building" as a way to define the standards that professionals in the field of architecture should uphold. Aspects that encompass this concept include the following:

- indoor air quality,
- thermal comfort,
- acoustic comfort,
- visual comfort, and
- practical comfort.

For a place to be considered for good indoor air quality, it must not contain any vapors, gases, germs, or radiation that could be harmful to people living or working there if levels exceed certain limits. Such an unwanted effect might be caused by mildew, various adhesives, paints, heating and electrical equipment, cosmetics, and more.

1. To achieve thermal comfort, one must make sure that the air has the right temperature, humidity, and air flow.
2. To achieve acoustic comfort, one must keep both the outside and inside noise levels below a certain threshold.
3. To achieve visual comfort, one must use appropriate lighting, both natural and artificial, and psychologically beneficial color effects.
4. Finally, to achieve functional comfort, one must ensure that the furnishings and spaces within are not obstructed. The second part of comfort is showing empathy by thinking about how people with disabilities live their lives, which is a big deal [6].

Modern data is driving engineers and architects to develop buildings using software technologies that outperform their predecessors in terms of functionality. Engineering and architectural professionals have easy access to information about site disturbance, energy efficiency, lighting quality, material properties, and hypothetical comparisons of new and remodeling projects in these resources. The industry has rebranded this approach to building design as BIM. Due to its variations from CAD software, BIM is used. BIM expands architects' architectural form research by eliminating the requirement for 2D drawings.

The transition from a 2D to a 3D environment, along with visualization, simulation, and analysis of the entire design process, can always lead to a decrease in the human, material, and financial expenses of the finished design. Therefore, different material costs can be looked at and chosen, various design concepts can be tried out and finally chosen, and various design flaws can be identified and fixed quickly. Many people in the construction industry associate BIM's innovations with green architecture. The capacity to forecast a design's performance, compare alternatives, foresee environmental consequences, and model the implementation process throughout the design development phase is essential to this idea for the prevention of damage and waste, the development of green resources, the execution of cleaner production, and the ongoing optimization and improvement of the ecological environment.

Sustainable design is becoming an increasingly important concept in almost all areas of design nowadays, from production and industrial manufacturing to urban planning and architecture [7, 8].

12.2.1 Environmental, Economical, and Social Considerations

The environmental, social, and economic impacts of existing building stock can be mitigated by green building practices, which are one of several suggested solutions. Sustainability in green buildings is often emphasized in isolation from its social, economic, and cultural components. Very few comprehensive empirical studies have addressed the significance of these unconventional components of green building assessment, despite numerous studies highlighting their importance. Green buildings have several benefits. Green buildings increase urban biodiversity and preserve ecosystems through conservation land use. Sustainable building design requires reducing construction and demolition waste. The environmental impact of building and demolition waste must be reduced by over 90% through recycling, including the use of repurposed and recycled materials in new construction.

When compared to regular buildings, green buildings are better for the earth because they use less water, energy, and carbon dioxide. If all new buildings in Seoul used Leadership in Energy and Environmental Design (LEED)-certified rating methods, carbon dioxide emissions might go down by a lot. According to the research, LEED certification has the greatest impact on reducing CO_2 emissions from commercial buildings, followed by public and residential structures.

From a life cycle perspective in particular, there is a correlation between cost reductions and improved building performance. As a result, the operational expense is minimized. Upgrading your construction budget by 4% and 10%, respectively, can help you obtain the Green Building Council of Australia (GBCA) Green 5 Star and 6 Star ratings. However, with the high energy price and carbon trade cost, it is as costly to not go green. Savings on operation and maintenance will help recover initial investment in green building elements.

Despite the wealth of literature on sustainability's economic and social implications, research is scarce. Social performance of green buildings needs more research. The technical and environmental components of green construction use life cycle evaluation, which will help social sustainability. The newest upgrades of LEED, Building Research Establishment Environmental Assessment Method (BREEAM), and GBCA Green Star have equivalent functionality, which is positive. The new grading tool needs more research to support its construction and help clients make judgments [9].

12.2.2 Climatic Study on Location

New ideas, technology, and materials for construction that enhance thermal comfort and energy efficiency are necessary to meet the sustainability targets set by the construction industry. There have been a number of studies looking at how different exterior building surface characteristics, such as color and reflective coating, affect thermal performance. Improving the energy efficiency of the buildings could be as simple as taking global measures to decrease heat leakage. Walls and roofs must be built airtight to achieve heating expense savings of up to 25%. Large thermal mass walls can collect solar radiation for passive heating and bulk insulation throughout the winter. Windows that are double-glazed, appropriately proportioned, and shaded provide the best insulation. Simple strategies to prevent heat gain exist in every environment. To reduce heat gain, install tiny windows that are properly oriented, shaded, and made of low-E glass.

Airflow aids sweat evaporation in humid situations. Hot and humid conditions require windows that face each other for cross ventilation, a long and narrow floor plan for bedroom ventilation, and an open floor plan with a high roof for air circulation and roof radiant heat reduction. The building is warmer at night with reflective foil insulation than with bulk insulation without any air conditioning (AC). Otherwise, use double-paned windows and strong insulation. Because there is a big temperature difference between day and night, bedrooms with low thermal mass cool the building swiftly at night by cross ventilation, whereas living areas with high thermal mass cool it during the day. Roof exhaust blowers at night promote thermal comfort.

Studying climate zones helps develop environmentally friendly, cost-effective buildings. Sustainable growth requires environmentally friendly construction materials and components. For best heat retention in any environment, use airtight construction, enough insulation, thermal mass in the walls and floor, solar radiation passive heating in winter, correctly sized, orientated, and shaded windows, and double-glazed windows for better insulation. Small east–west windows, low-emissivity glass, shading to reduce radiation absorption, lightweight construction, reflective insulation, light colors to reflect radiation, external wall shading from trees, high thermal mass in areas with frequent temperature swings, and enough ventilation to cool the building are general heat gain prevention measures [10].

12.3 STUDY OF PASSIVE AND ACTIVE STRATEGIES

There are many social, economic, and ecological aspects in which the building construction industry impacts society. The construction of buildings provides people with a safe place to live, as well as space for work, study, and play, all of which contribute to the expansion of economies and cultures around the world. On the other hand, the construction and occupancy of buildings can cause a deluge of pollution, trash, and carbon emissions that can overwhelm the ecosystem. Architects, engineers, researchers, and others involved in the building industry have created Green Building Rating Tools (GBRTs) and associated building environment assessment schemes to tackle the global issues of high-quality building design, construction, and operation, along with their societal impacts. Rating methods, building technology, environmental consequences, social and economic effects, and assessment programs have all been the subject of study since the initial voluntary evaluation program was implemented in the 1990s.

Modern building codes place stringent requirements on energy efficiency, cost-effectiveness, grid compatibility, occupant comfort, sustainability, and net-zero energy consumption. Thermal simulation models, sensor networks, and integrated control systems are needed to develop energy-efficient Nearly Zero Energy Buildings (nZEBs) and Positive Energy Buildings (PEBs). Renewable energy technologies and green energy credits help many construction projects approach net-zero energy performance. More precise definitions of net-zero and net-positive energy require new design concerns. By highlighting the building's impact on its environment and systems, net-positive energy redirects the focus back to maximizing energy performance at the system level. Beyond net-zero energy buildings, this approach highlights technical, behavioral, policy, and regulatory potential and problems, questioning the concept that all new buildings must reach net-zero performance and emphasizing the need for a more holistic energy assessment. Cities use so much energy and pollute that we must decarbonize to satisfy energy and climate goals. Promoting net-zero or positive energy buildings and localities helps cities become sustainable. Smart city projects highlight new low-carbon solutions, but their wide range of metropolitan conditions makes replication difficult.

To further minimize heat gain or loss over the building envelope, insulation is an essential component of passive design. Insulation acts as a barrier to heat transfer, lowering the interior temperature in the summer and increasing it in the winter. Air leaks and insufficient insulation are the two most common reasons for heat loss in buildings. Consequently, insulation is meticulously installed in the building's walls, floors, and ceilings to enhance thermal comfort and promote energy efficiency. Walls that face the sun can absorb a lot of heat, which can change the inside temperature and make people feel less comfortable. To increase the inside temperature and decrease heat loss, solar reflecting coatings can be applied. In addition to lowering the temperature inside, these covers also make cooling easier. According to study, solar reflecting coatings can lower temperatures by up to 41°C when put on gray or darker paint. Color choice is also crucial because cooler shades release more infrared radiation. Several factors, such as the material and the surrounding environment, affect the way light-colored paint is applied. It is possible to save energy, particularly during peak hours, by thoroughly assessing these factors. A more pleasant and energy-efficient indoor climate can be achieved with the help of this technique [11].

12.4 BUILDING ENVELOPE DESIGN

"Green buildings" use resources more efficiently and sustainably. Green construction approaches involve pollution, carbon dioxide emissions, non-toxic materials, water and energy conservation, and renewable energy. Sustainable exterior design is essential to green construction. Governments have prioritized energy usage, environmental laws, and carbon emission reduction measures like the Kyoto Protocol and Paris Agreement. Emission Trading Scheme (ETS) permitted prices are broadly endorsed as carbon pricing plans. South Korea, Switzerland, Germany, and the USA (California) have introduced ETS to price carbon emissions from buildings' heating and cooling energy needs. These factors emphasize the need for green building insulation. The Life Cycle Cost Analysis (LCCA) will consider thermal energy use, building envelope materials, and CO_2 emissions pricing.

Environmental factors and the building envelope are the two most important things to think about when planning a green building with thermal insulation in mind. It would not be smart to start a program for all temperature zones that requires all buildings to fit thermal insulation. An essential thermal approach that takes climate into account is the building envelope. The building's envelope, which typically consists of walls, insulation, and plaster, is what keeps the outside elements at bay. The thermal balance of the building, which affects CO_2 emissions and energy consumption, is primarily determined by the thickness of the building envelope and its thermal conductivity (λ_h) value, which varies among different climate zones. The seismic zone where the project will be developed is often overlooked while designing energy-efficient and green building envelopes for the LCCA. Because structures are prone to natural disasters like earthquakes, their structural deformation depends on location. These distortions reduce building envelope energy efficiency.

Engineering calculations, numerical simulations, and machine learning are the main building energy analysis methods. A rigorous mathematical solution strategy is needed to solve large and complicated structures, making the procedure unapproachable. The variety and qualities of the building envelope materials, however, have evolved substantially due to technological developments. To create an eco-friendly and energy-efficient structure in a fair length of time, it is practically impossible to choose the ideal building envelope design from the many alternatives available in simulation programs. To evaluate a wide variety of building envelope combinations, more complex methodologies are required than what is currently available in modeling systems. Machine learning allows for multiple assessments of building energy usage to be carried out more readily and rapidly than conventional approaches. Machine learning techniques have been widely used in building energy analysis for predicting future energy demand based on parameters such as population density, weather, and the amount of lighting, heating, and cooling that a building consumes on a monthly, daily, or even hourly basis. Producing output based on several input attributes manually could be quite challenging, particularly when dealing with building energy analysis challenges involving massive amounts of data. These analyses, however, can only be conducted once a certain quantity of data has been suitably trained using machine learning methods.

Building energy consumption trends, influencing factors, and emerging patterns are all identified in the course. By comparing the results from the training data with the results from the actual data, we may evaluate the method's performance. As a result, a system is established to assess the energy consumption of the structure. One common machine learning tool for creating more eco-friendly structures is artificial neural networks (ANNs). By using a few basic input attributes to establish the building envelope during the early stages of design, this tool can assist architects and engineers in estimating the energy performance of a project. Long-running simulations determine the full solution space, which ANN can then utilize to predict [12].

To compare the two scenarios, the projected ENVLOAD value is 87.45 ± 0.5 kW h/m$^{2\cdot}$ yr, which is lower than Taiwan's green building energy conservation criteria (<92 kW h/m^2). Few windows in the building envelope (low openness ratio (OR)) reduce heat gain, energy use, and construction cost, but architects and decision-makers loathe its appearance. I optimized envelope configurations to cut costs while ignoring OR. Scenario II optimized envelope configurations with the same window opening rate and ENVLOAD value as the first design. Structure material, sunshade type and length, and window number and length are optimum envelope factors. Optimization was performed for scenario II (Figure 12.2) to find the best design plans and illustrate OPOBEM's applicability as a design reference [13].

12.5 SITE PREPARATION AND PLANNING

To increase a wide range of ecosystem services, it is crucial to rehabilitate urban green spaces by planting native plants. Few details regarding the reintroduction of native ecological components, particularly forb species, into city parks have been revealed despite extensive study on ecological restoration, plant establishment, and land reclamation. Composting at a higher rate improved soil texture and nutrient levels; however, some nutrients degraded after a year, while others grew. The organic matter that decomposes in naturalized areas, such as leaf litter and woody debris, can increase the amount of water that plants can use. Numerous studies show that planting native forbs (wildflowers) among trees and shrubs is a relatively new practice that is being more and more adopted by urban planners and landscapers. When urban green spaces are restored with native forbs, native biodiversity is enhanced and beautiful flowering plants are generated for recreational enjoyment and educational purposes. Research on the employment of native forbs to restore urban green spaces is limited, despite their potential usefulness. Lack of knowledge about forbs as distinct species makes it difficult to use them in restoration efforts for urban green spaces. Visual attractiveness and plant material availability are currently the primary considerations in species selection. Restoring urban green spaces with non-native species requires plants that are robust and competitive.

When it came to reintroducing forb species to an urban park, site preparation procedures were ineffective compared to compost soil addition, which enhanced the growth and survival of planted forbs. There was an increase in cover, growth, species richness, and noxious weed cover in compost treatments compared to controls [14].

12.6 LIGHTING AND CONTROLS

Even though natural light has been around since the beginning of civilization, the Industrial Revolution saw the rise of artificial light as the preferred method of illumination due to its dependability, safety, and other advantages. After air conditioning, the artificial lighting system in Malaysian offices uses the most energy, accounting for over 20% of the commercial power consumption. As lighting fixtures are among the most energy-intensive components of tall buildings, researchers around the world have been looking into ways to lower their energy consumption. Possible solutions include rethinking fixture placement and shading, developing more effective control systems, and upgrading glazing and windows.

In 2008, the Green Building Index (GBI) was developed as a tool for environmentally conscious construction. It was specifically tailored to the social, economic, infrastructural, and tropical climate conditions of Malaysia. In an effort to reduce their environmental impact and save money, a growing number of local architects and engineers are utilizing the GBI evaluation tools to create more sustainable and

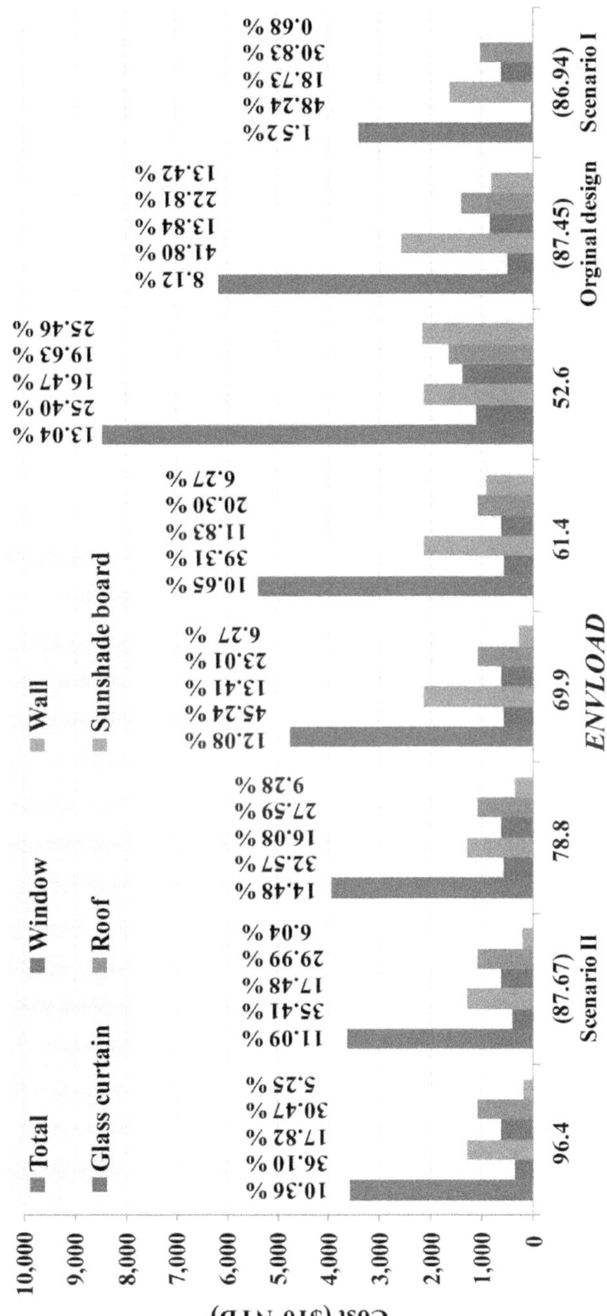

FIGURE 12.2 Building envelope costs in different *ENVLOAD* requests [13].

environmentally friendly structures. Attributes that enhance the aesthetic satisfaction of tenants receive 7 out of 21 points in the GBI evaluation criteria for non-residential existing buildings (NREB). More and more modern commercial buildings are opting to use natural light during the day instead of artificial lighting, and bigger windows are a common feature in these green-certified structures. This is especially important considering the fact that the impact of increased glazing to floor area ratios on interior illumination is still largely unknown. Energy costs for cooling and other purposes are increased due to solar heat transmission into buildings through fenestrations, which is typical in tropical areas.

Due to its fluctuation and flux, natural light is harder to regulate than artificial light, which raises concerns about its use in environmentally friendly buildings:

1. The open-plan office has solely natural light—how bright is it during work hours?
2. How much light makes a room cozy? Local renters' visual perceptions due to building natural light levels.

The visual comfort and illumination levels of office building occupants in constructions with big glass window sections to increase daylighting must be studied in field investigations. Post-occupancy comfort surveys (PCSs) on building interiors have shown their efficacy. Daylight Glare Index (DGI) and related glare indices affect sources with different illumination levels. These experiments also addressed daylight discomfort glare. These included the occupants' subjective estimate of glare severity and the source's brightness fluctuations.

To reduce sun glare, the building's automatic sunlight monitoring system should control the blind curtains at facade windows with excessive sunlight reflection. After that, an inherent manual override option would let nearby personnel change the blind curtains 24/7. Over-illuminated fenestrations may need outdoor shades and exterior louvers to improve visual clarity. To avoid glare, the glazing system can be updated. Sunlight-linked lighting management and solar film coatings also reduce illuminance and solar heat gain in high-illuminance buildings. Radiance software modeling shows that office buildings may save energy by carefully placing light shelves that vary in height, angle, and placement to exploit natural light. Current research suggests a cutting-edge glazing system that lowers space cooling load and glare, enhancing visual and thermal conditions. Because of this, it is important to understand how window-input solar heat affects tenant comfort. Even though these results are promising, repair and maintenance expenses should be addressed before installing new glazing or altering the current system [15].

REFERENCES

1. He, W., Zhang, Y., Li, S., Li, W., Wang, Z., Liu, P., & Kong, D. (2024). Reducing betrayal behavior in green building construction: A quantum game approach. *Journal of Cleaner Production*, *463*, 142760.
2. Silva, B. V., Holm-Nielsen, J. B., Sadrizadeh, S., Teles, M. P., Kiani-Moghaddam, M., & Arabkoohsar, A. (2024). Sustainable, green, or smart? Pathways for energy-efficient healthcare buildings. *Sustainable Cities and Society*, *100*, 105013.

3. Gao, X., & Pishdad-Bozorgi, P. (2019). BIM-enabled facilities operation and maintenance: A review. *Advanced Engineering Informatics*, *39*, 227–247.
4. Succar, B. (2009). Building information modelling framework: A research and delivery foundation for industry stakeholders. *Automation in Construction*, *18*(3), 357–375.
5. Tang, L., Chen, C., Tang, S., Wu, Z., & Trofimova, P. (2017). Building information modeling and building performance optimization. *Encyclopedia of Sustainable Technologies*, 2, 311–320.
6. Kunszt, G. (2003). Sustainable architecture, *Periodica Polytechnica Civil Engineering*, *47*(1), 5–10.
7. Li, S. (2009). Building information modeling leads to sustainable design. *China Computer User*, *9*, 42–43.
8. Wang, Y., Liu, S., Song, J., Zhou, J., Kong, Y., & Jiang, F. (2024, September). Sustainable architecture design. In *Proceedings of the 2024 7th International Symposium on Traffic Transportation and Civil Architecture (ISTTCA 2024)* (Vol. 241, p. 4). Springer Nature.
9. Zuo, J., & Zhao, Z. Y. (2014). Green building research–current status and future agenda: A review. *Renewable and Sustainable Energy Reviews*, *30*, 271–281.
10. Albatayneh, A., Alterman, D., Page, A., & Moghtaderi, B. (2018). The significance of building design for the climate. *Environmental and Climate Technologies*, *22*(1), 165–178.
11. Sarir, P., & Sharifzadeh, M. (2024). Application of passive and active scenarios to a residential building in a dry and hot climate to achieve a positive energy building (PEB). *Heliyon*, *10*(10), 1–15.
12. Himmetoğlu, S., Delice, Y., Aydoğan, E. K., & Uzal, B. (2022). Green building envelope designs in different climate and seismic zones: Multi-objective ANN-based genetic algorithm. *Sustainable Energy Technologies and Assessments*, *53*, 102505.
13. Lin, Y. H., Tsai, K. T., Lin, M. D., & Yang, M. D. (2016). Design optimization of office building envelope configurations for energy conservation. *Applied Energy*, *171*, 336–346.
14. Rojas, J. A., Dhar, A., & Naeth, M. A. (2022). Urban green spaces restoration using native forbs, site preparation and soil amendments — A case study. *Land*, *11*(4), 498.
15. Kwong, Q. J. (2020). Light level, visual comfort and lighting energy savings potential in a green-certified high-rise building. *Journal of Building Engineering*, *29*, 101198.

13 Future Trends and Innovations

Green building approaches have gained popularity in urban areas as an effective means of protecting both the environment and human health for future generations. In addition to ensuring rigorous quality assurance and control, smart green buildings provide efficiency, occupant comfort, accessibility, and systematic monitoring of building performance. Despite their potential benefits, there remains limited research into emerging technologies for sustainable construction. Although these technologies often have high initial costs, their affordability remains particularly crucial for developing countries, such as India and Iran. An essential area where to concentrate on is improving the workability and strength of eco-friendly materials. The reduction of environmental impacts remains the primary objective, emphasizing the need for improved recycling and composting methods for industrial waste. Integrating artificial intelligence (AI), 5G networks, and smart materials can significantly advance the smart green building sector, provided these technologies become economically viable.

According to the International Finance Corporation (IFC), approximately 70% of India's buildings are yet to be constructed. Adopting smart green construction practices in India would positively impact both the economy and the environment. While requiring higher initial investment, green smart buildings are substantially more cost-effective over their life cycle. It is therefore critical to emphasize the economic viability of sustainable construction, "sustainable construction is budget-friendly." Green buildings maximize the utilization of natural sunshine instead of relying on artificial lights. This is because sunshine calms down people's tensions and makes them feel more at peace [1].

Improving energy efficiency and conserving resources will also play a role in the advancement of sustainable construction methods, along with emerging trends and innovative materials. Improved management of natural resources necessitates the development of novel materials and technologies as well as the wise, long-term enhancement of current materials.

13.1 RETROFITTING EXISTING BUILDINGS

Energy efficiency retrofits in existing buildings significantly lower greenhouse gas emissions and energy consumption. However, additional non-energy impacts (NEIs),

DOI: 10.1201/9781003752837-13

such as improved occupant health and comfort, also influence retrofit decisions and enhance the overall value. Existing studies typically categorize NEIs into passive (e.g., building envelope improvement) and active (e.g., Heating, Ventilation, and Air Conditioning (HVAC) system and related components upgrades), yet this categorization often oversimplifies their interactive effects, hindering comprehensive retrofit strategies.

Effective retrofit plans that consider occupant needs are hindered by this category, which makes it hard to understand the actual mechanism(s) by which each NEI functions.

Residential energy usage is 16% of total energy use and approximately 20% of greenhouse gas (GHG) emissions in the USA. Passive approaches primarily enhance the building envelope's performance, while active methods improve energy systems. Residential energy retrofits rely on the following two methods.

1. The building envelope ensures interior conditions despite environmental changes as the principal barrier between inside and out.
2. Studies show that retrofitting a building's exterior can dramatically cut energy use and carbon emissions [2].

Retrofitting residential buildings in hot climates is crucial for improving indoor comfort and reducing energy consumption. In their pursuit of this goal, numerous strategies, including both proactive and reactive actions, have been investigated and considered. Effective strategies include installing high-quality thermal insulation, double-pane windows with low-emissivity coatings, and shading devices such as louvres and awnings.

Having proper thermal insulation is crucial in hot climates to keep cooling expenses down. Some buildings have well-insulated walls, roofs, and floors, which makes it harder for heat to get inside. Numerous studies have shown that insulating the outside of a building can greatly reduce the amount of energy it uses. It is possible, for instance, to reduce annual energy demand by 44% by installing high-efficiency thermal insulation materials. An Oman study found that by insulating the building shell and using other retrofitting procedures, the annual energy usage of a residential structure may be reduced by 42.5%. Quality insulation can help buildings in hot areas be more energy efficient, as shown by these results.

Building thermal mass is affected by window glass. For hot regions, double-pane windows with low-emissivity coatings reduce heat gain without sacrificing transparency. Energy-efficient glass reduces cooling loads and improves comfort. Researchers in Qatar found that enhanced glazing can dramatically cut cooling needs after studying retrofitting options, including window quality. Energy savings of up to 53% were found with retrofitting methods including improved window glazing.

By lowering the amount of heat that the sun gains, shading devices can also help cut down on the need for air conditioning. An external shade system can block the sun's rays from reaching a structure using louvres, overhangs, or awnings. Another option for reducing the interior temperature is to install internal shade, such as blinds and curtains. A case study in southern Spain showed that solar shading effectively

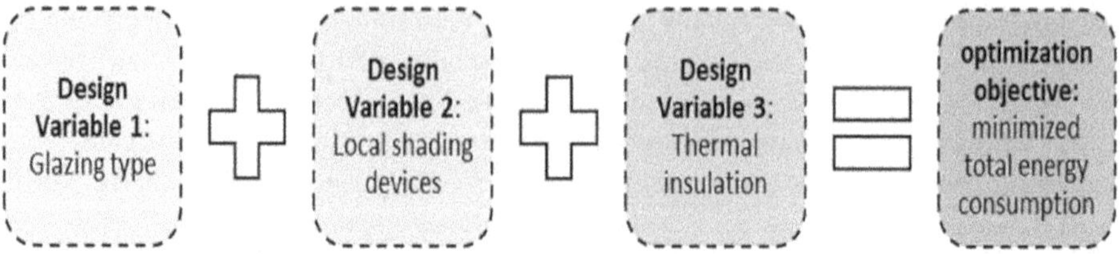

FIGURE 13.1 Process and design optimization goals [3].

reduced energy demand while simultaneously boosting indoor comfort. Including sun protection measures in the retrofitting intervention on the building envelope resulted in a 38% reduction in total energy consumption.

Integrating renewable energy technologies, such as solar photovoltaic (PV) systems, further enhances energy efficiency in retrofitted buildings. In regions with high solar radiation, such as Saudi Arabia, solar energy systems effectively reduce cooling-related energy demands. They are the way to go because they can turn solar energy into useful power.

Optimized combinations of glazing type, shading mechanisms, and insulation materials provide the best outcomes for energy-efficient residential design. In ref. [3], the authors reported on the optimization strategy used to reduce residential villa energy consumption (Figure 13.1). The optimization strategy focused on glass type, local shading mechanism type, and outer-shell insulation type and thickness. The worst-case scenario included one pane of clear glass, no local shade, and insufficient insulation. Permutations of these design components were considered to find the best energy-efficient layout.

13.2 LOW-INCOME HOUSING

Modern times have seen a surge in the popularity of green building practices. Building eco-friendly structures is becoming increasingly important; thus, the industry must find strategies to ensure they perform as intended once operational. Green buildings are being constructed in several countries and can offer win–win scenarios, but there are still some things standing in the way of their broad adoption. The growing emphasis on green buildings must extend to low-income housing, especially in countries like India, where the construction sector is critical to economic growth. Indeed, construction industry as the second-largest and second-most important sector of India's economy, after agriculture, employs a massive workforce. Because of its fast-growing population, a country like India experiences ever-increasing needs. There has been a recent uptick in the demand for real estate such as houses, shopping malls, hotels, and office buildings. The concept of "green" housing is emerging as a viable alternative. When it comes to green construction and greenhouse programs in India, the Indian Green Building Council (IGBC, 2014) is in charge of everything. Some studies have looked at greening and environmental conservation, but few have examined how low-income and marginalized populations may help. Even now, most of these people are still living in filthy, isolated villages. In the long run, addressing the housing needs of marginalized communities through sustainable solutions can significantly improve living standards and overall economic stability. Essential components of sustainable low-income housing include energy-efficient designs, effective water management solutions like rainwater harvesting, and sustainable, locally sourced construction materials.

To become more energy-independent, one of the most important things we can do is make sustainable homes use less energy. It is easy and quick to put buildings into use that use no net energy. This greatly reduces the use of fossil fuels. Decreased water use and maintenance of water quality are primary concerns when building a sustainable home. Two options for water optimization were considered; groundwater and a simple, cheap rainwater collection system. With this system, water from rain is collected in a holding tank by means of a roof and gutters constructed of sheet material or plastic. Before settling on eco-friendly construction supplies, it is important to weigh a number of criteria, such as the following:

1. How close are they to your installation site?
2. How easy are to recycle?
3. How long do they last?
4. How can hazardous gases enter into the air?
5. How long have they been obtained sustainably?
6. How long are they reusable?

Deciding to optimize a green building during the early stages of design is essential for achieving the structure's full green potential, minimizing redesign requirements, and guaranteeing the project's overall viability and success. Although the specifics vary by region, country, and even city, similar constraints do exist [4].

13.3 WATER CONSERVATION METHODS

In comparison to more traditional methods, most water conservation and energy-saving strategies for commercial and residential buildings demand a larger initial investment. The greater outlay of capital is likely to be more than offset by the benefits of these "green" building practices in the end. Because of variables including regional rainfall patterns and the unpredictability of market-related expenses and advantages, the use of green roof systems to cool homes is fraught with uncertainty. Research on green energy efficiency has mostly concentrated on smart grids, better insulation, and decreased emissions of greenhouse gases. The water–energy nexus in sustainable development has overlooked the complementary impact of water efficiency and reduced energy use, despite its importance in building environmentally friendly and low-carbon urban areas. Thus, including water conservation and energy savings in green building design strategies is a growing objective on a global level.

A green building that uses these principles would typically have a greater initial capital cost compared to a conventional residence.

13.4 CARBON NEUTRALITY

The construction sector, responsible for over 40% of global energy consumption and carbon emissions, faces growing demands to reduce its environmental impact. Green building rating systems have become essential tools to encourage and measure sustainable construction practices. Governments, businesses, and the public worldwide are prioritizing construction sustainability and lowering buildings' environmental impact. Many green building assessment standards and methods have assisted the construction industry's sustainable development. In addressing climate change, carbon neutrality entails compensating for carbon emissions to achieve a net-zero emission status. However, "Carbon Peak" refers to the point at which an entity's emissions are expected to decline. As governments and organizations worldwide reduce emissions and promote sustainability, these ideas are becoming mainstream in the fight against climate change. To align with Carbon Neutrality and Carbon Peak assessment standards, evaluation standards should be rebalanced to emphasize carbon emission reduction. Smart technology can streamline and improve reviews by real-time recording and analyzing building energy use, material choices, and other parameters. With this, Carbon Neutrality and Carbon Peak are closer [5].

13.5 GREEN OR LIVING ROOFS

Due to the great quality of life in big cities, more and more people are moving to urban areas, which brings with it a host of serious problems, such as noise pollution, poor water and air quality, and climate change as well as the urban heat island (UHI) effect, which is produced by an absence of green spaces. It is logical to suppose that buildings' thermal comfort is negatively affected by the rapid increase in ambient temperature in places with minimal vegetation. Effective long-term solutions include integrating vegetation management strategies like green roofs and living walls. These systems enhance both indoor and outdoor environmental quality, mitigate UHI, and improve building efficiency.

Green roofs and walls are common landscape elements that are considered in building plans. The efficiency of a structure and the quality of its indoor and outdoor areas are both enhanced by these systems, and their adoption is anticipated to rise in tandem with the growing environmental consciousness of the general public.

Figure 13.2 displays a variety of green façade. The modular trellis, the basic green façade, and the double-skin green façade (DSGF) are the two main varieties of green façades, each with three subtypes. Instead of using soil or another growing medium below ground, plants on traditional green façades depend on the façade itself for structural support. In DSGFs, constructed on surface-running double-skin frameworks, an air gap separates the wall from a vertical support component. Encircling the structure are planters designed to fit the roofs of modular trellises [6].

13.6 LIVING WALLS OR VERTICAL GARDEN

An artificial or purposefully produced vegetal element that can serve many functions and give intended environmental advantages to its built surroundings is called a living

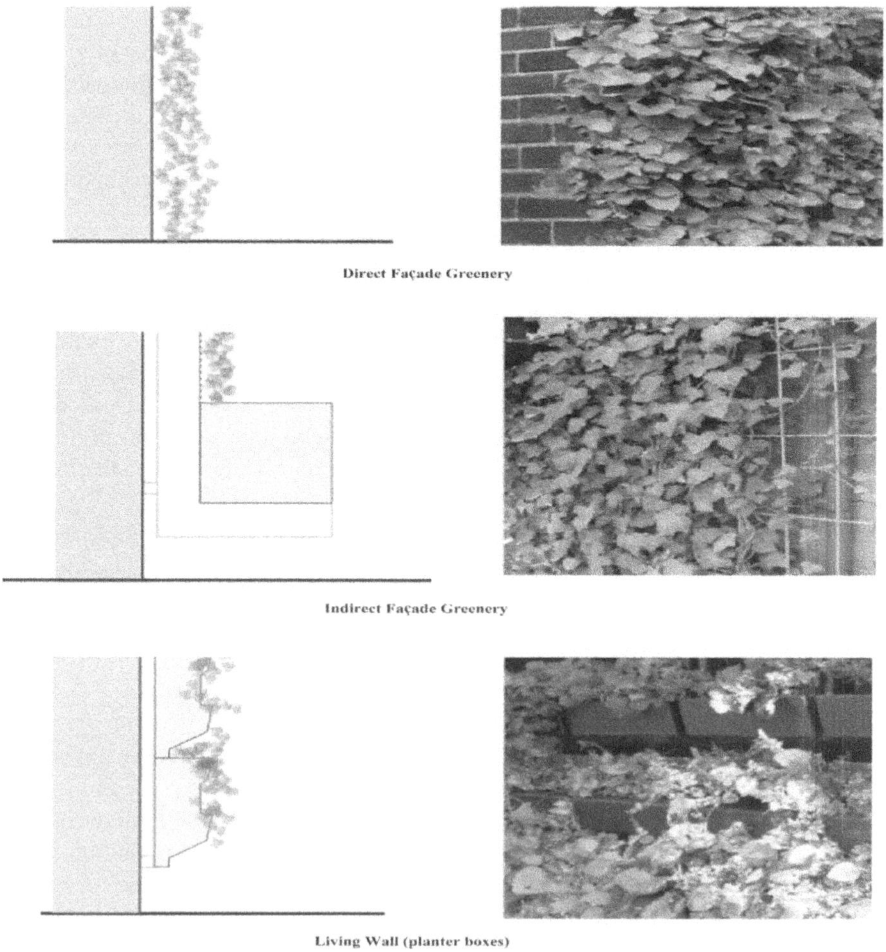

FIGURE 13.2 Types of green façades [6].

wall. “Living walls” or “green walls” differ from “green facades” as they are self-sustaining structures that may be affixed to either the exterior or interior of a building. The plants here establish roots in a structural framework that is firmly affixed to the wall, distinguishing them from other types of green façades, such as ivy walls. Because they can get all the water and nutrients they need from the vertical support, the plants do not even need soil fertilizers. Potentially, Vertical Greenery Systems (VGS) benefits might be felt at the city and building levels simultaneously. Several large-scale benefits are better health, better envelope protection, less noise inside, higher property value, better indoor environmental quality, better indoor air quality, and higher energy economy. Another benefit that has not been fully tested experimentally is the chance that VGS could make the air better. Even while many European municipalities employ VGS to try to clean the air, there is not enough conclusive scientific evidence to say whether or not it really works. The optimal plant species for air cleaning, soil pollution removal, and dust capture has been the subject of very little research. It is clear that more research is needed to determine how various VGS construction types can improve air quality. This research should consider the supporting structures, soil types, and plant species of each type. Additionally, these kinds of research should determine which VGS varieties work best in various climate zones and other specific settings.

Similar problems arise in empirical studies that attempt to put a price on the advantages to VGS hydrology and noise mitigation. Much of the data is presented descriptively and based on assumptions made about the similarities with green roofs.

Gaining a better understanding of how other people perceive different elements of VGS design—including the type of plant, color, scent, height, placement on the wall, and tactile experience—would be advantageous as the number of empirical research evaluating VGS's visual effects keeps growing. Insights into occupant preferences and the reasons behind the superior performance of certain VGS in particular building types could be uncovered in this way.

VGS definitely does not get enough credit for its social and educational benefits. A few studies have found that VGS might be good for society because it could improve health and lower crime. Future studies should focus on whether or not VGS can improve other social benefits, such as the growth of a sense of community [7].

13.7 BIOMIMICRY

Innovation in agriculture must be guided by a commitment to environmental sustainability, which is crucial in the battle against climate change.

It comprises looking at the natural environment and the architectural design process side by side and contrasting and analyzing the main similarities and motivating factors that impact each. The building's skin is a thin membrane that encases the skeleton and defines the building's structural, mechanical, plumbing, and electrical systems as well as the interior regions. The building's skin, with its numerous layers and filters, can detect air, light, humidity, sound, and heat, like real skin. One commonality of natural skin is its ability to control internal conditions in reaction to its function. Similar to real skin, the building skin separates the controlled and uncontrolled areas. The results that are influenced by causes both within and outside the organization are arranged in this way. They both play the role of filters, regulating what can enter and what can exit [8].

13.8 CRADLE-TO-CRADLE DESIGN

Unlike zero-emission and eco-efficiency programs, cradle-to-cradle design and eco-effectiveness provide an alternative perspective on production and style. An alternative to more traditional eco-efficiency strategies, eco-effectiveness focuses on making products and systems that are both safe and environmentally benign. Reducing the detrimental impacts of economic activity on ecological systems without compromising or lowering the value of economic output is the aim of eco-efficiency strategies.

Maximizing economic value without negatively impacting the environment is the goal of zero-emission, the highest form of eco-efficiency. The goal is to create a clear demarcation between the two areas.

Eco-efficiency requires a one-way, linear flow of materials through industrial processes, from natural resource reclamation to disposal. Environmentally friendly practices can only lessen the material flow system's volume, velocity, and toxicity; they cannot change its linear evolution. Some materials are only recycled when they have served their final functional purpose as recycling was not their initial intention. Instead of recycling, what is taking place here is downcycling. Decreasing material quality decreases material usability and maintains a linear material flow system from cradle to grave [9].

In contrast to traditional eco-efficiency approaches, eco-effectiveness envisions industries as beneficial rather than harmful, enhancing ecosystems while fostering sustainable economic growth (Figure 13.3). This is where the idea of the triple top line first came from. Businesses are increasingly using the "triple bottom line" strategies to reduce their negative effects on society and the environment. They integrate social, ecological, and economic values while fostering synergies that lead to comprehensive sustainability outcomes. Trifecta growth firms, on the other hand, value social and natural capital separately, spread the word about how important they are, and enjoy the synergies that can result from putting all three priorities together [10].

The future of sustainable construction and green building innovations will significantly depend on integrating advanced technologies, sustainable high-performing materials, and strategic design practices. Emphasis on retrofitting existing structures, optimizing water conservation, achieving carbon neutrality, and utilizing innovative vegetative solutions such as green roofs and vertical gardens will continue to

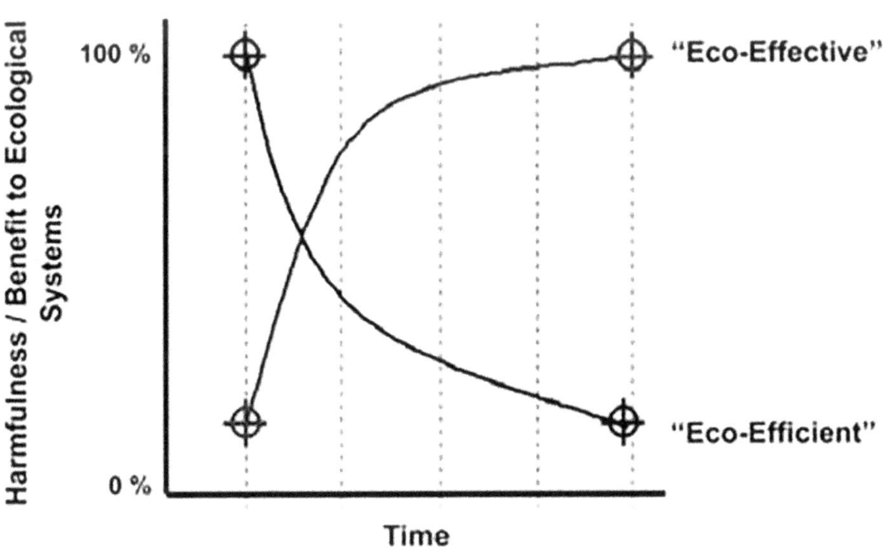

FIGURE 13.3 Eco-effectiveness seeks to benefit ecological systems 100% [9].

shape the built environment. Approaches like biomimicry and cradle-to-cradle design further underline the transformative potential of sustainable practices, promoting ecological regeneration and long-term economic viability. As global sustainability targets become increasingly ambitious, the strategic combination of these trends and innovations will be critical in fostering resilient, sustainable urban environments.

REFERENCES

1. Patil, M., Boraste, S., & Minde, P. (2022). A comprehensive review on emerging trends in smart green building technologies and sustainable materials. *Materials Today: Proceedings*, *65*, 1813–1822.
2. Lozinsky, C. H., Casquero-Modrego, N., & Walker, I. S. (2025). The health and indoor environmental quality impacts of residential building envelope retrofits: A literature review. *Building and Environment*, 270, 112568.
3. Aloshan, M. (2025). Achieving near-zero energy in hot climates: Retrofitting building envelopes for existing homes. *Heliyon*, 11(3), 1–21.
4. Uddin, M. N., Selvam, A. M., Shahoonda, J., & Prasanth, R. (2018). Optimization of green building for low-income people at Pondicherry. *Civil Engineering and Architecture*, *6*(6), 283–292.
5. Mao, J., Yuan, H., Xiong, L., & Huang, B. (2024). Research review of green building rating system under the background of carbon peak and carbon neutrality. *Buildings*, *14*(5), 1257.
6. Besir, A. B., & Cuce, E. (2018). Green roofs and facades: A comprehensive review. *Renewable and Sustainable Energy Reviews*, *82*, 915–939.
7. Radić, M., Brković Dodig, M., & Auer, T. (2019). Green facades and living walls—A review establishing the classification of construction types and mapping the benefits. *Sustainability*, *11*(17), 4579.
8. Allahverdiev, E. G., Baysaeva, M., & Chigvintseva, I. (2024). Impact of biomimicry on the economics of green construction: Cost reductions and increased sustainability. In *BIO Web of Conferences* (Vol. 140, p. 03002). EDP Sciences.
9. Braungart, M., McDonough, W., & Bollinger, A. (2007). Cradle-to-cradle design: Creating healthy emissions–A strategy for eco-effective product and system design. *Journal of Cleaner Production*, *15*(13–14), 1337–1348.
10. Reddy, V. S. (2016). The emerging trends of green building construction practice. *International Journal of Scientific & Engineering Research*, *7*(10), 1564–1569.

Index

For Product Safety Concerns and Information please contact our EU representative GPSR@taylorandfrancis.com
Taylor & Francis Verlag GmbH, Kaufingerstraße 24, 80331 München, Germany

www.ingramcontent.com/pod-product-compliance
Ingram Content Group UK Ltd.
Pitfield, Milton Keynes, MK11 3LW, UK
UKHW022220190926
475702UK00003B/464

* 9 7 8 1 0 3 2 8 1 8 3 2 0 *